T0093232

DIAGNOSING SOCIAL PATHOLOGY

Can a human society suffer from illness like a living thing? And if so, how does such a malaise manifest itself? In this thought-provoking book, Frederick Neuhouser explains and defends the idea of social pathology, demonstrating what it means to describe societies as "ill," or "sick," and why we are so often drawn to conceiving of social problems as ailments or maladies. He shows how Rousseau, Hegel, Marx, and Durkheim – four key philosophers who are seldom taken to constitute a "tradition" – deploy the idea of social pathology in comparable ways, and then explores the connections between societal illnesses and the phenomena those thinkers made famous: alienation, anomie, ideology, and social dysfunction. His book is a rich and compelling illumination of both the idea of social disease and the importance it has had, and continues to have, for philosophical views of society.

FREDERICK NEUHOUSER is Professor of Philosophy at Barnard College, Columbia University, and Permanent Fellow at the Center for Humanities and Social Change in Berlin. His books include *Fichte's Theory of Subjectivity* (Cambridge University Press, 1990), *Foundations of Hegel's Social Theory* (2000), *Rousseau's Theodicy of Self-Love* (2008), and *Rousseau's Critique of Inequality* (Cambridge University Press, 2014).

DIAGNOSING SOCIAL PATHOLOGY

Rousseau, Hegel, Marx, and Durkheim

FREDERICK NEUHOUSER

Barnard College, Columbia University

CAMBRIDGE
UNIVERSITY PRESS

University Printing House, Cambridge CB2 8BS, United Kingdom

One Liberty Plaza, 20th Floor, New York, NY 10006, USA

477 Williamstown Road, Port Melbourne, VIC 3207, Australia

314–321, 3rd Floor, Plot 3, Splendor Forum, Jasola District Centre, New Delhi – 110025, India

103 Penang Road, #05–06/07, Visioncrest Commercial, Singapore 238467

Cambridge University Press is part of the University of Cambridge.

It furthers the University's mission by disseminating knowledge in the pursuit of education, learning, and research at the highest international levels of excellence.

www.cambridge.org
Information on this title: www.cambridge.org/9781009235037
DOI: 10.1017/9781009235020

First published 2023

Printed in the United Kingdom by TJ Books Limited, Padstow, Cornwall

A catalogue record for this publication is available from the British Library.

Library of Congress Cataloging-in-Publication Data
NAMES: Neuhouser, Frederick, author.
TITLE: Diagnosing social pathology : Rousseau, Hegel, Marx, and Durkheim / Frederick Neuhouser.
DESCRIPTION: 1 Edition. | New York, NY : Cambridge University Press, 2022. | Includes bibliographical references and index.
IDENTIFIERS: LCCN 2022009563 | ISBN 9781009235037 (hardback) | ISBN 9781009235020 (ebook)
SUBJECTS: LCSH: Social problems. | Sociology – Philosophy.
CLASSIFICATION: LCC HN18 .N478 2022 | DDC 361–dc23/eng/20220224
LC record available at https://lccn.loc.gov/2022009563

ISBN 978-1-009-23503-7 Hardback

Für Gene, das Gute in leiblicher und geistiger Gestalt

Contents

Preface

When I began this project ten years ago, my plan was straightforward – and, as I now see, naive: I wanted to demonstrate the indispensability of the concept of social pathology for normative social philosophy and, by drawing mostly on resources from European thinkers from Rousseau onward, to articulate more precisely than had been done before how illness in the domain of the social ought to be conceived. My initial strategy for defending this concept was to consult contemporary philosophy of biology and medicine with the aim of finding a generally accepted account of sickness and health in biological organisms that would serve as the basis for arguing that analogous features of social life justified applying the concepts of health and illness to social phenomena. Two discoveries led me to revise my plan. The first was that contemporary philosophy of biology and medicine, no less discordant than other fields of philosophy, offered no uncontroversial account of health or illness that I could simply avail myself of in defending the idea of pathology in the social domain. Moreover, the controversies only increased when turning from purely physiological conceptions of health, applicable to nonhuman organisms, to conceptions of health appropriate to human beings. It is not only that in the human realm a new category appears – that of mental health – but also, and more interestingly, that, in contrast to the case of veterinary medicine, no full account of bodily health for humans can be given that abstracts from what I call (and explain below) the "spiritual" aspects of human beings. Although the account of social pathology I provide in this book is informed by ideas deriving from the philosophy of biology and medicine, I have had to decide for myself which aspects of the views on offer there belong to the best account of illness in human beings (and other animals) and shed the most light on what illness in the social domain might consist in.

The second discovery that led me to change my plan for this book was that, as I soon found when presenting my ideas in academic contexts – among contemporary philosophers, political theorists, and sociologists – resistance to the concept of social pathology is so entrenched that framing my task as an outright defense of that concept generated premature dismissals of my claims and unfruitful disputes that tended to reenact long-familiar academic debates rather than shed light on the phenomena that interested me. At the same time, when I described my project to nonacademics, I was greeted with a degree of enthusiasm I had never before experienced when trying to explain to "laypeople" what I wrote about as a philosopher. In earlier years my attempts to say something, when asked, about the self-positing subject, moral autonomy, or the conditions under which the will of a citizenry could be considered "general" and therefore binding for all had elicited mostly polite responses and not so lightly veiled (but understandable) attempts to change the topic of conversation. Suddenly, I discovered, nonacademics had some sense of what I was talking about and were eager to volunteer their own examples of ways in which society appeared to be ill. (Not surprisingly, this tendency increased dramatically in the United States after 2016, even if the political events that evoked this response merely made it no longer possible to ignore pathological conditions that had been developing for decades. Now the question is no longer whether that society is sick but whether it – especially its version of liberal democracy – is dying.) What, I asked myself, did the responses of my fellow citizens, if not my fellow academics, say about the relevance of the concept of social illness?

These experiences changed my conception of my project. I decided to start from a fact about the discourse of social pathology that seems incontrovertible: beginning at least with Plato there appears to be an irresistible propensity among philosophers, social theorists, cultural critics, and journalists of very different outlooks – and not only in the West – to conceive of social problems and their solutions in terms of the vocabulary of illness, health, and cure. From Plato's fevered polis (Plato 1992: 369–74e) to Shakespeare's "something … rotten in the state of Denmark"[1] to Machiavelli's "hectic fevers" (or wasting disease) of the state[2] to Hobbes's infirmities of "the body politic" (boils, scabs, bulimia, rabies, epilepsy, parasitic worms[3]) to Frank Lloyd Wright's description of "landlordism" as a social disease (Wright and Pfeiffer 2008: 418), thinkers of various epochs seem irrepressibly drawn to "illness as metaphor" in the domain of social philosophy (Sontag 1978: 74, 77, 78). Moreover, despite the fact

[1] Shakespeare, *Hamlet*, Act I, scene iv.
[2] Machiavelli [1532] 1950: 11.
[3] Hobbes [1651] 1994: ch. xxix.

that conceptions of physiological illness have varied greatly over the centuries, the tendency to think of social problems on the model of illness has remained strikingly resilient.[4] My project, then, is an attempt to take seriously the indisputably widespread impulse to think of societies as susceptible to falling ill by trying to understand the powerful pull it exerts on our imagination and to assess what value the concept of social pathology might have, even when we reject the idea that human societies are simply biological organisms writ large.

With this starting point in view, a slightly different question came into focus: What does the irresistible urge to think in terms of illness when thinking about social problems say about the nature of human society? Or, posed in "ontological" terms: What kind of thing must human society be if it is vulnerable to falling ill? This book, then, is about the connection between social ontology[5] and the discourse of social pathology, and its central claim is that we can learn something important about human social life by taking seriously past and present attempts to understand and criticize society using a vocabulary borrowed from medicine. Establishing this constitutes a limited defense of the concept of social pathology by showing that theorists who employ that concept have good reasons for turning to the language of illness; that their doing so is motivated not by a priori philosophical commitments but by empirical inquiry into the real phenomena of social life; and that conceiving of social problems as pathologies enables one to discover and think productively about aspects of social life that cannot be grasped by discourses confined to the categories of legitimacy, justice, or moral rightness, as typically (and narrowly) construed by most Anglo-American political philosophy.

As a result of these revisions of my project, its central argument is no longer that critical social philosophy *must* employ the language of social pathology but, more modestly, that there are good reasons for doing so and that an outright dismissal of that theoretical framework risks losing sight of important social phenomena that purely moral or political approaches to social life cannot capture. It is true that many of the critical concepts of

[4] The metaphor of social illness is so widespread that one can find it in nearly every issue of a serious newspaper or treatise devoted to social issues. Two examples are Krugman (2019) and Mau (2019), which analyzes the problems of contemporary eastern Germany using the analogy of a bone fracture.

[5] A caveat: social ontology as pursued here is a less abstract project than many contemporary analytic philosophers take it to be, e.g., Gilbert (1989), Tuomela (2013), and Searle (1995). A good discussion of analytical social ontology can be found in Stahl 2013: ch. 4. My project is continuous with but still broader than the accounts of social reality offered by Searle (1995 and 2010) and Descombes (2014). Anthony Giddens uses the term in roughly the sense in which I use it (Giddens 1984: xx).

social theory that we should continue to take seriously have been, or can be, formulated without explicit reference to pathology. Among them are alienation, ideology, reification, colonization of the life-world (Habermas 1987: 232; Hedrick 2018: ch. 5), and the tendency of capitalism to generate recurring crises. And yet even when the language of pathology is absent, most of these critiques have been formulated within the framework of a conception of social reality that places the idea of *life* at its core. It is noteworthy that almost every thinker who can be read as a theorist of social pathology – including Hegel, Marx, Nietzsche, Freud, Durkheim, Dewey, and Habermas – understands social reality as *social life*, where the term "life" is just as important as "social." That social reality is social life is the main ontological claim I investigate here.

No doubt this book has many defects, but one is especially worthy of mention upfront: the accounts of social pathology reconstructed here take the nation-state as the basic unit of social analysis despite the increasingly globalized nature of social life nearly everywhere. Globalization does not mean that the nation-state has become irrelevant to social theory and critique – numerous decisions of import are still made at that level – but it does mean that many determinants of social life within nation-states are inseparable from processes and developments that extend far beyond their national borders. To varying degrees, Hegel, Marx, and Durkheim are aware of the increasingly international character of modern life, but their diagnoses of social pathology, and my reconstructions of them, do not for the most part reflect this awareness. (Such awareness does appear, among other places, in Hegel's allusions to the "necessity" – internal to civil society – of colonial expansion [PhR: §248], and Marx, more than any thinker covered here, provides resources for considering how, under capitalism, international social life might exhibit pathologies not visible from a merely national perspective.) Contemporary critical theorists urgently need to think further about whether – and if so, how – classical accounts of social pathology can be expanded to take account of global interdependence. At the very least, there are surely distinct pathologies that afflict formerly colonized societies that do not show up if one focuses only on states in Western Europe and North America. Even more important, the social dynamics between the global North and South, or between former colonizers and colonized, can be expected to exhibit patterns of "functioning" not visible from a merely national perspective. The question is: Is the category of pathology useful for understanding these dynamics? Is dysfunction a relevant concept, given that the "healthy" functioning of former colonizing nations relies on ethically objectionable relations to their counterparts? Or are the interactions between these two poles (and among the nations that somehow lie between them) so asymmetric and oppressive that thinking of them as composing a single "organism" that might or might not function well obscures rather than illuminates our contemporary condition? (Is there, even in biology, an example of an organism whose functioning depends so thoroughly on "higher" organs living at the expense of the "lower"?) These are important questions that I have not addressed here.

Although the chapters of this book are devoted mostly to specific thinkers, its structure is unusual and calls for explanation. Most obviously, I do not discuss the figures I treat in chronological order. Instead, chapters are arranged conceptually, beginning with less complex conceptions of social pathology and social ontology and proceeding to increasingly richer accounts of what human societies are and of the illnesses to which they are susceptible. As readers will quickly discover, this scheme yields only a loose form of organization. For the most part, I do not offer a developmental argument that proceeds by revealing defects in the theories covered in earlier chapters and showing them to be remedied by the theories that come after them. I in no way want to suggest that the specific pathologies discussed in earlier chapters are less deserving of our attention than those examined later. Still, the general account of society and social pathology found in the final three chapters is more sophisticated and theoretically adequate than those treated earlier. In this sense, then, Hegel is the hero of this book, although I do not take this to mean that Hegelians have nothing to learn from Marx, for example, the first figure discussed in detail. Perhaps I am trying to say that much of what we can learn from Marx can be integrated into Hegel's framework, whereas important possibilities for pathology rendered visible by Hegel would go undetected if we restricted ourselves to Marx's understanding of (capitalist) society. It follows that my approach here, like that of my past work, is more syncretic and conciliatory than many readers find appropriate. I will not defend this approach beyond saying that, as with most ways of doing things, it has its advantages and disadvantages. I hope that some of those advantages come across to readers of this book despite its less than perfect structure.

The initial two chapters of this book are introductory. The first explores the concept of social pathology in general, distinguishing five interpretations of that idea from the conception of social illness I adopt here. It also discusses various advantages and disadvantages of the concept of social pathology, especially the circumstance that diagnosing a society as ill allows one to thematize defects in social life that the narrower category of injustice cannot capture. Chapter 2 investigates the ways in which theories of social pathology do and do not rely on a picture of human societies as akin to biological organisms and argues for a limited version of that analogy.

In Chapters 3 and 4, after a brief survey of conceptions of social pathology that can be found in Marx, I focus on those bound up with his account in *Capital* of the formula for the circulation of capital, which distinguishes money that is capital from money that is merely money (and, so, provides a definition of capitalism) in terms of the *function* money plays in each case. Marx's biological language makes it plausible to interpret the dysfunctions of capitalism he points to here as social pathologies. One of his contributions to theories of social pathology is to bring to light an ambiguity in the concepts "functional" and "dysfunctional": the same phenomenon that appears functional from the perspective of what capital requires to function can appear dysfunctional from a broader perspective that takes into account the good of capitalism's participants. Exploring this point requires us to introduce the idea of a distinctively spiritual aspect of human existence, an idea that will accompany us throughout this book. Indeed, the chapters' main thesis is that Marx regards social life as spiritual in the same sense I attribute later to Hegel's social theory, namely, as informed by the aspiration of social members to unite in their social activity the ends of life with those of freedom. Related to this is the claim that capitalism's failure to allow for the unity of life and freedom constitutes its principal defect for Marx and the core of the most important conception of social pathology I ascribe to him. Finally, I argue that Marx's conception of human society leaves out certain elements of the spiritual aspect of social life that theories explored in later chapters enable us to incorporate into a more adequate social ontology.

Chapters 5 and 6 focus on Plato and Rousseau in order to explore the ontological thesis that thinking of societies as *functionally organized systems* that are *artificial*, or humanmade, is crucial to understanding how theories of social pathology can ascribe nonarbitrary standards of healthy functioning to social institutions. The first of these points is set out in Plato's *Republic*, and it is appropriated by many later social philosophers, including Hegel, Comte, Spencer, and Durkheim. Rousseau is the first modern philosopher to elaborate the second point in a form that is promising for contemporary social thought. The most important sense in which social institutions are made by us is expressed in Rousseau's claim that institutions are grounded in *conventions*. The upshot of this claim is that a kind of self-consciousness, or subjectivity, is intrinsic to social life, namely, a collective acceptance of the authority of the rules governing social institutions, which, in the most fundamental of these, includes a shared conception of the good that explains their "point," part of which consists in (some version of) the freedom of social members. Because acting in accordance with such a conception is constitutive of the activity in which institutional life consists, the *functions* of institutions – including a conception of their healthy functioning – are accessible, if imperfectly, to the agents on whose activity those functions depend.

Chapter 7 briefly treats two social philosophers who directly influenced Durkheim: Auguste Comte and Herbert Spencer. Although Durkheim's position is more sophisticated and compelling than his predecessors', examining their theories serves to introduce several themes that play a key role in his social philosophy. Looking at his predecessors' use of the society–organism analogy, their versions of functional explanation and functional analysis, and their conceptions of what a scientific sociology must be like will help us to understand not only the content of Durkheim's positions but also some of the arguments – formulated in conversation with Comte and Spencer – behind his claims. This chapter distinguishes three types of functional explanation employed by Comte that shape the discussion of Durkheim's method in the following chapter, especially with regard to what I call functional analysis. Perhaps most important, the chapter argues that Comte and Spencer rely too heavily on the society–organism analogy, leading to an overly biologistic understanding of the types of normative critique available to social pathologists.

Chapter 8, the first of three devoted to Durkheim, examines his version of functionalism in social theory. I reconstruct his position with an eye to defending it as far as possible and to determining which aspects of it are worth retaining for a contemporary understanding of social pathology, including, most fundamentally, the functionally organized character of human societies. Focusing on his claims regarding the moral function of the division of labor, I examine the tortuous epistemological issues bound up with his ascriptions of functions to specific features of society, including the relation between functional explanation and functional analysis. I argue that the method underlying his functional analysis is best understood as a complex form of holism whose claims depend less on single facts and individual arguments than on the plausibility of the whole picture of society that emerges from a variety of mutually reinforcing arguments, empirical facts, interpretive suggestions, and analogies. In this respect Durkheim's method for ascribing functions to social phenomena bears similarities to other *interpretive* enterprises, from the reading of texts to (even) the construction of theories in the natural sciences.

Chapter 9 continues my discussion of Durkheim by explaining his understanding of moral facts and the conception of social solidarity at the core of his account of the division of labor's function in organized societies. These views expand on Rousseau's understanding of human society as normatively constituted – governed by rules accepted as authoritative by social members – and also introduces the idea, of central importance to my account, that healthy social institutions serve moral and not merely "useful" social functions. On the basis of these discussions I lay out the resources Durkheim has for conceiving of social pathology and examine in detail the modern pathology most important to him, anomie. Finally, I reconstruct his understanding of what is bad about social pathology – why social members should care about whether their society is ill.

Chapter 10 reconstructs Durkheim's conception of sociology as a science of morality, which bears similarities to Marx's historical materialist account of morality but in addition claims to legitimize the moral systems whose existence it explains. I distinguish three tasks of Durkheim's science of morality and conclude that Durkheim does not adequately explain how historically specific moral systems can claim a *moral* authority that does not reduce to the narrowly functional value they have for social reproduction. Finally, looking ahead, I suggest that Hegel's conception of spirit offers more promising resources for doing precisely this by understanding morality and social reproduction as inextricable aspects of social life, neither of which can be reduced to the other, and by conceiving of the moral ideals of later societies as rational responses to crises, both functional and moral, encountered by earlier societies, where the idea of a rational response to such a crisis plays a key role in justifying the later moral ideals under question.

Chapter 11 begins my discussion of Hegel's social philosophy with an extended examination of objective spirit. This concept is central to Hegel's social ontology because it specifies the kind of being, or reality, characteristic of the social world and distinguishes it from other domains of reality, such as nature and subjective spirit (or mind). I first explain what objective spirit means for the most compelling exponent of such a view of social reality today, Vincent Descombes – borrowing as well, but to a lesser extent, from Durkheim and John Searle. I end with an overview of some respects in which Hegel's view goes beyond Descombes's. This chapter distinguishes four claims espoused in some version by all of the major positions I consider in this book: (i) There is a form of mindedness that exists outside the consciousness of individual social members; (ii) the externally existing mind, embodied in social institutions, is in some sense prior to the individuals whose lives are ruled by them; (iii) social reality depends on a collective acceptance of its institutions' normative rules; and (iv) such rules constrain what social members do but also expand their practical possibilities and hence enrich their agency.

Chapter 12 continues the discussion of Hegel by examining his characterization of human society as the "living good." As the term "living" suggests, this aspect of his position can be read as his distinctive take on the familiar analogy between human societies and living organisms. For him that analogy implies that human societies both incorporate the processes of life (in carrying out the activities necessary for material reproduction) and mirror them in the sense that social and biological life exhibit a similar structure. The latter claim appeals not only to the functional specialization of human societies emphasized by Plato and Durkheim but also to the idiosyncratic Hegelian thesis that the processes of biological and social life are like those of "subjects," the principal characteristic of which is to maintain itself by positing "contradictions" internal to itself and then negotiating them in a way that establishes its own identity without completely abolishing the internal differences it has posited. (What this means will become clearer in the chapter itself.) If Hegel emphasizes the continuity between life and social being, he also insists on the differences between mere animal organisms and spiritual beings (including human societies): most importantly, the presence of self-consciousness and the capacity for freedom. The realm of (objective) spirit, then, consists in cooperative processes of life that are imbued with ethical significance deriving from their potential to be consciously self-determined, a potential that the life processes of mere animal organisms lack; thus, spiritual activities are life processes that simultaneously aim at realizing the freedom of those who carry them out. At the end of this chapter these ideas are fleshed out by examining how Hegel's famous master–slave dialectic brings to light the fundamental elements and structure of any set of human relations that count as a society.

Chapter 13 examines Hegel's understanding of both animal and mental illness and, making use of the concepts of objective spirit and the living good, extrapolates an account of various conceptions of social pathology that his social philosophy licenses. It argues that for Hegel social pathologies should be understood not only in terms of impaired functioning generally or as imbalances among specialized functional spheres, but also as ways in which society fails in the spiritual task of enabling its members to relate to life in the mode of freedom. Specific forms of such pathologies include (but are not exhausted by) social practices losing their spiritual features and becoming indistinguishable from processes of mere life; social impediments to the realization of practical selfhood, such as inadequate sources of recognition or the generation of infinite, unsatisfiable desires; forms of ideology involving a mismatch between what social members do and what they take themselves to be doing in their social practices; the failure of social life to bring together the ends of life and those of freedom, including cases where social participation becomes merely a means for staying alive rather than a site of freedom; and socially caused impairments of individuals' ability to reconcile themselves to the fact of death. The final paragraphs of this chapter point to the form of immanent critique found in Hegel's account of bondsman and lord as a promising solution to one problem encountered by Durkheim's science of morality, namely, its inability to provide an *ethical* justification of social norms that avoids reducing the point of morality to its mere functionality for social reproduction.

I am indebted to many individuals and institutions who have aided me in the writing of this book. I am sure that I have failed to mention some of them here, and to them I apologize. Most recently, I had the good fortune to receive extensive anonymous comments from two attentive and insightful readers of the manuscript whose help was enlisted by Cambridge University Press. I have also benefited from philosophical conversations with many individuals: Mark Alznauer, Barbara Carnevali, Maeve Cooke, Mattia Gallotti, Amanda Greene, Axel Honneth, Rahel Jaeggi, Jan Kandiyali, Bruno Karsenti, Philip Kitcher, Richard Moran, Karen Ng, Andreja Novakovic, Lea-Riccarda Prix, Eva von Redecker, Isette Schumacher, Achille Varzi – and undoubtedly many others.

Two institutions provided the generous support of research stays that greatly enhanced my ability to make progress on this project: the Center for Humanities and Social Change, Humboldt University (Berlin) and School for Advanced Studies in the Social Sciences (EHESS-Paris). In addition, many academic audiences patiently endured lectures in which, in early phases of the project, I tried to work out ideas not yet fully formed, including at Boston University, City University of New York, Colgate University, Columbia University, DePaul University, School for Advanced Studies in the Social Sciences (EHESS-Paris), Georg August University (Göttingen), Georgetown University, Goethe University (Frankfurt), Harvard University, Humboldt University (Berlin), Indiana University, Institute for Social Research (Frankfurt), International Hegel Conference (Göttingen), International Hegel Congress (Stuttgart), Kansas State University, New School for Social Research, New York University Law

School, North Carolina State University, Oxford University, Princeton University, School of Higher Economics (Moscow), Stockholm University, Technical University (Darmstadt), Metropolitan Autonomous University (Mexico City), Catholic University (Lima), University Ca' Foscari (Venice), University of Lucerne, University of Lausanne, University of Leiden, University of California–Riverside, University of Cambridge, University of Essex, University of Georgia, University of Helsinki, University of Pittsburgh, University of Sydney, University of Toronto, University of Wisconsin–Milwaukee, and York University.

Hilary Gaskin, as always, provided invaluable advice throughout the writing and editing process. Finally, Eugene O'Keefe continually pushed me to make my prose more concrete and, more importantly, provided support of a nonacademic sort without which I might never have finished the book.

Note on Citations

Although I refer throughout to English translations of primary texts, *I have revised many quotations, often substantially*, in order to render the original passages more accurately.

I have made use of the following abbreviations in citing the texts I refer to most frequently:

Vincent Descombes

IM *The Institutions of Meaning* (1994)

Émile Durkheim

DLS *The Division of Labor in Society* (1984)
[cited by page number, followed by page number in French]
II "Individualism and the Intellectuals" (1969)
[cited by page number, followed by page number in French]
ME *Moral Education* (1961)
[cited by page number, followed by page number in French]
RSM *Rules of Sociological Method* (1982)
[cited by page number, followed by page number in French]
S *Suicide: A Study in Sociology* (1997)
[cited by page number, followed by page number in French]
SP *Sociology and Philosophy* (2014)

Georg Wilhelm Friedrich Hegel

EL *Encyclopedia Logic* (1991)
[cited by section number, where *A* refers to the Addition to the cited section]

PhG	*The Phenomenology of Spirit* (2018) [cited by paragraph number (e.g., *PhG*: ¶182)]
PhM	*Philosophy of Mind: Part Three of the Encyclopedia of the Philosophical Sciences* (1971) [cited by section number, where *A* refers to the Addition to the cited section]
PhN	*Philosophy of Nature: Part Two of the Encyclopedia of the Philosophical Sciences* (2004) [cited by section number, where *A* refers to the Addition to the cited section]
PhR	*Elements of the Philosophy of Right* (1991) [cited by section number (e.g., *PhR*: §270A), where *A* refers to the Addition to the cited section]
VPR	*Die Philosophie des Rechts: Die Mitschriften Wannenmann* (1983)

Immanuel Kant

CJ *Critique of Judgment* (1987)

Karl Marx

Cap.	*Capital*, Vol. 1 (1992)
MER	*The Marx-Engels Reader* (1978)
MEGA	*Karl Marx, Friedrich Engels Gesamtausgabe* (1975) [cited by volume and page number (e.g., *MEGA*: XXIII.529)]

Friedrich Nietzsche

GM *On the Genealogy of Morals* (1989)
[cited by essay and section number (e.g., *GM*: III.13)]

Jean-Jacques Rousseau

DI	*Discourse on the Origin of Inequality* (1997a) [cited by English page numbers followed by French]
OC	*Oeuvres Complètes*, Vol. 3 (1964)

PE *Discourse on Political Economy* (1997b)
[cited by English page numbers followed by French]
SC *The Social Contract* (1997c)
[cited by book, chapter, and paragraph numbers (e.g., I.3.ii)]

John Searle

CSR *The Construction of Social Reality* (1995)
MSW *Making the Social World* (2010)

CHAPTER 1

Can Societies Be Ill?

This chapter offers some preliminary thoughts about the general concept of social pathology and its usefulness for social philosophy. The first section distinguishes five conceptions of social illness that differ from the one I endorse in this book. Following that, I discuss various advantages and disadvantages of the concept of social pathology. Finally, analyzing a little-known example of Rousseau's, I illustrate various possible features of a sick society that help to illustrate how that concept aids us in understanding and evaluating social reality.

Rival Conceptions of Social Pathology

It is necessary to say something first about the sense in which I speak of social pathology in this book since the term has various meanings – I note here five – that I want to distinguish from my own usage. There is, for example, a very simple conception of social pathology, according to which (i) *a society is "ill" whenever a significant number of its members is ill.*[1] (I will often place "ill" and its synonyms in quotation marks to remind us that those terms are used metaphorically.) One might claim to find this conception of social pathology in Richard G. Wilkinson's *Unhealthy Societies: The Afflictions of Inequality*, which marshals empirical evidence for the claim that developed societies with extensive economic inequality tend to have higher rates of illness among their members – including among the better off – than more egalitarian societies (Wilkinson 1996: 3). In this case the unhealthy societies mentioned in the book's title are simply those with large numbers of physiologically unhealthy members.

Two features of this simple conception of social pathology distinguish it from the one I employ here. First, it makes illness in the social

[1] Honneth (2014a: 684) mentions and rejects this conception.

realm derivative of nonsocial illness. In Wilkinson's case the illnesses in question are physiological, but one could devise a similar conception of social pathology that referred to mental illness instead.[2] Such a conception does not countenance a distinct way in which societies can be ill that is not traceable back to the physiological or mental illnesses of its members. Second, this conception of social pathology requires no understanding of human society as anything more than a collection of individuals. It offers an aggregative picture of social pathology, in which a society's being ill consists simply in the logically prior fact that many of its members are ill. Such a view does not depend on any specifically *social* understanding of human society, of how it is structured, or of what its distinctive functions consist in.

In fact, however, Wilkinson's unhealthy societies come closer to exemplifying social pathology in the sense I defend here than the previous paragraph suggests. For his thesis is not only that unhealthy societies are those that contain many physiologically ill individuals but also that such societies foster those illnesses. According to this second, more complex conception of social pathology, (ii) *societies are ill just in case they play a substantial causal role in making a large number of their members physiologically or mentally ill*; that is, the illnesses of social members indicate social pathology only if those illnesses have social causes. In Wilkinson's case the relevant cause of the illnesses found in the individuals of an unhealthy society is a social phenomenon – economic inequality – which both *consists in* social relations (not in properties individuals possess on their own) and *is explained by* supraindividual features of society, such as its economic structure or its laws and practices.

I have no objection to regarding Wilkinson's unhealthy societies as pathological since the illnesses of their members result from distinctly social factors. A society that makes many of its members ill qualifies as pathological, but its specifically social pathology resides in the social dynamics that produce the illnesses of its members. As I argue throughout this book, the socially pathological character of such a society lies in its *dysfunctional* dynamics, not in its members' physiological or mental illnesses. To see why Wilkinson's examples count as social pathologies, compare them with a society with inordinately high rates of physiological or mental illness where this is due to (nonsocially caused) environmental

[2] One might interpret Freud as suggesting a conception of social or cultural pathology of this type: late nineteenth-century Viennese society was sick because of a high incidence of neurosis among its (especially female) population.

changes. As this case suggests, the mere fact many persons in a society are ill does make that society ill.

Wilkinson's unhealthy societies are not, however, paradigm cases of sick societies, for they suggest too narrow a picture of social pathology. This is because, like the first conception of social pathology, his account presupposes a picture of what health in individuals consists in and then speaks of social pathology whenever social conditions negatively affect the health of social members so conceived. As I treat the concept here, social pathology does not require that individual social members themselves be ill, either mentally or physically. To see this point, consider Marx's account of alienation. Workers are alienated in capitalism, but this does not imply that they are ill. What makes capitalist alienation a social pathology is that it is a systematic result of capitalism's class structure and its mode of organizing production and accumulation. As noted above, social pathology consists in a dysfunction at the level of social structure or in a society's constitutive dynamics. Whereas anything that counts as a social pathology must be bad for (at least some) social members, the way in which individuals are negatively affected generally does not take the form of illness. To take another example from Marx: the inherent tendency of capitalism to produce recurring crises qualifies as a social pathology but not because those crises make individuals sick (even if illness might be one effect of the unemployment produced by crises). Such crises produce alienation and poverty for many – and would not constitute a social pathology if they did not have some such negative effects – but they are relevant to a theory of social pathology because they are signs of dysfunction in social processes, not because they make individuals ill.

My claim that social pathologies must be bad for at least some social members does not, however, imply that socially caused *suffering* is either a necessary or a sufficient feature of social pathology. Some philosophers of medicine place suffering at the center of their accounts of illness, and certain social theorists likewise regard (iii) *"social suffering" as the hallmark of social illness.*[3] My objection to this conception is not that suffering is a rare feature of social pathologies – it is not – but that it blinds us to the fact that felt suffering need not be present in them. (And, as Durkheim notes in making the same point, plenty of organic illnesses do not involve suffering either [RSM: 87/50].) Here, again, think of Marx's claim that although

[3] Honneth (2007: 686) rejects this conception as well. For a treatment of social suffering and its relation to social pathology, see Renault (2017).

the bourgeoisie is alienated in capitalism, it "feels confirmed and at ease in its alienation, experiencing it as its own power" (MER: 133/MEGA: II.37). Conversely, taking social suffering as a sufficient indicator of social pathology yields false positives: when I walk in my neighborhood in Berlin and see passersby's expressions of pained outrage at seeing so many black faces or hijab-wearing women in "their" streets, I am reminded that not all social suffering is a response to genuine social ills. (Moreover, as Nietzsche reminds us, some suffering, as in pregnancy, signifies growth and new possibilities rather than degeneration or dysfunction.)

None of this implies that theorists of social pathology should avoid taking the expressed suffering of social members seriously but only that one cannot assume that experienced suffering in the social domain by itself indicates social pathology. Rather, social theorists (some of whom themselves belong to suffering groups) must interpret suffering and judge it in light of normative criteria – those implicit in some inclusive version of the idea of a good human life – that go beyond experienced physical or psychic pain. This may sound harsh or paternalistic, but it is nevertheless true that social suffering must be articulated and made comprehensible to those who do not suffer from it if social transformation is to occur. In the twentieth century this was seen, by the "sufferers" themselves, as a principal task of the Civil Rights movement in the United States, of women's liberation, of unionization drives, and of lesbian and gay politics. In all these cases those suffering from the relevant injustices and pathologies succeeded in translating their suffering into terms that those not directly afflicted came to understand as ethically compelling. The mere expression of social suffering that was previously invisible to others played a major role in the progress achieved by these social movements, but even more important was their ability to articulate the meaning of their suffering to others, showing it to be an injustice or, more broadly, a grievous impediment to living a good human life.

A fourth conception of social pathology conceives of society on the model of an individual human being and of its illnesses as (iv) *large-scale versions of physiological or mental illnesses that afflict individual humans.*[4] Plato's positing of an isomorphism between polis and soul moves in this direction, as does the famous frontispiece of Hobbes's *Leviathan* depicting the commonwealth as a super-sized human individual. Although strange, it is not unheard of to depict social pathologies as near-literal analogues

[4] Honneth (2014a: 684) describes this conception as "the collective understood as a macro-subject."

to the physiological illnesses suffered by human bodies, as does the *Leviathan*'s notorious (and entertaining) Chapter XXIX. It is more common and more plausible, however, to think of social pathologies conceived of on this model as collective forms of mental illness, as in popular invocations of collective psychosis, when describing contemporary political discourse in the United States or in Carl Jung's conception of collective neurosis (Jung 1964: 85). On this model of social pathology, human society is conceived of as an organism (or collective subject) and is taken to be ill when it falls short of the standards of health applicable to individual organisms (or subjects). Although I do not take a stand on the plausibility of accounts of collective mental illness, I distance myself from the general model of social illness that informs them. None of the classical instances of social pathology rely on that model, and, more important, thinking of the relevant functions of social life as close analogues to those of human bodies or minds diverts our attention from real social phenomena and from the specifically social dysfunctions in which, according to my account, social pathologies consist.

Finally, there is a loosely defined conception of social pathology that once enjoyed wide currency in academic circles, especially in the United States. Developed by sociologists in the first half of the twentieth century, it achieved popularity in the 1960s and 1970s outside academia, when it became common in the press and among social workers to refer to (v) *a haphazardly collected group of "social problems"* – including poverty, crime, drug abuse, "promiscuity," and racism – as social pathologies (Mills 1943: 165–80).[5] Pathology in this conception is defined primarily as a deviation from prevailing social norms and is often understood as a failure on the part of individuals to "adapt" or "adjust" to those norms. The norms in question include such "virtues" as thrift, sobriety, heterosexual monogamy, discipline in work, and commitment to family, all of which are regarded as essential to social order and as conditions to be reproduced rather than called into question.

Even if such pathologies are condemned primarily because of the threat they pose to the smooth functioning of society, they are also taken to be bad for the afflicted individuals, and the social work inspired by this conception of social pathology was surely motivated more by the desire to improve the lives of individuals than by the larger aim of putting society in order. For this reason, this conception of social pathology goes beyond the

[5] See, for example, the "social disease" about which the Jets sing in "Hey, Officer Krupke!" in *West Side Story*.

alternatives discussed above in expanding the idea of social illness beyond a narrowly physiological or psychiatric definition to include social phenomena that appear as pathological only in the light of broader normative criteria for a good or flourishing human life. Poverty, for example, counts as pathological for this view not because it makes the poor sick, physically or mentally, but because (as Hegel pointed out) it prevents them from achieving basic goods – satisfying work, self-esteem, forms of relaxation and enjoyment – available to the better off in their society. In relying, however implicitly, on some idea of the human good and of how social conditions can promote or hinder it, this conception of social pathology takes us a step away from the original idea of illness. This, however, is not its defect. (The fact that homosexuality, for example, was long taken to be one of these pathologies should remind us, however, of the dangers inherent in judging social conditions to be pathological according to prevailing criteria for a good human life, as well as of the fallibility of our present judgments as to what fails to meet those criteria.) For, as noted above, a central claim of this book is that an adequate conception of pathology in the social domain must be normative in respects that go beyond the narrower ideas of physiological and mental illness.

I do not deny that some of the problems picked out by this conception of social pathology are indicative of pathologies, but I reject the loose conception of social illness it employs. One respect in which that conception is insufficiently social shows itself in the fact that it was often interpreted as attributing responsibility, even moral blame, for social pathologies to the "maladjusted" and undisciplined individuals whose (for example) "weak ego structures" (Rigdon 1988: 113) prevented them from complying with social norms. The tendency of this conception of social pathology to moralize social problems is surely connected to its deeper theoretical deficiency, namely, the implicit assumption that a list of "social problems" constitutes an account of social illness. My chief objection to this once-popular way of regarding crime, poverty, and drug abuse, then, is not that it is excessively normative but that it lacks a sufficiently complex conception of human social life to grasp the problems it concerns itself with as social pathologies in the more robust sense in which I employ the concept here. For on this conception, diagnoses of social pathology are made independently of any specific understanding of a society's structure or basic functions. The problem with this approach is not that it regards alcoholism or crime or high rates of suicide as social ills but that, in the absence of an account of a society's structure and basic functions, its diagnostic procedures amount to little more than compiling a list of diverse problems, the causes of which it then locates in a variety of social conditions. As C. Wright Mills pointed out, this way of conceiving of social pathology is capable only of "collecting and dealing in a fragmentary way with scattered problems" (Mills 1943: 166). Mills can be taken to imply that an enlightening conception of social pathology must do more than uncover a diversity of ways in which individuals are afflicted by social conditions; it should also be able to say how a society in its basic structure is deformed or unbalanced, or how its essential functions are disrupted, impaired, or misaligned. One way of expressing this point is to say that a high rate of alcoholism might indeed be a symptom of social pathology (rather than, say, the effect of cold, dark winters) but that the pathology resides in a deeper functional deficiency of society that explains the high incidence of alcoholism and reveals its connections to other social ills bound up with the same pathology.

Disadvantages of the Concept of Social Pathology

It is important to acknowledge that there are good reasons – conceptual, rhetorical, and ideological – for approaching the idea of social pathology with skepticism. There can be no doubt, for example, that the concept has been used, by philosophers and politicians alike, in the service of projects that are intellectually and morally objectionable. To cite the most egregious instance: Nazi ideology made extensive use of the idea of a diseased society to impress on its adherents the need to attack with violence those parts of the body politic – those groups of human beings – in which society's malady, imagined varyingly as syphilis, cancer, or tuberculosis, was thought to reside (Sontag 1978: 82–3).[6] Along with the notion that sickness must be treated by violent means, the baffling idea that agents of disease are morally culpable, and therefore deserving of punishment, appears to be a persistent element of our (mostly unconscious) attitude to the basic fact of our vulnerability to illness, and this creates a standing potential for corrupting our responses to whatever phenomena we diagnose as social pathologies. Some of the oddness of our moralistic attitude to illness is evident in the fact that it often coexists with its precise opposite: sometimes a condition's being regarded as an illness functions to shelter the ill from moral condemnation, as in the thought, regarded not long ago as progressive, that homosexuals are not morally depraved but sick, implying

[6] In an interesting example of how metaphors can travel in the reverse direction, plague was understood in the Middle Ages as a sign of moral pollution, which, requiring a scapegoat, led to massacres of Jews.

that medical treatment rather than persecution is the appropriate response to "abnormal" sexual orientation. In any case, the line between regarding persons as ill and perceiving them as depraved is easily traversed, as can be seen in the fact that academic sociologists in the United States today are most likely to associate the discourse of social pathology with "social deviance," a term whose problematic connotations hardly need to be pointed out. The fact that perceiving some condition as an illness has the potential to engender hostility, disgust, and condemnation is an important reason for theorists of social pathology to be scrupulous in insisting that societies or institutions, not individuals, are the bearers of the illnesses their theories diagnose.

Beyond this, the popularity of social Darwinism in the latter decades of the nineteenth century surely accounts for some of the disfavor into which the concept of social pathology and the society-as-organism analogy has fallen today. Social Darwinists gave a distinctive ideological twist to the idea that society is akin to a living organism. For them, this analogy implied that the key to establishing a science of society lay in appropriating the methods and outlook of the newest advance in biological science: Darwin's theory of evolution. Although the founder of this school, Herbert Spencer, developed his main ideas before the publication of *The Origin of Species*, much of the influence social Darwinism enjoyed in subsequent decades depended on the mistaken perception that he and his followers were applying the principles of Darwinian science to social life.

Spencer's social theory is based on the idea that, despite superficial differences, biological and social "organisms" are subject to the same laws of evolutionary development. The details of this theory need not concern us, apart from the most significant respect in which Spencer's conception of the evolutionary process diverges from Darwin's: in the case of the social organism, Spencer posits an end-point – a state of "equilibration" (Spencer 1969: 141) – at which perfect adaptation has been achieved and no impetus for further development is present. Although this end-state is supposed to be one of social integration in which peaceful, industrial activity becomes society's chief occupation, it must be preceded by an ongoing "struggle for existence" among antagonistic social units. Unfortunately, it was this idea that most captured the imagination of Spencer's followers, who elevated it into a full-blooded ideology in support of laissez-faire social policies they took to serve the "survival of the fittest" (a term coined by Spencer, not Darwin). Government aid to the poor, state-financed education, public health measures, even the regulation of commerce – all were regarded as misguided attempts to interfere with the natural workings of society,[7] which, if left to its own devices, would eliminate the least "fit" of society's members (Spencer 1969: 379). (The pseudoscientific character of social Darwinism is especially visible in its interpretation of the "fittest." Spencer's most prominent American follower, William Graham Sumner, took the fittest to be, as one commentator puts it, the frugal, tax-paying, middle-class man who "went quietly about his business, providing for himself and his family without making demands upon the state" [Hofstadter 1959: 64].[8])

There are also philosophical reasons for avoiding the concept of social pathology, and they, too, must be acknowledged upfront. The most obvious of these reasons is ontological: the concept of illness belongs to the study of animal organisms, and societies – so the objection – are not animal organisms nor sufficiently like such organisms that the concept of illness could be meaningfully applied to them. Of course, that societies are not animal organisms is plain enough – and recognized by all serious theorists of social pathology – but this alone does not settle the issue of whether the two are so dissimilar that categories applicable to the latter have no value for understanding the former.[9] Here, too, it is worth noting that the figures I rely on here reflect extensively on the differences between organisms and societies and go to great lengths, if not always successfully, to do justice to them. The widespread assumption that the entire tradition of theories of social pathology can be dismissed by the observation that human societies are not animal organisms betrays either a penchant for easy philosophizing or an ignorance of how and why such theories have employed the concept of social illness.

It is instructive that some of the figures mentioned in the Preface – Machiavelli, Marx, and Habermas, for example – appear unable to dispense with the idea of social illness even when they generally avoid conceiving of human society on the model of an organism. Of course, the more one distances oneself from the organism analogy, the more difficult it becomes to see why one should employ the idea of social illness at all, especially if one wants to avoid emptying it of all content by referring to whatever one disapproves of as "sick" (Sontag 1978: 74). In line with this thought, I will argue that there are good reasons to employ the concept of social pathology

[7] The implications for international relations are no less severe: "Progress of … nature is everywhere manifested in the subjugation of weaker tribes by stronger ones" (Spencer 1969: 316).

[8] For Darwin, fitness could be defined only in terms of success in biological reproduction (making, perhaps, the sexually "promiscuous" fitter than the chaste).

[9] For a nuanced critique of the society-as-organism analogy, see Laitinen and Särkelä (2019).

and that doing so commits one to thinking of human societies, not exactly as animal organisms, but as "alive" – as "living" beings – in some nonliteral but still meaningful sense. Or, formulated differently, an important part of my task is to articulate in which respects societies are like living organisms (and in which respects they are not), since in the absence of any such similarities, speaking of sick societies would indeed be empty talk (Honneth 2014a: 701). Before addressing this issue further (in Chapter 2), it is important to consider why, despite its disadvantages, the concept of social pathology has an important role to play in social philosophy.

Advantages of the Concept of Social Pathology

Diagnosing a society as ill involves claiming sometimes *more than*, sometimes something *other than*, that it is unjust. This gives theories of social pathology two advantages: First, they can draw our attention to social phenomena worthy of critique to which theories focused exclusively on justice are blind; and, second, they have at their disposal critical resources beyond those employed by most liberal political and social philosophy. It is not that the category of injustice is always irrelevant to diagnoses of social pathology but rather that, even when it is relevant, conceiving of the social deficiencies in question only as injustices underdescribes them. Consider the example of global warming. It is certainly appropriate to regard human-caused global warming as an injustice (to future generations, or to contemporaries who only suffer its effects while others profit from it). It would be odd, however, to take this as an exhaustive description of the problem. It is hard to avoid the impression that there is something sick – or perverse, or gravely awry, but in any case something more than unjust – about social practices for which we are responsible that systematically thwart fundamental human ends, in this case ends as simple as those deriving from our biological nature. The appearance of illness becomes only stronger when one considers that, after becoming aware of global warming, we continue, and even intensify, the very practices that threaten our species' survival.

One way of bringing this issue into focus is to abstract from those to whom injustice is done – future generations and the global poor – thereby removing injustice from our picture of why global warming is worthy of critique. What is left then could be described as a systematic thwarting of the – in this case, self-preservative – ends of the very agents whose activities produce global warming or, alternatively, as a systematic undermining of the conditions of those agents' good and, ultimately, of their very agency and lives. What makes human-caused global warming a pathology and not only an injustice is that it is the result of a *social dynamic* – a self-reproducing nexus of collective practices – that, apart from its connection to injustice, *diminishes the good* (or impedes the ends) of those who participate in those practices.[10]

The first of these points implies that theories of social pathology take as their object not isolated human actions but ongoing social processes that constitute a dynamic with a coherent logic and point. This is the idea behind Max Weber's distinction between isolated, sporadic acts of profit-seeking and "capitalistic enterprise," the distinguishing feature of which is that it is a continually repeated series of actions unified by a specific aim (the maximization of profit) and exhibiting a certain structure (economic efficiency, as determined by market-oriented calculations of profitability) (Weber 1992 [1905]: xxxi–xxxii). Social processes such as these have not only ends; they also exhibit a characteristic logic or dynamic that may or may not be consciously apprehended by those whose activities sustain it. For this reason, diagnoses of social pathology rely on a dynamic understanding of how social processes work (or function) and how they reproduce or transform themselves over time; in other words, social critique on this model requires social theory, and the primary objects of such critique are not individual actions but social practices and institutions. The social forces behind contemporary global warming, for example, cannot be understood as the aggregative result of independently undertaken actions on the part of thoughtless or greedy individuals. Rather, global warming is the outcome of a system of production and consumption that follows a logic of its own – bound up with the aim of capitalist accumulation – that cannot be countered without substantially reforming institutions.

The second point (that, apart from considerations of justice, social pathologies diminish the good or impede social members' ends) suggests that in most cases such disorders are – to employ a familiar but not fully transparent distinction – failures in realizing the good, broadly construed, rather than in achieving the right. Examples of such failures include but are not exhausted by the following phenomena: felt estrangement from

[10] One possible response to this aspect of theories of social pathology is to claim that such theories *expand* our conception of justice (Honneth 2014b: 3–19). In my view, it is preferable, and truer to ordinary usage, to retain a relatively narrow concept of justice – bound up with ideas of "mine and thine" and of what we owe to one another – and to embrace a broader range of (different) ethical values. Making "justice" mean many things decreases the precision of critique. Hegel appears to agree: In the context of domestic right he speaks of justice (*Gerechtigkeit*) only in relation to "the administration of justice," which is limited to abstract right and civil society (PhR: §§99A, 214).

social institutions; having a purely instrumental attitude to one's social activity; failing to realize the distinctive goods available through participation in collective enterprises; missing out on satisfying forms of work and self-esteem; and embracing values and ends the pursuit of which is self-undermining or inimical to one's own good. In general, then, diagnosing a society, practice, or institution as ill involves ascribing to it specific ends bound up with conceptions of human flourishing that are somewhat "thicker" than those typically admitted by liberal theories of justice.[11] A social theory that makes pathology a part of its critical arsenal relies on a vision of the social, according to which human societies cannot be adequately grasped or evaluated without attributing ends to their practices and institutions that – because connected with ideals of human flourishing – are broadly ethical in nature. For this reason the concept of social pathology is less restricted to "ends set by nature" than is the concept of illness in the case of merely animal beings;[12] a diagnosis of social pathology is always in part an ethical critique.

To show that the concept of social pathology can illuminate deficiencies in social life not capturable by discourses confined to the categories of legitimacy, justice, or moral rightness, it will be helpful to return to Marx's critique of capitalism (which is inseparable from an ambitious theoretical account of how capitalism works as a system, reproducing and transforming itself in accordance with its own logic.) Of the types of critique attributable to Marx – that capitalism is alienating, exploitative, self-undermining, and that it ultimately fails to develop human productive forces as well as other practically available forms of society would – only one has a natural home in contemporary Anglo-American social or political philosophy: only the claim that capitalism requires the systematic exploitation of workers lends itself to reformulation in the language of justice that contemporary liberalism takes as the central category of social

[11] Among liberal political philosophers, Rawls comes closest to grasping some of the phenomena I understand as social pathologies. In emphasizing the social bases of self-respect; in applying standards of justice to nonpolitical institutions; in relying on some ("thin") conception of the human good; Rawls approximates some of the normative criteria employed by theories of social pathology. Even so, certain topics important to more comprehensive social theories remain untheorized by Rawls, for example, how labor should be organized so as to avoid alienation; how certain injustices are systematically reproduced by ongoing social dynamics; and how the ideals of free citizenship relate to different values realizable in nonpolitical social spheres.

[12] Throughout this book the "merely" in "merely animal" means "only" rather than implying "lower than." Of course, the contrast between the merely animal and the human that runs throughout this book attributes a "higher" form of being to the latter. Nothing in this implies, however, that animals may be treated "as mere means" to satisfying humans' needs or desires.

critique. One of the strengths of Marxist theory is that it sheds light on broadly ethical problems of modern society typically ignored, even today, by justice-oriented political philosophies, where topics such as alienation, reification, self-defeating social dynamics, and justice in nonpolitical social spheres go largely unmentioned. To take only the first example, the problem with alienated labor is not primarily that it is unjust.[13] The problem, rather, is that the conditions under which such labor is carried out make it impossible for laborers to realize spiritual goods – recognition, self-esteem, successful execution of complex tasks, the satisfaction of producing for others, and so on – that can be had from labor in societies with highly developed productive forces.

Merely to decry such conditions as unjust not only says too little about what is problematic about them; it also fails to grasp how they are grounded in the structure and logic of existing institutions rather than being sporadic or contingent. This points to a further respect in which theories of social pathology go beyond mainstream social and political philosophy: in addition to employing a broader set of normative standards, such theories aspire to uncover the social dynamics that explain why the pathologies they diagnose are more than accidental. One might say that theories of social pathology aspire to distinguish symptoms from underlying pathologies and that their diagnoses go beyond mere classification to include an account of the social forces or underlying structural conditions responsible for producing the symptoms at issue. This explanatory aspiration means that a diagnosis of pathology typically carries implications about the treatment likely to eradicate or ameliorate the diagnosed condition; for the social pathologist, as for the physician, diagnosis goes hand in hand with practical orientation.

Finally, theories of social pathology differ from much (but not all[14]) contemporary political philosophy in eschewing a priori justification of the critical norms they employ, seeking them instead *within* the social practices they investigate. This point raises the tricky question of how

[13] Durkheim can be read as suggesting that injustice is a necessary condition of alienation, even if the two are not identical (DLS: 407/403). I suspect, however, that that claim is false.

[14] Here, too, Rawls's theory of justice is closer to the theories I endorse than other examples of contemporary political philosophy, insofar as it reconstructs the norms informing an already existing tradition of political liberalism rather than proceeding foundationally. The device of the original position can look like an attempt to find a free-standing foundation for standards of justice, but only if one forgets the role played by reflective equilibrium, which allows features of the original position to be revised if the results following from it diverge too much from the considered judgments of actual participants in the practices whose logic is being reconstructed (Rawls 1999: 18–19, 42–5).

the diagnosis of social pathologies relates to what has come to be called immanent critique, as well as – since some notion of *contradiction* is central to such critique – to what extent social pathologies involve contradictions in the sense in which Hegel and Marx employ that concept. These questions would be easier to answer if immanent critique and contradiction were univocal concepts.

The *locus classicus* of the method of immanent critique is Hegel's *Phenomenology of Spirit*, but, in a different form, it plays a role in his *Philosophy of Right* as well: his critique of widespread poverty there, for example, rests on the idea that basic features of civil society impede its ability fully to realize the ideals – such as self-reliance and finding meaning in one's work – that animate its own workings. In Marx the closest analogue to this form of immanent critique is *Capital*'s account of how the ideals that justify the wage–labor relation and the appropriation of surplus value – "freedom, equality, property, and Bentham" (Cap. 280/MEGA: XXIII.189) – are necessarily realized only one-sidedly in capitalism. In both cases, one could speak of contradictions – between normative aspirations that inform institutional life and the internal impossibility of fully achieving them – but for Hegel and Marx "contradiction" tends to imply something more as well, namely, an internal potential for transformation, or the presence of real forces that have the capacity to resolve the contradiction at issue and, in doing so, produce a new, "higher" social configuration in place of the old. (This dimension of "contradiction" has an analogue in the method of immanent critique employed in the *Phenomenology*, but it is not part of the *Philosophy of Right*'s account of poverty.) This conception of contradiction in the social domain finds its clearest expression in Marx's account of the "contradictions of material life" – the conflicts between the forces of production and the relations of production – that explain epochal change in his vision of historical materialism:

> No social order ever perishes before all the productive forces for which there is room in it have developed; and new, higher relations of production never appear before the material conditions of their existence have matured in the womb of the old society itself. Therefore mankind always sets itself only such tasks as it can solve; since … the task itself arises only when the material conditions for its solution already exist or are at least in the process of formation. (MER: 5/MEGA: XIII.9)

Diagnoses of social pathology, as I conceive of them here, depend on a form of immanent critique but do not necessarily regard pathologies as contradictions in this sense: uncovering a pathology in social life does not imply the presence of forces that will or could resolve the relevant dysfunction and lead to radical transformation "at a higher level." In other words, the metaphysical commitments of theories of social pathology are more modest (and realistic) than views of history that emphasize contradiction-based progress, and this in two ways: in rejecting the idea that dysfunctions carry within themselves the resources for overcoming them, and in denying that only radical transformation – revolution rather than reform – is genuine progress. In both respects theories of social pathology are closer to the diagnosis of medical illness than to "dialectical" social critique, for they presume nothing about the likelihood or potential for overcoming the illnesses they diagnose. Thus, illness in its social form, like physiological pathology, is sometimes cured; it sometimes leads to change or death; and sometimes it merely persists indefinitely. On this point, the view of social pathology I defend here is closer to the positions of Rousseau and Durkheim than to Marx's or Hegel's. The metaphysical modesty of such a view may be disappointing, but it also avoids the embarrassment of predicting social transformations that never come about.[15] (At the same time, forward-looking critical social theories ought to *attempt* the kind of analysis Marx provides in *Capital* of real social and material developments that might have the potential to remedy contemporary dysfunctions and to transform the social world in ways that make their recurrence less likely; although such analyses are important for guiding political action, they cannot be regarded as predictive or determinable with the precision of natural science.)

There is another aspect of the uncovering of social contradictions that, at least for Marx, has implications for the form social progress must take: *Capital*'s method of uncovering the "contradictory movement of capitalist society" – primarily in its account of the periodic cycles of capitalism that culminate in "universal crisis" – is said to be "revolutionary in its essence" (Cap.: 103/MEGA: XXIII.27–28). In other words, social contradictions in the sense at issue here are taken to apply not to superficial features of societies but to their deep structure,[16] which means that resolving those contradictions is closer to killing off the extant social "organism" than to curing it.[17] In this respect – in insisting on the revolutionary character

[15] The views expressed in this paragraph are heavily indebted to comments made by an anonymous reader of an earlier version of this chapter.

[16] I take these points from another anonymous reader of the book's manuscript.

[17] One could ask whether the contradictions in modes of knowing uncovered by Hegel's *Phenomenology* are revolutionary in a similar sense. Although they pertain to the deep structure of the modes of knowing considered, resolving their contradictions always involves incorporating elements of the

of social critique and of the political actions implied by it – Marx fits less comfortably into the social pathology tradition than other figures covered in this book. Nevertheless, as I argue in Chapters 3 and 4, both the language and concepts of social pathology critique figure prominently in *Capital*'s accounts of the nature of capitalist production and its inherent defects.

In any case, drawing too close an analogy between physiological and social illness can make diagnosis of the latter appear more conservative than it need be. Normally we take cures of physiological illness to aim at restoring the sick organism to its previous healthy state without transforming it or giving it new powers. If the same were always true of diagnoses of social pathologies, they would indeed be inherently conservative in an objectionable sense. Yet no major theorist of social pathology takes the concept to be restricted in this way – certainly not Durkheim, who, because of being associated (unfairly) with sociological functionalism,[18] has seemed to many the social pathologist most vulnerable to this charge. No doubt this difference between biological and social "organisms" can also be traced back to ontological differences between what Hegel calls spirit and mere life, especially to the intrinsically historical character of spiritual beings due to their self-conscious nature.[19]

While restoring society to a previous state of health is in principle a possible aim of theories of social pathology, it in fact plays little, if any, role in the theories I take most seriously here. Instead, there is a general presumption in these theories that remedying social pathologies requires real change that typically falls short of "revolution" but does not for that reason amount to "mere reform" that leaves the underlying causes of dysfunctions unaddressed. Like medical approaches to disease, which distinguish symptoms from underlying causes, theories of social pathology are not opposed to grasping social problems "at their root," but they do not assume that an adequate response to such problems requires complete extraction of the roots in question. A satisfactory response to poverty or economic inequality *might* involve abolishing the market economy and replacing it with production organized on a different basis, but it might also be remedied by substantial changes to the market economy, including revised conceptions

earlier stage rather than wiping the slate clean and beginning from scratch. Is this revolution or (substantive) reform?

[18] See my discussion of sociological functionalism in Chapter 2.

[19] I leave this thought undeveloped here, though resources for exploring it further can be found in Chapter 12 and, in more elaborated form, in Brandom (2007: 127–9).

of property and of the rights belonging to its owners,[20] that should not be dismissed as "mere reform." In truth, we do not (yet) *know* whether in this case revolution or reform is required. For his part, Durkheim calls his vision of the social order that has overcome the pathologies of modernity "socialism," and the fact that Marx does not use the term in this way should not prevent us from entertaining the possibility that Durkheim's socialism might represent real growth and substantial change – something beyond "mere reform."

Finally, it is important to forestall a further possible misunderstanding regarding the sort of immanent critique theories of social pathology rely on that may have its source in the circumstance that the *locus classicus* of this method is Hegel's *Phenomenology* rather than his social philosophy.[21] The source of this potential confusion is that the object of inquiry in that text is not real social formations but configurations of *consciousness* (or modes of knowing), the contradictions of which consist in mismatches between the norms for knowledge espoused in those configurations and what subjects who subscribe to those norms actually do when they attempt to know the world in accordance with them. This aspect of the *Phenomenology* might suggest that immanent social critique always involves revealing how consciously held norms of social members fail to be realized in the society they inhabit. On this issue, the form of critique employed by theories of social pathology is more like the immanent critique one might ascribe to medical diagnoses of illness, where "immanent" means not "internal to a form of consciousness" but immanent to the form of life of the relevant species (Thompson 2008: 81). In the case of physiological illnesses the defect picked out by "immanent critique" involves (but is not exhausted by) dysfunctions defined relative to the species' normal functioning.[22] Obviously, such diagnoses do not appeal to consciously held norms but to standards of the well-functioning of the species. As we shall see in later chapters, in the social domain consciously held norms play an important role in constituting social reality, and a systematic failure to realize such norms is relevant to, but not exhaustive of, diagnoses of social pathology. The more fundamental thought underlying immanent social critique is

[20] Using concepts of "social," "public," and "partial" ownership, Thomas Piketty proposes changes of this sort as a response to the massive inequalities in contemporary Western societies (Piketty 2020: 493–8, 508–10, 611, 972–5, 989–90). Rawls's alternative to capitalism, property-owning democracy, might also be regarded as a proposal of this type (Rawls 2001: 135–40, 158–62).

[21] For an extended discussion of immanent critique, see Jaeggi (2019: 190–214).

[22] According to Michael Thompson, what I am calling normal functioning is defined relative to an animal's "form or kind and the natural history that pertains to it" (Thompson 2008: 81).

that social life is made up of goal-directed practices informed by internal criteria for their own success. In this respect social life bears a resemblance to the processes of animal life – the circulation of blood, the digestion of food, the production of sweat – with the difference that in the former, practice-immanent norms are in part (but not always) *consciously* known and followed by the humans whose activities constitute the practices in question. In other words, the standards enabling the social pathologist to diagnose dysfunctionality in social life are already present – and therefore already partially realized – in the institutions under investigation, even if many questions remain concerning how the specific functions carried out in social life are to be determined. On this conception, even Marx's account of the recurrent crises of capitalism counts as a form of immanent critique, insofar as they impede the function (or work against the "point") of economic cooperation.

An implication of this feature of social life is that the projects of understanding and critique are more interdependent for the social pathologist than they are normally taken to be in contemporary moral and political philosophy. In some form the interdependence, and therefore inseparability, of the two projects is a dominant theme in much of post-Kantian European philosophy – in Hegel, Marx, Nietzsche, Freud, and Durkheim, for example – but it is not found only there. It is also present in the best social theories outside this tradition – in Adam Smith, who predated the tradition, and in Max Weber, who, although a member of it, did his best in his methodological reflections (but, to his credit, not always in his empirical studies) to embrace as his "official" normative position a rigorous, neo-Kantian version of the separation of fact and value.

Let us consider how these issues play out in Smith's social theory. Nothing distinguishes *The Wealth of Nations* more from the social and political philosophy of Smith's contemporary Rousseau than this methodological point. The normative logic of the *Social Contract* begins with an abstract, relatively a priori account of the basic interests all humans share and proceeds from there to deduce the fundamental principles of legitimate political association that allegedly apply at all times and places (SC: I.6.v). The contrast to Smith is even more apparent in Rousseau's *Discourse on Political Economy*, which, like *The Wealth of Nations*, discusses trade, the division of labor, and the proper balance between commerce and agriculture but does so from the same a priori normative perspective that guides the *Social Contract*. This is evident in the *Discourse*'s opening paragraph, where political economy is defined as the science dealing with the wise administration of a state (PE: 3/OC: 241), after which it proceeds as though the rudiments of sound economic policy could be derived directly from the principles of the *Social Contract*, assuming that both follow from a single normative principle, that of the general will.

Smith's treatment of commercial society could hardly be more different. In fact, it is an example of social theory that engages in what Hegel calls "comprehending what is." In this context comprehending involves understanding both how an existing economic system, "commercial society," works and how its functioning nonaccidentally realizes certain ends that for this reason can be thought of as inscribed in that system itself, even if some of them may not be immediately apparent to its participants. Smith's first aim, in other words, is to understand how a part of the existing social world actually functions rather than to evaluate it from an a priori normative perspective or to construct superior institutions from scratch. That is, Smith takes as his object an already existing economic system and asks not how it ought to function but how it (or an appropriately idealized version of it) in fact does: Which factors determine the prices of commodities and the wages of labor? How do increases in the rate of profit affect the various classes of society and the economy as a whole? What is the source of a nation's wealth? Finding systematic answers to these questions enables Smith to see the *good* realized by commercial society, which turns out to consist, as it does for Hegel, in some combination of well-being and freedom. The latter value is not imposed on commercial society by Smith's own normative commitments. Rather, freedom, conceived in a certain way, is intrinsic to the functioning of commercial society, insofar as the latter relies on the normative status of laborers as free beings who, independently of others' wills, enter into wage-labor contracts with their employers, without which production in commercial society would not take place. Smith's defense of the free-market economy derives not from a priori arguments about which institutions are ideally suited to human beings given their essential nature or interests[23] but from a comprehensive account of how actual institutions function and of what they can accomplish under favorable but realistic conditions. Similarly, his prescriptions for commercial society are limited to measures that would fine-tune an

[23] I do not deny that Smith has a normative conception of human nature, but it plays less of a role in justifying the market economy than his analysis of how commercial society works and of its systematic consequences for human freedom and well-being.

already functioning system rather than replace what exists with a more perfect, wholly invented ideal.

Examples of Social Pathology: Rousseau

An illustration of some of the claims made above about the concept of social pathology can be found in a passage from Rousseau's *Discourse on Political Economy*, which, along with his *Discourse on Inequality*, is a founding text of the modern tradition of social pathology critique:

> A time may come when citizens, no longer seeing themselves as having an interest in the common cause, cease being the defenders of the fatherland, and the magistrates prefer to command mercenaries rather than free men, if only in order … to use the former to better subjugate the latter. Such was the state of Rome at the end of the republic and under the emperors; for all the victories of the first Romans … had been won by courageous citizens who were ready to shed their blood for the fatherland when necessary, but who never sold it. Marius was the first … to dishonor the legions by introducing … mercenaries into them. The tyrants, having become the enemies of the peoples whose happiness was their responsibility, established standing armies, in appearance to contain foreigners, and in fact to oppress the local population. In order to raise these armies, tillers had to be taken off the land; the shortage of them lowered the quality of the produce; and the armies' upkeep introduced taxes that raised its price. This first disorder caused the people to grumble: In order to repress them, the number of troops had to be increased, and, in turn, the misery; and the more despair increased, the greater the need to increase it still more in order to avoid its consequences. On the other hand, those mercenaries, whose worth could be judged by the price at which they sold themselves …, despising the laws … and their brothers whose bread they ate, believed it brought them more honor to be Caesar's henchmen than the defenders of Rome, and … they held the dagger raised over their fellow citizens, ready to slaughter at the first signal. (PE: 28–9/OC: 268–9)

This tale contains several elements relevant to social pathology in imperial Rome. One is that citizens fail to "see themselves as having an interest in the common cause," implying a lack of the civic unity characteristic of a well-functioning society. This deficiency – the absence of solidarity among citizens – manifests itself in their failure to recognize a convergence among their interests and those of fellow citizens. As described here, this failure is said to reside in how citizens perceive themselves and their relation to society. This itself might be taken as an indication of social pathology, but in this case the problem is deeper, for the absence of civic unity is not merely a matter of false consciousness. Rather, the perceptions of citizens track something real, namely, that their society is divided into factions – citizens, magistrates, paid henchmen – whose members, because of these divisions, pursue interests so opposed that consensus on what is best for all becomes impossible: The magistrates are interested in subjugating the citizens; the latter are interested in having enough to eat and in avoiding subjugation; and the mercenaries are interested only in the money they receive for defending Rome.

It is important that this problem is not merely one of injustice, for example, the citizens being subjugated by their rulers. Rather, it is that something like a tear in the social fabric makes healthy social life impossible. Civic unity is important for Rousseau because it is necessary if individuals are to be "morally free" *qua* citizens (SC: I.8.iii). There is an important difference between being subjugated by others and being unfree in the sense of being obligated to obey laws that one does not endorse or see as expressing one's will. Although both forms of unfreedom are problematic, it is the latter – a forerunner of what Hegel calls alienation – that is at issue here. In Rousseau's tale the inability of citizens to regard laws as proceeding from their own wills might well be bound up with the fact that those laws produce their own subjugation, but there could also be instances where the absence of civic unity is not due to injustice – for example, when a society is basically just but individuals do not identify with it because their institutions fail to foster in them the values or self-conceptions necessary for doing so, or where divergences among interests pose insurmountable obstacles to the formation of a general will without those divergences depending on injustice.[24] In such cases, too, the absence of civic unity would constitute an ethical deficiency that has its source in nonaccidental social conditions; it would be, in other words, a social pathology.

A second element of Rousseau's tale pointing to social pathology is the magistrates' hiring of mercenaries or, more broadly, the inappropriate introduction of money into social relations. We should not saddle Rousseau with the view that it is bad for social relations of any kind to be mediated by money; a more plausible claim is that certain types of social relations are ruined once money comes to serve as their organizing principle. One of Rousseau's complaints is that individuals who carry out the duties of citizens only because they are paid to do so are easily

[24] This may be true for Smith of the three "ranks" that compose commercial society.

manipulated by those in power in ways that undermine the proper ends of political life. (Mercenaries let themselves be used to subjugate citizens rather than to secure their freedom and promote their good.) A further point is that "citizens for hire" tend to assume an instrumental attitude to their associates and to their own social activity. Adopting a purely instrumental attitude to what we do, and to those with whom we do it, may be consistent with acting freely (voluntarily), but when this attitude extends to much of what we do, the result is estrangement from others and from our own activities. Again, the idea is not that it is always inappropriate to relate to others as a self-interested calculator of gain but rather that there are types of social activity in which cooperation mediated by money is incompatible with the kinds of bonds and cooperation that constitute the very point of those activities. Here, again, the problem is not injustice but the failure of individuals to realize certain goods in principle available to them as social members.

Rousseau's characterization of the hiring of mercenaries as dishonoring the legions suggests a further respect in which his description of Rome is relevant to an account of social pathology, namely, in its depiction of what has come to be called, by interpreters of Axel Honneth's work, pathologies of misrecognition.[25] This aspect of Rousseau's tale points to a specific good, social recognition, that plays an important role in many accounts of social pathology and serves to illustrate one of the ethical deficiencies such accounts are concerned with. At the same time, Rousseau's description of the consequences of the hiring of mercenaries brings out a more formal point about what can make a social problem, loosely conceived, a pathology. For his treatment of misrecognition does more than establish that large numbers of Romans failed to find recognition from their fellow citizens; it also diagnoses a *pathology* of misrecognition, insofar as it reveals the dynamic underlying those recognitive failures, enabling us to understand the social forces that perpetuate them.

The import of Rousseau's remark that the legions were dishonored when money replaced civic allegiance as the reason to serve the fatherland is that this event, itself a form of misrecognition (of those who had served in the legions out of conviction and attachment), undermined other relations of recognition in society, generating a self-perpetuating system of recognition gone awry. One aspect of Rousseau's account concerns the effect the hiring of mercenaries had on their self-conceptions and their sense of their

[25] For Rousseau's versions of this idea, see Neuhouser (2010: chs. 2–4).

own worth: the mercenaries judged their own value only "by the price at which they sold themselves." This point is familiar to members of societies in which the market plays a dominant role: when money becomes the main principle around which social relations are organized, individuals are subject to a nearly irresistible tendency to judge their own worth in terms of the monetary value the market places on them and their services. One problem with this is that prices determined by supply and demand are ethically arbitrary. In the same way that a market responds not to need but to effective demand – production is organized not by what humans need in order to live a good life but according to what those with the greatest resources desire – it is accidental whether the measure of worth employed by the market tracks the characteristics that make for good societies and individuals.

Closely related is the fact that markets measure value according to a quantitative, one-dimensional metric: price expressed in terms of money. This encourages social members to value themselves quantitatively rather than qualitatively, less for the intrinsic merits of their qualities and achievements than for the (numerically measured) extent to which they happen to be in demand by those with the resources to purchase their services. This, too, could be described as alienation from who one is and what one does. In such circumstances we tend to value our socially beneficial qualities not because they contribute to a human good but because they serve to increase our value as determined by the price we command on the market. Moreover, measuring one's worth by a quantitative metric makes it harder to satisfy the desire to have value in one's own eyes and others'. In contrast to qualitatively defined ideals – being a good nurse or parent or citizen – measuring one's worth quantitatively admits of no natural stopping point in the quest for confirmation of one's value. Money prices can in principle always be improved on, and once this way of measuring one's worth has been internalized, satisfaction tends to become thin and unstable – there is no reason not to seek ever larger sums of what one seeks and hence no reason to place bounds on what Hegel calls a dynamic of "bad infinity."

Rousseau's tale also suggests that a distorted sense of where one's own value lies translates into a distorted picture of the value of others; a faulty evaluation of self goes hand in hand with the misrecognition of others. As Rousseau puts the point, once the hired security forces began to judge their own worth by the price at which they sold themselves – once they "believed it brought them more honor to be Caesar's henchmen than the defenders of Rome" – they also lost respect for the state, its laws, and their fellow citizens. The most vivid sign of this is that losing respect for others allowed them to hold "the dagger over their fellow citizens, ready to slaughter at the first signal." Yet this misrecognition is but a consequence of a more fundamental one: the fact that the magistrates no longer cared about the happiness of their subjects and, in "preferring to command mercenaries rather than free men," ceased to recognize their subjects as citizens, whose fundamental right is to remain free from the arbitrary wills of others.

This account of pathologies of misrecognition in imperial Rome points to an idea that frequently underlies diagnoses of social pathology, that of a self-perpetuating dynamic that makes a bad situation worse and, once initiated, is extremely difficult to break. This idea cannot be illustrated more clearly than Rousseau himself does in describing another pathological dynamic in which Rome found itself caught up:

> In order to raise ... armies, tillers had to be taken off the land; the shortage of them lowered the quality of the produce; and the armies' upkeep introduced taxes that raised its price. This first disorder caused the people to grumble: In order to repress them, the number of troops had to be increased, and, in turn, the misery; and the more despair increased, the greater the need to increase it still more.

The idea of a self-perpetuating dynamic represents another way in which a diagnosis of social pathology goes beyond merely uncovering social problems and how grasping certain phenomena as pathologies can provide a better understanding of social reality than theoretically less ambitious alternatives, in this case because doing so reveals the social forces at work that explain the persistence of the social ill in question. In Rousseau's example, the cycle of increasing poverty, taxation, inflation, and militarization makes it hard to avoid the impression that the society described, beyond the respects in which it is unjust, is internally dysfunctional in a way reminiscent of certain physiological illnesses. Apart from the diagnosis of impaired functioning, the idea of a self-perpetuating social dynamic points to another respect in which the analogy of illness seems not to be out of place in social theory: to locate the source of dysfunction in a social dynamic is to regard society as something like an autonomous, "living" system of forces, where one function affects and is affected by others and where their interaction acquires a life of its own not directly dependent on the will or consciousness of those whose activity constitutes those forces.

A similar point about dynamics can be seen in Marx's treatment of what is now called structural unemployment. Merely establishing that capitalist societies exhibit high rates of unemployment, and that they have done so for a long time, falls short of locating a social pathology. Marx's analysis of the industrial reserve army, in contrast, diagnoses a pathology of unemployment because it shows how the phenomenon is required by the capitalist system itself, which is to say, it shows how capitalist accumulation necessarily produces (and reproduces) unemployment and, moreover, how doing so serves the interest of the dominant, surplus value-appropriating class. Understood in this way, unemployment is more than merely a social problem; it points to a pathology because it is the nonaccidental result of an ongoing dynamic, a remedy for which requires a realignment of social structure and a transformation of institutions, rather than piecemeal attempts to stimulate a lagging economy or to provide welfare aid to those who cannot find work.

Before concluding this chapter, I want to point out two significant disanalogies between animal and social illness. The first is relevant to social ontology because it marks a fundamental distinction between the kinds of being characteristic of animal and social "organisms." I first develop this point in Chapter 6 in conjunction with Rousseau's claim that human societies, in contrast to products of nature, are artificial, or humanmade. The point underlying this claim is that human societies are normatively constituted entities whose workings depend on the agency, including some sort of "acceptance" of social norms, of their human members. Moreover, because the functioning of social institutions depends on the (free) agency of their participants, there is an important sense in which the continued existence of institutions, as well as potential transformations of them, is "up to them," that is, up to those whose activities and attitudes sustain them. In contrast to the life-sustaining components of animal organisms, it is in principle within the capacity of human social members to transform at least parts of the social world they inherit from previous generations.

This means that the diagnosis of social pathologies contains a moment of *critique* lacking in the diagnosis of animal illness. Although even the latter involves revealing defects of a certain sort – the sick organism falls short of the standards of well-functioning appropriate to its species – in the former case diagnosis implies criticism of a more robust sort. It is appropriate to speak here of critique, and not merely diagnosis, first, because the diagnosis at issue is reflexive, carried out both on and by the very same being, much like in Kant's critiques reason is both the subject and object of critical inquiry; and, second, because diagnoses of social pathology ascribe to their objects a kind of responsibility (to be explained in Chapter 6) and imply a practical imperative addressed to the human wills on which the illness of those objects depends.

It is important, however, to distinguish the critique involved in diagnoses of social pathology from moral critique,[26] if the latter is taken to involve ascriptions of moral blame or praise for individually imputable actions. Although diagnoses of social pathology typically invoke ethical standards beyond that of efficient functioning narrowly understood, appealing to some understanding of the human good, this does not mean that the critiques delivered by such diagnoses imply moral culpability. (This point is exemplified by Rousseau's *Discourse on Inequality*, which offers an ethical critique of modern society and yet, while regarding society and its ills as humanmade, does not ascribe moral culpability to those whose free actions, unintentionally and without foresight, produced the ills in question.) The main reason that diagnoses of social pathology do not imply moral culpability is that although the dysfunctions characteristic of them involve failures to realize the good, such failures result from social dynamics the persistence of which is independent of the intentions of specific individuals; ascribing blame for them to individuals is therefore highly problematic.

Even if it were possible to apportion moral responsibility for social pathologies, it is unclear what relevance moral critique would have to the main aims of critical social theories, namely, to understand, evaluate, and, when appropriate, indicate directions for transforming social life. One might put this point by saying that, whereas the attribution of moral blame is largely backward looking – concerned to establish whose actions are causally responsible for a certain state of affairs and, when appropriate, to attach moral blame to those responsible – theories of social pathology are forward looking in attributing responsibility to social actors (in the sense of: "It is up to us to transform the world our actions maintain"). Such theories ask not: "Whose actions created our condition, and what good or evil intentions do they express?" but instead: "How are existing institutions bad for us, what social forces maintain them, and how can we collectively transform them?" Theories of social pathology can criticize institutions for being unjust and for being less good than they can be, giving us reasons to seek alternatives, but figuring out whom to blame for those institutions is not a principal concern.

To take the classic Marxist example: the "moral capitalist" who comes to regard the wage–labor relation as unjust and therefore refuses to participate further in the system – perhaps even giving away his

[26] I am indebted to Macalester Bell for raising this question.

accumulated capital for relief of the unemployed – may be in some way morally laudable, but his actions, because they do nothing to alter the systematic injustice that motivated his withdrawal from the system, remain irrelevant from the perspective of social theory. Or consider cruel or environmentally harmful agricultural practices (which, the social theorist will remind us, became pervasive only after family farms were ruined by entities ruled by the logic of "agribusiness"). While there might be room somewhere for finding individuals morally guilty of inhumanely enclosing livestock or of overusing chemical fertilizers, doing so is of little concern to social theory. Apart from the fact that the agents of such practices are largely corporate entities rather than individuals, whatever moral badness those practices contain does not explain why they exist (and persist). Again, understanding, evaluating, and transforming social practices depends not on assigning moral responsibility but on understanding the social forces and conditions – the logic of capitalist accumulation but also the absurd, manipulated "preferences" and habits of consumers – crucial to explaining why the agents involved act as they do.

The other disanalogy between animal and social illness poses a more significant challenge to my project since it can appear to cast doubt on the wisdom of attempting to rehabilitate the concept of social pathology. This issue came to my attention in discussions of that concept with nonacademics, when the question frequently arose, "Have there ever been human societies that were not sick?" This initially startling question, to which the answer is immediately obvious, reminded me of the greatest of all diagnosticians of *cultural* pathologies, Nietzsche, and of his haunting remark – a fitting epigram for Rousseau's *Discourse on Inequality* – that "the human being is more sick, more uncertain, more changeable, and more unsettled than any other animal; of this there is no doubt – the human being is *the* sick animal" (GM: III.13). The import of Nietzsche's remark is twofold. First, it raises the possibility that no human reality – and hence no human society – can be completely void of illness, so that to wish illness out of the social world is to wish away its humanness. Second, the context of Nietzsche's remark suggests that it is precisely this tendency to fall ill that not only makes human existence "interesting" (GM: I.6) but also creates the conditions that make possible great *spiritual* health (GM: II.20). Taking these points seriously appears to render the social pathologist's aspirations doomed to failure, naïve about the conditions of the human good, and, worst of all, inimical to the optimal development of the human species.

What, if Nietzsche is correct, are we to make of a social theory whose chief aim is the diagnosis and cure of social pathologies? An answer to this question can be found in Nietzsche's thought itself. For, as he recognizes, the fact (if it is one) that human societies are always somehow ill implies neither that all are equally ill nor that all illnesses are equally dangerous. Nietzsche makes this point in characterizing the cultural world bequeathed by Christianity to nineteenth-century Europe as plagued by "the most terrible sickness yet to have raged in the human being" (GM: II.22), an illness that risks exhausting the very source of human vitality and bringing about "the great nausea, the will to nothingness, nihilism" (GM: II.24). Or, to put the point differently, there are healthier and less healthy ways of dealing with a condition in which illness is unavoidable. (The ancient Greeks' conceptions of their gods and what is owed them is an example of the former.) In the first place, the value Nietzsche attaches to human self-transparency, if not unconditional or overriding, is sufficient to convince him that, apart from all prospects for a cure, there is a point to diagnosing human illnesses – revealing them for what they are – rather than "suffering" them, unspiritually, in self-ignorance. Beyond this, however, he recognizes that the impossibility of eradicating illness is not a reason to let it simply run its course without seeking to meliorate it – including the needless or unproductive forms of suffering it causes – or to mitigate the dangers it poses to human well-being. Indeed, no one (with the possible exception of Freud [Whitebook 2011: 120–42]) deserves the title "diagnostician and treater of spiritual pathologies" – or "philosophical physician" – more than Nietzsche. All this confirms, it seems to me, that diagnosing and combating social pathologies remains a legitimate aim of social theory, even if there has never been, and never will be, human social life free from all illness.

CHAPTER 2

Society as Organism?

In the previous chapter I illustrated how diagnoses of social pathology go beyond pointing out injustices in social life. The examples I considered there are also useful in considering the extent to which a theory of social pathology is committed to thinking of human societies on the model of animal organisms. Many thinkers, even in modernity, avail themselves of this analogy in more substantial ways than do the conceptions of social pathology I reconstruct here. That Rousseau in the *Social Contract* describes the just republic as an "organized, living" body animated by a common self and possessing a unitary life and will (PE: 6/ OC, 244–5; SC: I.6.ix–x) is common knowledge, but Kant, too, endorses a robust version of the analogy between state and organism (conceived of as a *Naturzweck*) (Sedgwick 2001: 171–2, 174). According to this analogy, the "body politic" – or at least the ideal behind it[1] – is appropriately thought of as an organic totality in a very strong sense, in which "each member [is] not merely a means but also an end and, in working together [with other parts] so as to make the whole possible, is in turn determined with respect to its position and function by the idea of the whole" (CJ: §65n). In an organic totality of this perfect type – both teleologically organized and self-organizing – no part is "gratuitous [or] purposeless" (CJ: §66), and there is nothing accidental in how its parts are constituted or in terms of the relations in which they stand to one another. Complete teleological harmony, then – where all parts work together to maintain the organism's stability and cohesiveness as perfectly as a well-tuned watch – is for Kant the hallmark of organic totality, whether social or biological.

After Darwin, we might question whether this conception of organic totality applies even in the biological case, but, conceived this perfectly,

[1] That Kant regards *Naturzweck* as an idea of reason with regulative rather than constitutive import does not affect my claim here: that the tradition is rife with overly perfect conceptions of organic unity that, if transferred to the social realm, cast suspicion on the society–organism analogy.

organic totality is emphatically not a feature of human societies nor an ideal by which they should be measured. Critics of the society–organism analogy often have this conception of the biological organism in mind when they reject it as both theoretically false and, as an ideal, repressive and reactionary. Of the social philosophers I most rely on, however, only Hegel could reasonably be taken to employ something approximating this idea when describing and evaluating modern societies. Yet the aspects of Hegel's view I appropriate here do not depend on the assumption that perfect teleological harmony is an actual or desirable feature of human societies; indeed, the best readings of Hegel's social philosophy emphasize that even the (somewhat idealized) social world that he thinks we have reason to affirm as good is marked by internally necessary ruptures in such harmony, which show up in such phenomena as divorce, poverty, and war.

At the same time, there is a more modest conception of social totality that finds its way into the defensible version of the society–organism analogy that theories of social pathology are committed to. This conception of totality is less a normative ideal – though it remains one to a limited extent – than a principle for understanding social phenomena. On this conception, to regard human societies as totalities, or "social wholes," is to reject a species of methodological abstraction common in contemporary political philosophy. What is rejected, more precisely, is the seductive conceit that one can understand or evaluate one social sphere or institution in isolation from other spheres and institutions. Rejecting this assumption entails that no adequate political philosophy, for example, can be carried out in abstraction from a social theory that investigates nonpolitical social life and the respects in which political institutions are intertwined with the practices of social life more generally. Hegel and Marx are the most consistent upholders of social totality in this sense. It appears, for example, in Hegel's insistence that one can neither understand the functioning of the modern nuclear family nor assess its goodness without taking into account how it interacts with – both supports and poses threats to – the other spheres of the modern social order. The same methodological orientation, if not the same normative conclusions, characterizes Marx's historical materialism and, even more illuminatingly, his earlier views in *On the Jewish Question* about the function and value of liberal democracy when coupled with an egoistically organized civil society. This construal of totality implies a weak form of holism that also applies to our understanding of biological organisms: adequate knowledge of both how the heart works and what its proper functions are requires considering it in relation to other organs and systems of the same animal body.

The limited respect in which the idea of totality retains normative significance in theories of social pathology is expressed, then, in the modest principle that the differentiated parts of a healthy organism must cooperate so as to achieve the ends of the whole and avoid excessive, but not necessarily all, internal discord and working at cross purposes. The Nietzschean thought that certain kinds of internal discord are consistent with both animal and social health is especially visible in Durkheim's treatments of suicide and crime: Socrates, after all, was both a criminal and a promoter of positive moral change (RSM: 97–103/65–72). In any case, the normative implications of viewing societies as totalities in this deflated sense – with its theoretical directive "Do not view parts in abstraction from others!" – fall far short of those bound up with Kant's stronger conception of organic totality.

My references to the *functions* of *differentiated* social institutions point to three further respects in which theories of social pathology presuppose a weak analogy between human societies and living organisms. First, like organisms, societies are functional beings in the sense that how they are constituted and how their parts act and interact cannot be understood without ascribing ends to them, as well as to the whole they make up. Of the philosophers studied here, Hegel is the most unequivocal on this point, citing Aristotle's description of an animal organism as "acting in accordance with ends" (PhN: §360) and explicitly ascribing the same characteristic to what he takes to be "spiritually alive" beings, such as human societies. This is of great importance for conceptions of social pathology since, without this claim, the idea at the heart of that concept, dysfunctionality, would make no sense. Of course, ascribing functions to societies and to their constituent parts raises thorny theoretical issues that accounts of social pathology cannot ignore. Later chapters of this book address various aspects of those problems, including, in Chapters 7 and 8, the vexed issue of functional explanation.

Second, theories of social pathology posit an analogy between living organisms and human societies in regarding the latter not merely as functional but as *functionally organized*, which is to say, as beings that carry out their characteristic functions via specialized and coordinated functional subsystems (or "social spheres"). It is tempting to take specialization of this sort as primarily a characteristic of modern societies – Durkheim's distinction between segmental and organically structured societies encourages this view – but the idea that societies function by means of the coordinated actions of specialized spheres is already part of Plato's account of the healthy polis in the *Republic*, one of the first moves of which is to introduce into social life a specialized division of labor necessary for even the most rudimentary of societies, including those derided by Glaucon as cities of pigs. Here, again, it is important to avoid exaggerating the degree of complementarity among a society's parts, as well as the intricacy of their coordination. Yet, with these cautions noted, it is hard to see why one should dispute the claim that functional specialization is a basic feature of human societies.

Third, *reproduction* is a central task of human societies. Every society, in order to survive, must produce the human bodies that will make up future generations as well as the goods that allow for the maintenance of those bodies. In other words, biology places *some* constraints on the health of human societies, and there is no reason to expect that the production of adequate food, shelter, clothing, and medicine will ever cease to be a central concern of social organization. Nevertheless, this fundamental fact about social life is subject to two qualifications, both prominent themes in the chapters that follow. First, for humans, material reproduction is never simply a matter of responding to purely biological needs. Material reproduction is always at the same time spiritual reproduction, which means not only that, for example, the food we aim to produce is always food we take to be *appropriate to humans* (varied, appealing, tasty – something in any case better than the slop we feed to pigs[2]), but also that we aspire to make the activities through which we produce what our bodies need as *free* as they can be. Expressed slightly differently, reproduction must be understood as reproduction *as the specific kind of being* something is, and to conceive of social reproduction in purely biological terms is to leave out of sight the spiritual – the self-conscious and self-determining – nature of human beings. Second, reproduction, though perennially a major concern of healthy social life, is never its exclusive, or even preeminent, end. There is no reason to assume – and none of the thinkers appealed to here do – that the healthy reproduction of human societies involves eternally reproducing the status quo. To do so would be, again, to fail to grasp how human life differs from that of mere animals (even though the latter, too, evolve over time as species).

Avoiding "Sociological Functionalism"

Many of the points just made can be reformulated by saying that regarding human societies as functionally organized beings that aim at their own reproduction need not imply the view commonly known, and widely dismissed,

[2] This is the point behind the distinction German makes between *essen* (human eating) and *fressen* (the eating nonhuman animals engage in).

as sociological functionalism (Radcliffe-Brown 1935: 394–402; and 1957).[3] More precisely, since there are numerous versions of functionalism, theories of social pathology need not embrace many of the dubious doctrines that have been ascribed to it and that account for its contemporary unpopularity. Of course, any theory that places function and dysfunction at its center can be regarded as functionalist in some sense. One will see the precise respects in which the views endorsed in this book are functionalist only after having read all its chapters, especially those on Durkheim and Hegel. Still, it may be helpful to note in advance a number of theses frequently associated with functionalism that do *not* belong to the theoretical perspective I endorse here:

(i) *Stability* and *harmony* are the principal marks of healthy societies. (This book argues instead, stability and harmony are generally good, though not in all cases, and they are never the most important element of social health; not all conflicts indicate social dysfunction.)

(ii) Healthy societies exhibit a *tendency toward homeostasis*, or equilibrium, and achieving that state, or returning to "normality," is a leading ideal of social health. (This book argues instead, social change, and the disequilibrium that generates it, can be a sign of vitality rather than dysfunction, and "the normal" is not an important ideal or diagnostic concept.)

(iii) *Survival* – and, even less so, success in the "struggle for existence" – *is the overarching end* defining social health. (This book argues instead, reproduction is a central function of human societies, but it is never purely material and always incorporates spiritual, including ethical, ends; expressed in ontological language: healthy reproduction is always reproduction of the kind of being something is, which in the case of human society comprises both material and spiritual qualities.)

(iv) *All enduring aspects of human societies serve a purpose* (or have a function) relating to their long-term survival. (This book argues instead, it is always reasonable to search for a – material or spiritual – function that enduring features of a society serve, but there is no guarantee that such a function can always be found.)

(v) *Ascribing a function* to a feature of society *explains its origin*. (This book argues instead, the cause of a given feature of society must

[3] I do not distinguish between sociological and anthropological functionalism (Turner and Maryanski 1979: chs. 1–4).

be distinguished from its ascribed function; to use language developed in Chapter 7, functional analysis does not imply (existential) functional explanation.)

(vi) *Adaptation*, especially the one-way adjustment of individuals to existing social norms, is a central function of human society and a hallmark of its health. The adaptation of a whole society in response to changes in its natural and social environment is likewise a central function of society. (This book argues instead, spiritually healthy processes of internalizing social norms are compatible with, and even require, reflective distance from, and critical assessment of, prevailing norms; moreover, "adaptation" is too passive and biological a term to capture how healthy societies respond to external threats.)

(vii) A central task of social theory is to *determine and classify the functions any human society must fulfill* – for example, education and socialization; social control; economic adaptation; political authority (Turner and Maryanski 1988: 114) – and then to use such a scheme as a guide for empirical inquiry.[4] (This book argues instead, because of the great variation among them, a universally valid list of the functions of human societies is either unattainable or so indeterminate that its empirical usefulness is negligible.)

(viii) *Phenomena that are similar in kind* across different societies – religious practices, for example – *serve similar functions* in each. (This book argues instead, phenomena that appear to be similar across different societies may have different meanings and functions.)

Given this picture of sociological functionalism, it is no surprise that theories emphasizing the functionality of social institutions are often criticized as intrinsically conservative. One form this charge takes is that such theories employ narrowly functionalistic criteria in evaluating social life (and hence, in a further sense, exaggerate the similarities between animal organisms and human societies). The most frequent version of this objection – voiced nearly every time the topic of social pathology

[4] Honneth (2014a: 698) appears to endorse a version of this project in naming what the tradition, from Marx to Parsons, regards as "the challenges that societies have to cope with in order to preserve themselves for a certain period" – namely, "confrontation with external nature; social shaping of inner nature; and regulation of inter-human relations."

is raised among academic philosophers – is that human societies can be highly functional and morally repugnant at the same time and that in such cases theories of the sort defended here cannot avoid deeming the morally repugnant healthy. There is something about the unreflective certainty with which this objection is typically raised that should lead us to pause and ask whether it is well founded or instead reflects an entrenched prejudice of moral philosophy: Why does it seem so obvious (to us moderns?) that the principles of good living can be separated from considerations of what works humanly well in social life? Where are the supposedly abundant examples of well-functioning but unjust or immoral societies? North Korea? The former Soviet Union? Saudi Arabia? Israel? The antebellum United States? The United States in 2022? (China would be an interesting example to consider, but one would have to think about it long and hard and rely on more empirical knowledge than philosophers usually avail themselves of.) In any case, it is my intention in this book to broaden the concepts of functional and dysfunctional so as to take account of the fact that the good of human beings is more complex than that of nonhuman living organisms and for that reason social arrangements that work humanly well will also be responsive to ethical standards. A large group of the most insightful social theorists of Western modernity – among them Smith, Hegel, Marx, Durkheim, and Weber – tacitly or explicitly accept some version of this principle, and one aim of this book is to show it to be more philosophically compelling than it is usually taken to be.

There is, however, a more interesting version of the conservatism charge that deserves consideration. Theories that emphasize the functional character of social life are primed, as it were, to seek functionality in whatever social arrangements they investigate, even if they also must refrain from assuming it can always be found. For such theories there is a defeasible presumption that enduring social formations "work" (or worked in the past) in some way, even when we can no longer regard them as ethically desirable. Thus, theories of the sort I endorse run the risk of finding the good wherever they look, even in deeply immoral societies, and hence of being ideological in the pejorative sense. There is some truth behind this worry, and it must be taken seriously. Yet it is worth noting that Hegel, Marx, and Durkheim all subscribe to versions of the functionalist view I defend and that Marx, for one, can hardly be suspected of sharing the conservative tendencies thought to be implied by it. (Hegel and Durkheim are best understood as reformers – with Durkheim, the advocate of a kind of socialism, the more radical of the two – but they are decidedly not reactionary thinkers who want to hold on at all costs to whatever "has worked in the past.") To take the most extreme examples, both Hegel and Marx viewed ancient slaved-based societies and medieval feudalism (but not slavery in the antebellum U. S.) as generally functional social worlds whose ways of accomplishing the fundamental tasks faced by human societies exhibited a degree of rationality – constrained by their historical circumstances – and as best regarded from the perspective of sociological understanding rather than moral condemnation.[5] I argue here that functionally oriented social theories do indeed run the risk of undue conservatism but that they also have an important advantage over theories that approach social phenomena only from abstract moral principles.

I will attempt to make these points clearer by considering Hegel's account and defense of the modern nuclear family (PhR: §§158–80). The version of the family Hegel presents as rational is indisputably patriarchal, and that feature of it, far from being accidental, is essential to its functioning. Hegel endorses an especially stark (and depressingly familiar) view of sexual difference – there is no room here for distinguishing sex from gender – as grounded in the different biological natures of men and women, where natural differences translate into spiritual differences, the most important of which is that women are beings dominated by feeling who for that reason have limited rational capacities. Since the family is the social sphere bound together by a type of feeling (love), it is the main province of women, while their more independent and rational husbands also lead lives in civil society – where they work in order to provide for their families – and in the state. Although officially some type of equality between husband and wife – reflecting their equal status qua human beings and persons – is supposed to obtain internally to the family, the husband is clearly its head and representative in society at large (raising the question of what remains of intra-family equality once this condition is in place).

Where, one might wonder, is the rationality in all this? As we will see in later chapters, the core of Hegel's complex answer to this question lies in the idea of functional specialization within the context of a collective project guided by the aim of satisfying human needs while, in that very function, realizing freedom (and its conditions). The family is a functionally specialized institution in relation to society at large that, to simplify, produces the bodies of future generations; socializes children into beings who, as adults, are capable of love, trust, and various forms of freedom;

[5] Piketty takes a similar approach when treating premodern *sociétés ternaires*, made up of the three traditional estates – clergy, nobility, and laborers (plus merchants) – which, emphasizing the specialized functional character of such societies, he calls *ordres trifonctionnels* (Piketty 2020: 71–83).

and creates the framework for a satisfying, identity-constituting form of life ("family life") in which, in contrast to other social spheres, participants lead a common life grounded in relations of love and trust (PhR: §157).

What makes the internal (patriarchal) structure of the family rational for Hegel is the same principle, namely, functional specialization in the service of need satisfaction and the promotion of freedom. The basic idea is that the tasks the modern family is called on to solve are complex and diverse, requiring different skills and dispositions among its members. Raising children requires love of an especially precarious sort – one that (somehow) brings together both unconditional love and constructive forms of discipline, and at the same time fosters both attachment and independence – and the talents required for this are different from those required in a market economy, where competition is pervasive and individual participants (mostly fathers) carry a weighty responsibility for the well-being of those at home who depend on them. Moreover, the challenges of making a living where there is no guarantee of success create a further task for the spouse responsible for emotional affairs: ensuring that the family offers the haven of love Dad needs when he comes home from the dog-eat-dog world of work. The modern, patriarchal family, Hegel thinks, exhibits a substantial degree of rationality insofar as it exploits allegedly natural differences (and the natural passions of sexual attraction and love for one's offspring) in order to carry out the various domestic tasks necessary for society's material and spiritual reproduction.

This abbreviated account of Hegel's view of the modern family may not seem well suited to warding off the charge of conservatism I mean to counter. It is clear, two hundred years later – but notice how much historical experience was necessary to arrive at this point – that this version of family life, though functional in significant respects, is also beset by grave spiritual defects, not least that it drastically curtails the spiritual possibilities (including the freedom) of mothers and wives, who, in the name of a naturalized essentialism, are not able to assume the same range of social roles accorded to men, which Hegel himself takes to be essential to self-realization in modernity. A social institution that for a long time "worked," even in the eyes of many mothers and wives, became at some point, and for good reason, no longer "habitable"[6] (Pinkard 2012: 8) for

[6] "Inhabitable" belongs to the same register occupied by "pathological" and "dysfunctional," indicating more than that the institution in question is unjust.

large numbers of women (as well as for a smaller number of men who came to see that satisfying relations with the women they loved were not achievable under patriarchal conditions).

The most common and seductive response to the historical development described is to condemn the patriarchal nuclear family as unjust and to demand its reform or abolition. In itself there is nothing wrong with this response, but moral critique alone cannot guide effective social reform precisely because it ignores the (imperfectly) functional character of the existing institutions to be reformed. In other words, to condemn an institution as immoral, however justified the charge, is not to diagnose it as pathological – it is not to grasp its dysfunctional character because it has not yet been grasped as a functional phenomenon – and moral condemnation by itself provides little orientation for thinking about what is to be done. The advantage of Hegel's account of the – as we now see, merely partial – rationality of the patriarchal nuclear family is that it reveals what the old structure accomplished and therefore alerts us to the array of tasks any replacement will need to carry out. Hegel's functional account should lead us to expect that merely introducing gender equality into the family will produce a host of problems the new institution must somehow solve. If husbands and wives have equal claims to self-realization, how are conflicts between them to be resolved in a way that preserves the unity of purpose the nuclear family is supposed to embody? (Think of the "two-body" problem in academia and elsewhere and the disruptions to relationships and families it produces.) If both spouses pursue professional aspirations in civil society, what happens to the rearing of children? Is that affectively complex task compatible with the commodification of child care that seems to be required when both parents pursue careers? Does something get lost when childrearing (or care for the aged) becomes another service-for-pay like dry cleaning or hair styling? What happens to the character of street play in cities when there are no mothers at home to keep watch and maintain neighborhood order when children come home from school (and when, because the extended family has been left behind, no grandparents are around to do that job either)?

It should be clear that these and other problems are not reasons to wish the old family back into existence. Nor is there reason to think such problems are unsolvable. But something like Hegel's functional understanding of the old institutions must be kept in view if we are to have a realistic picture of the challenges that arise when we respond to the injustices (and therefore dysfunctionalities[7]) of existing institutions. In other words, the danger of a certain tendency toward conservatism is indeed intrinsic to functionally oriented social theories; their practitioners should acknowledge this tendency and work hard to resist it. At the same time, one should not lose sight of the radical potential of such theories (their potential for grasping social problems "at their roots"). This, too, can be seen in Hegel's treatment of the nuclear family: the fact that Hegel sees not only the family but also society as a whole as functionally organized – where features of one institution complement and adjust to features of the others, and vice versa – has the virtue of alerting us to the likelihood that responding adequately to patriarchal inequality in the family will require more than just educating boys and girls to think of themselves as equals; in addition, it is likely to require far-reaching changes in other institutions, especially in the workings of economic life. Fixing the family requires, at a minimum, reforming civil society.

That all of this lies in the spirit of Hegel's thought (and of theories of social pathology more generally) is confirmed by the strategy he employs in exhibiting the rational advantages of the modern nuclear family. For his account of these advantages is articulated against the background of the type of family the modern form has replaced, and it highlights both how the later institution can be seen as responding to (ethical) deficiencies of the earlier and how the disappearance of the traditional, extended family poses new problems for social organization since certain of its functions can no longer be carried out by the family in its modern form. For Hegel, the ethical advantages of the nuclear family are relatively clear: it fosters forms of individual freedom and identity for which the traditional family had no place. Most obviously, women and men in the modern world are free to choose their spouses in accordance with their personal longings (PhR: §§162, 164, 168) rather than have those choices determined by others based on considerations of what best serves "the Capulets" or "the Montagues." More important, the family's nuclear structure – with its implication that offspring are mostly released from family obligations

[7] "Therefore" does not imply that all dysfunctionalities result from injustices; rather, it marks a point emphasized by Durkheim: when significant numbers of social members see their roles within institutions as imposing unjust burdens on them, those institutions become dysfunctional (DLS: 384/377–8). Applied to the case at hand: abstracting from the most important point – that women who feel overly constrained by their roles in the family suffer unnecessary and unjust spiritual deprivation – such women also do not make the best mothers and wives, even when making heroic efforts to do so. A further question is whether unjust institutions have dysfunctional tendencies even when not perceived as unjust. Piketty's remarks on the increased productivity of free, socially recognized laborers, as compared to slaves and serfs, suggests an affirmative answer (Piketty 2020: 92–4).

in adulthood – affords children (or sons) a nearly unlimited range of options for structuring their adult lives, including the career to which they dedicate their lives as members of civil society. The very structure of the modern family – in ensuring that one's life activities are unconstrained by the needs of grandparents, uncles, aunts, and cousins (PhR: §172) – gives expression to the values of independent personhood (PhR: §159), and these values are both reflected and reinforced in modern childrearing practices aimed at producing not only economically self-sufficient adults but also autonomously thinking moral subjects possessing the self-mastery and discipline required to pursue their conceptions of their good and the ethically good more generally. Moreover, beyond promoting individual freedom in these straightforward senses, the nuclear family "liberates" by introducing new possibilities for self-realization. In being grounded in love rather than (exclusively) in economic necessity or family prestige, the nuclear family makes possible hitherto unknown forms and degrees of intimacy (Honneth 2014b: 141–54); of being known, by both oneself and others, as a unique and unconditionally valued personality; and of knowing others in the same way (PhR: §§158, 167, 175).

The main point here is that Hegel's functional understanding of social life allows him to see not only the advantages of the modern family but also the functional gaps left behind when the traditional, extended family ceases to exist. Here again his analysis exhibits the strength of regarding the family as inseparably intertwined with other major institutions. There is an obvious sense in which the modern family is well-suited to complement another distinctly modern institution, the market-governed economy, which counts on participants regarding themselves as independent individuals who aspire to make their own choices and to pursue their particular life projects. At the same time, the thoroughgoing individualism of modern civil society, together with the unforeseeable results of its anarchic dealings, brings to light one function served by the traditional family that its modern counterpart cannot carry out: protecting individuals from the contingencies of economic fortune (PhR: §238), vulnerability to which is exacerbated by the nature of modern civil society, with the result that social forces – the workings of the market – replace natural forces as the principal causes of economic ruin.

It is this point Hegel has in mind in saying that, because of the disappearance of the traditional family, modern civil society must assume the role of a "*second* family" (PhR: §252), protecting its members from the contingencies of the market economy. In a move that anticipates Durkheim's later intervention to an uncanny degree, Hegel spots a functional gap in modern society that needs to be filled; he looks at the existing world to see where a potential for filling it might be found; and he hits on the idea of retooling an earlier institution, the medieval corporation, so as to make it suited to carrying out the needed function in the modern world. Although for Durkheim corporations are meant to solve the problem of anomie rather than to protect individuals from economic misfortune, the difference is less than it initially seems since Hegel, too, attributes integrative functions to the corporations, insofar as they form their members subjectively in ways that endow them with the capacity to engage in "conscious activity for a common purpose" (PhR: §254).

Two points about this move are worthy of note (and apply to Durkheim as much as to Hegel). First, Hegel's appeal to corporations is not an attempt to remedievalize the modern world. Although some of the functions he attributes to corporations approximate those of their medieval forerunners, he seeks the foundations of the new corporations – the source of their members' bonds – in modern conditions of civil society. Although the modern economy atomizes its participants to a degree unthinkable in medieval times – individuals act on their own initiative, rather than collectively, and are motivated by their particular interests – Hegel sees that *widely shared particular interests* have the potential to unite self-interested individuals into shared projects of certain kinds, and this is precisely what corporations – associations of similarly positioned members of civil society with shared interests – can help accomplish (PhR: §251). To take an anachronistic but not irrelevant example, a high degree of overlap in the particular interests of wage-laborers in the auto industry has the potential to motivate collective action, even if they enter the economic sphere with only self-regarding, particular ends. The need of each to be protected from the market's contingencies and the exploitation of their employers makes it reasonable for them to work together to bargain for better working conditions or to establish unemployment funds or forms of insurance that benefit all corporation members. At the same time, the process of working together to promote overlapping particular interests has – as Marx pointed out (MER: 99/MEGA: XL:553–4) – the tendency to create more substantial bonds among corporation members, ones that eventually surpass narrow self-interest and generate a genuinely collective, if not yet universal, "we," thereby addressing not only the problem of economic contingency but also that of social integration. This sort of transformation is what Hegel has in mind in describing corporations as having the potential to bring "what is common (*Gemeinsames*) into existence in the form of a collective (*Genossenschaft*)" (PhR: §251). If we conceive of some of the corporations Hegel recommended on the model of modern labor unions, and if we take seriously contemporary evidence of how severely disintegrated a union-poor social world is, then Hegel's proposed remedy for the functional gaps left in modern society by the demise of the traditional family appears sound and remarkably prescient.

The second point has to do with how theories of social pathology relate to practical orientation. Hegel's treatment of the corporations, institutions not fully developed in the social world he found before him, reveals how his project of "comprehending what is" goes beyond the "what is," aspiring to complete it, as it were, on the basis of what he has discovered about what the existent social world appears to be set up to accomplish. For Hegel, to comprehend the functional character of existing reality is at the same time to see where it falls short of full functionality, putting the social pathologist in a position to grasp or anticipate how it might become more completely what it already is – or, in Hegel's idiom, how the existent might be made to correspond more fully to its own "concept" (Jaeggi 2019: 120–2, 155–9). Although the details differ, something similar to Hegel's approach guides Marx's account of the transition from capitalism to socialism, insofar as the latter is taken to have the potential to realize aspirations and possibilities generated by the former that it itself cannot realize.

Social Pathology as Misdevelopment?

The foregoing paragraphs do not exhaust the ways in which social theorists have exaggerated or misused the analogy between human societies and living organisms. Before abandoning this topic, however, I want to note a cluster of respects in which even the thinkers I take most seriously here have preserved that analogy but from which I depart. This cluster of ideas concerns the development or evolution[8] of human societies, and it includes the following:

(1) It is theoretically important to classify human societies into "species" (and even genera) distinguished by species-specific, characteristic forms, corresponding functions, and characteristic patterns of, or possibilities for, development (Durkheim).
(2) Human societies, or their types, follow a more or less fixed pattern of development analogous to the life-stages of biological organisms,

[8] As I note in Chapter 7, Comte and Spencer exploit the analogy with biological evolution more egregiously than the possibilities I mention here.

where such stages are either species-specific or common to all societies (Rousseau, Comte, and perhaps Marx).

(3) All human societies belong to the same species in the sense that together they constitute a single, universal process of "human" development, spanning multiple historical epochs, in which no specific pattern of life-stages applies to each society; rather, specific societies, usually corresponding to different historical epochs, represent different life-stages in the development of the human species (Hegel, Marx).

In this book I reject all three possibilities for thinking about the development of human societies for the simple reason that each seems to me empirically untenable. I presuppose, in other words, no specific philosophy of history or overarching scheme of social development. It is in this respect that I diverge most fundamentally from the tradition of theories of social pathology I reconstruct here, and the importance of this divergence should not be underestimated. As I understand them, even "post-metaphysical" social philosophers of this and the previous centuries – Weber, Habermas, and Axel Honneth, for example – have found it necessary to preserve traces of a normatively tinged philosophy of history in attempting to secure a basis on which to distinguish healthy from ill societies. None of them espouses a teleological philosophy of history in the strong sense found in Hegel and Marx, yet each relies on a weakly teleological understanding of human history similar to Weber's: broadly speaking, human history is characterized by an enduring and (barring exceptional events) irreversible pattern of development (or "learning processes") that, though complex and characterized by stops and starts, is a process of increasing rationality, which, despite the dangers it brings with it, is, in its broadest outlines, civilizationally and ethically progressive.

The theoretical consequences of rejecting even weakly teleological conceptions of history are not negligible. First, without them, there can be no thought of relying on a vision of the general trend of history in order to find normative criteria for distinguishing healthy from ill societies, depending on the extent to which they have undergone the developments in question or incorporated the values implicit in them – "communicative rationality," for example (Habermas 1987: 62–76, 120–35, 140–8, 179–87, 318–26, 383–91) – into their institutions and practices. Second, one category of social pathology prominent in the work of Axel Honneth – that of a *Fehlentwicklung*, or misdevelopment (Honneth 2007: 4; 2014a: 700) – is, at least in its stronger incarnations, no longer available to theorists who abandon even a weak version of progressive universal history. Both of these points raise the question of whether a theory of social pathology shorn of all foundations in universal history or in a general account of social development has at its disposal sufficient normative resources to distinguish healthy from pathological social phenomena. This book as a whole can be read as an extended argument for answering that question affirmatively. One aspect of this answer, addressed in the discussion of Durkheim in Chapter 8, is that the category of misdevelopment (or "a wrong step forward") is not completely out of place in describing dysfunctional reactions to social crises, or "problem-solving" deficits in social life (Jaeggi 2019: 133–72). (Even here, however, I avoid the term "misdevelopment" because of its associations with deviations in standard, predeterminable patterns of development in organisms or, for Freud, human psyches; in accounts of *social* pathology "development" is a potentially misleading term because of its biological connotations.) As Jaeggi (2019) appreciates, a theoretical orientation of this kind cannot dispense with some fairly robust account of immanent critique, in which standards immanent to the object studied – ideals already partially realized – supply (at least some of) the normative criteria in terms of which the health or illness of that object is assessed.

CHAPTER 3

Marx: Pathologies of Capitalist Society

This chapter examines a variety of problems Marx takes to be inherent in capitalist societies that can fruitfully be regarded as social pathologies. Its aim is to clarify how dysfunction is to be understood if the most sophisticated aspects of his critique of capitalism are to be grasped. After a brief survey of some examples of social pathology to be found in Marx, I turn to those specifically bound up with his account of the formula for the circulation of capital, which distinguishes money that is capital from money that is merely money in terms of the *function* played by money in each. As we will see, Marx's biological language in these passages makes it plausible to interpret the various dysfunctions of capitalism he points to there as pathologies. At the same time, it is impossible to grasp the dysfunctions of capitalism without taking into account his (often implicit) conception of the distinctively spiritual aspects of human social being. I argue that Marx regards social life as spiritual in the same general sense I ascribe later to Hegel's view – namely, as informed by the aspiration of social members to unite in their social activity the ends of life with those of freedom – and that capitalism's failure to allow for this unity counts for him as its principal defect.

Organic Analogies and Various Conceptions of Social Pathology

My claim in this chapter is not that the idea of social pathology is the main element in Marx's critical repertoire. Concepts such as "contradiction" and "crisis" play a more central role for him (although note that in English "crisis," too, originally denoted a medical condition).[1] Rather, my claim is

[1] For Marx, crises and contradictions consist in fundamental tensions between productive forces and relations of production that produce social dysfunctions of such gravity that eliminating them requires radical social transformation. (Note, however, that dysfunction characterizes crises and contradictions as well as pathologies.) While remedying social pathologies, as I conceive of them,

that although Marx does not use the term, his critique of capitalism is rife with descriptions of phenomena that can plausibly be regarded as social pathologies. At one point in *Capital* he refers to "industrial pathology," but he means by this not a specifically social illness but the collection of various medical illnesses – the "crippling of spirit and body" (Cap.: 484/MEGA: XXIII.384) – that workers suffer from as a result of laboring in capitalist industry. Similarly, "illness" (*Krankheit*) appears in many places in *Capital*, but here, too, it nearly always refers to the physiological illnesses of workers. If we take seriously the preliminary account of social pathology given in *Chapter 1*, then these phenomena, too, deserve to be thought of as social illnesses since they consist in physiological illnesses caused by capitalist social conditions. As such, industrial pathology would belong to the same category as the unhealthy societies analyzed by Richard Wilkinson (1996), even if what causes physiological illness in the latter is economic inequality, whereas in the former, illness is due to social dynamics bound up with the conditions under which production is organized in capitalism, most prominently, its tendency to demand ever greater rates of exploitation, the health-endangering effects of which Marx portrays in detail. Expressed in the language of Chapter 1, capitalism is a form of society that nonaccidentally makes large numbers of its members physiologically ill.

What Marx calls industrial pathology is not representative, however, of the most prominent dysfunctions he takes to be internal to capitalist society. Some evidence that the concept of pathology is intimately linked to Marx's conception of the kind of thing (capitalist) society is can be found in his statements to the effect that his "scientific" treatment of capitalism regards its object as a "social organism" (Cap.: 101/MEGA: XXIII.27).[2] It is interesting that, whereas Marx is not generally drawn to organicist analogies when thinking about society, he frequently treats capitalist society as sharing important features with biological organisms or species.[3] What Marx often has in mind when invoking this analogy is that capitalism follows a distinctive "natural" course of development – a life-cycle, as it

might sometimes require radical transformation, they are more typically taken to imply reform rather than revolution; the latter might be thought of as aiming not to make an existing society healthier but to replace it with an "organism" of a new type.

[2] Marx approvingly cites a review of *Capital* that refers to *soziale Organismen* (Cap.: 101/MEGA: XXIII.26). Other references to the capitalist mode of production as an organism include Cap.: 93, 102, 448, 466, 479, 518, 531/MEGA: XXIII.16, 27, 352, 367, 379, 407, 416.

[3] At Cap.: 90/MEGA: XXIII.12, Marx refers to "cells" that must first be studied in order to grasp the "anatomy" of the capitalist mode of production, considered as a "fully developed body."

were, with determinate stages characteristic of all capitalist societies, much as the oak tree and the octopus have species-specific patterns of development. Discerning the natural life-cycle of a type of society, Marx avers, enables one to find in nineteenth-century England the broad outlines of the futures of less advanced capitalist societies.[4] Although the idea that a given type of society has a predetermined life-cycle might ground an account of social pathology as a species of misdevelopment, it is not used in this way by Marx, nor would it make much sense to do so – as if France would be ill if its course of development diverged from that of the classical case, England. (In the context of Marx's theory of historical materialism, however, one might plausibly speak of misdevelopment when a stagnant feudal society fails, or takes too long – however this is determined – to make the transition to capitalism or, more relevantly, when socialist relations of production fail to emerge as a response to the dysfunctions of late capitalism.)

When Marx describes capitalist societies as organisms in *Capital*, his preferred expression is "social organism of production" (*gesellschaftlicher Produktionsorganismus*) (Cap.: 172, 175, 201, 508/MEGA: XXIII.93, 96, 121, 407), by which he typically refers to a feature of capitalism deriving from its character as generalized commodity production, where goods are produced with the aim of exchanging them on the market.[5] The feature in question is the apparently anarchic and atomistic nature of capitalist production or, more precisely, the fact that it proceeds in the absence of an overall plan and appears to be the contingent result of countless independent decisions made by individual producers. In truth, however, capitalism is a "social organism of production" in the sense that the success of individual producers – their ability to find buyers on the market willing to pay the full value of their product – depends on an intricate set of market conditions they do not control and are only imperfectly aware of. (If too many shoes are produced, they cannot be sold at their full value, or if produced using outmoded techniques, more labor time is invested than is socially necessary, and their producer loses out in the competition for profit, und ultimately for survival, to more efficient rivals.) Even if

[4] "The industrially more developed country merely reveals to the less developed country a picture of its own development" (Cap.: 91/MEGA: XXIII.12).

[5] In discussing the organism-like structure of the division of labor, Marx distinguishes capitalist from pre-capitalist forms, but uses "organism" in both cases, where its defining feature is that "individual labor" carries out "the function of a member (*Glied*) of the social organism" (MEGA: XIII.21). Marx also refers to the cooperation internal to the capitalist factory as "organic" (Cap.: 465/MEGA: XXIII.366).

production appears to rest on the initiative and decisions of independent individuals, it is in fact the process of an "organism" – of a systematic whole organized from below – in the sense that the various parts of society's productive apparatus succeed and survive only if they stand in specific functional relations to other parts. A capitalist society, on this view, is an organism (or totality) in only the weak and unobjectionable sense that its overall production is carried out by functionally specialized units, the well-functioning of which depends on a host of intricate relations each has to the activities of all others: "It is as cooperators, as members (*Glieder*) of a laboring organism, that [workers] become … a mode of capital's existence" (Cap.: 451/MEGA: XXIII.352–3).

It is possible to ground certain diagnoses of social pathology in precisely this feature of capitalist production. On the one hand, social dysfunction might be located in a falling out of sync of the specialized activities in question, as, for example, in the recurring crises Marx took to be intrinsic to capitalist economies. On the other, one might – as Marx did – take this aspect of the normal functioning of capitalist production itself to constitute a dysfunction. For the way in which capitalist production is organized hides from its members the socially mediated character of their own apparently individually determined activities. This mode of organization obscures to its own participants the thoroughly cooperative and interdependent nature of their multifarious activities, giving rise to the "fetishism of commodities" (Cap.: 163–77/MEGA: XXIII.85–98).

These brief examples provide a glimpse of two points about the nature of the social pathologies I treat below. The first is that the most interesting forms of social pathology rendered visible by Marx require that we expand our conception of social functions to include what he, following Hegel, regards as their spiritual aspects. Whereas the falling out of sync of specialized productive activities can be seen as dysfunctional on a completely economistic construal of that concept, in abstraction from the subjective relations producers have to their productive activity, the systematic non-transparency of social members' activities looks dysfunctional only if we think of the nature of (free) human agency as placing certain spiritual constraints – deriving, in this case, from the conditions of self-transparency in what one does – on the success of those activities: commodity fetishism appears as a pathology only from a normatively expanded perspective on what successful social life requires.

The second point, distinctive to Marx and one of his most interesting contributions to a theory of social pathology, is that what from one perspective appears as functional can appear from a different, spiritual point of view as dysfunctional. For the same feature of capitalist production that renders economic activity nontransparent to its agents can also appear as an advantage from the general standpoint of economic efficiency, or from the standpoint of the "aims" of capital and its owners. What Marx criticizes as commodity fetishism is intrinsic to what enables capitalism to function and is in that sense functional. As we will see below, however, what is functional from the standpoint of capital and its owners appears dysfunctional when the conception of a well-functioning economy is expanded to include the interests, spiritual and otherwise, of those whose labor sustains it.

The aspects of Marx's social ontology that ground the conceptions of pathology I emphasize in this chapter derive less from capitalism's nature as a "social organism of production" – or as an organism of any sort – than from the various respects in which he conceives of economic activities as processes of *life*.[6] As noted, Marx relies heavily on analogies with life processes in explaining what capitalism is, how it functions, and what problems it inevitably generates, and this biological framework invites us to view many of the problems he finds in capitalism on the model of illness. An especially vivid example of this is his claim, articulated below, that the form of circulation that distinguishes capital from mere money produces cancer-like growth in the social organism (Postone 1996: 75–76). This, however, is but one example of a pathology that derives from a more general respect in which Marx views capitalist processes as life-like, namely, insofar as they are driven by social forces that take on a dynamic (or life) of their own that runs its course independently of the will and consciousness of the individual human agents whose activities sustain such processes.

Two further examples of pathologies that exhibit self-perpetuating dynamics are deserving of mention, although I will not examine them in detail. The first is Marx's claim regarding the necessity of recurring and increasingly destructive economic crises in capitalism, according to which continued capital accumulation, which brings with it the rising organic composition of capital,[7] tends to depress the rate of profit and – because reinvestment ceases to be profitable – result in crises of stagnation

[6] This theme is developed differently in Ng 2015: 393–404.

[7] The organic composition of capital is the ratio of capital invested in means of production, raw materials, and instruments of production to capital invested in labor power. Marx believes there is a tendency for this ratio to rise with continued accumulation because in industrial capitalism the drive to increase the productivity of labor takes the form of replacing laborers with machines and other forms of fixed capital, which requires greater expenditure on the means of production relative to that on labor power (Cap.: 762–81/MEGA: XXIII.640–57).

(produced, then, entirely by forces internal to the process of accumulation). These crises throw laborers out of work and result in bankruptcy for capitalist firms, which both lowers the price of labor power and decreases competition among capitalists, eventually making it profitable once again for the surviving owners of capital to reinvest. Production then resumes, until a declining rate of profit again produces stagnation, and so on, with the result that as time progresses, crises become ever more severe. One should note that although it involves a self-correcting mechanism, this dynamic is dysfunctional from *all* points of view, spiritual or otherwise: it produces general social upheaval as well as significant suffering not only for the masses of workers but also for the unlucky capitalists who cease to exist as such. Even from the perspective of surviving capitalists, such crises have a dysfunctional dimension because they paralyze, for longer or shorter periods, the process through which surplus value is created and accumulated, even threatening the very survival of the system. What we observe in this case is a self-undermining dynamic internal to the nature of capitalism; in this system of production, normal (standard) functioning is itself dysfunctional.

The second example is Marx's account of the dynamic at play in the production and utilization of the industrial reserve army, or the structurally unemployed. Here, too, the increasing organic composition of capital accompanying accumulation is explanatorily fundamental: because ever smaller proportions of total capital outlay go toward purchasing labor power, as opposed to machines, factory buildings, and so on, renewed production requires fewer laborers, resulting in both a downward pressure on wages and increased unemployment, each of which affects production. In the first case, a downward pressure on wages ameliorates to some extent the falling rate of profit (because labor power is cheap), and hence also the tendency toward crises caused by that falling rate. In the second, increased unemployment creates a standing reserve of laborers that can readily and cheaply be drawn upon whenever capital enters a phase of expansion and therefore needs, at least temporarily, more labor power than in the previous cycle. When stagnation sets in again, as it inevitably does, the ranks of this reserve army swell once more, recreating the stock of idle laborers that can again be called on at little expense in future phases of expansion, ensuring that, despite inevitable fluctuations in the need for labor power, capital will always have an abundant supply of cheap labor at its disposal. The parallel to certain physiological phenomena is obvious: the industrial reserve army is the central element of a self-regulating system that cushions capitalist production against the shocks it inevitably incurs due to normal fluctuations in business cycles. As such, the army of the unemployed carries out a function essential to capitalism's reproduction, and in a manner reminiscent of the self-regulating, self-maintaining processes of biological organisms. If there is dysfunction here, it is not a disturbance in the process through which capital achieves its characteristic end.

The described fluctuations in capitalist business cycles are normal only in the sense that they are standardly produced by forces internal to the system itself. For Marx, this does not imply that they are healthy or functional (all things considered), although their dysfunctional character comes into view only when regarded from a point of view other than that of capital's characteristic aim – one that takes into account something beyond the ends of capital, namely, the good of laborers, the other class of participants on which the system relies. For this class, the industrial reserve army represents not healthy functioning but poverty, unchosen idleness, and increased exploitation. This means that dysfunctionality for Marx will be more complex than the straightforward conception of it entertained above. It will require conceiving of the healthy functioning of an economic order from a position that is both more encompassing and more robustly normative than the one implied by the specific, class-relative end – the accumulation of surplus value – that drives capitalist production. Whatever the dysfunctions that constitute social pathology turn out to be for Marx, they will be inseparable from some conception of the good of the human beings whose activity makes capitalist production possible.

In both of these examples of self-perpetuating dynamics, capitalist production appears as life-like because it depends on processes that are organized as if to achieve a specific end, the accumulation of surplus value that reproduces capital and renews the production of surplus value. As we will see below, this end, like the ends of living beings, is species-characteristic in the sense that it is essential to what capital is, how it operates, and how it reproduces itself. That capital's motions exhibit a self-perpetuating dynamic depends on their being determined by its specific end – an end that can be achieved only insofar as it motivates certain individuals (capitalists) – such that it makes sense to speak of capitalism as a whole as being driven by ends internal to its nature, and it is this end-directed nature of capital's motions that makes it possible to ascribe a life-like dynamic to them. There can be *tendencies* and a *dynamic of social forces* in capitalist processes only because, by virtue of the kind of thing it is, capital reacts in predictable ways (predictable because of its own determining end) to the conditions, themselves nonaccidental features of capitalism, of competition and advances in technological know-how that make constant and rapid improvements in the productivity of the labor process both possible and necessary.

Although the most straightforward way of conceiving of dysfunctionality in capitalism is to think of it as a disturbance in the self-perpetuating dynamic through which capital reproduces itself and achieves its characteristic end of accumulation, it would be a mistake to take this conception of dysfunction as the only or most interesting model for social pathology in Marx's account of capitalism. As suggested above, many of the most provocative instances of social pathology can be recognized as such only by appealing to notions of health more complex than those used to assess the health of biological processes.

The remainder of this chapter illustrates this point by focusing on five criticisms of capitalism deriving from Marx's definition of capital, set out in *Capital* well before his discussions of crises and the industrial reserve army, that can be plausibly understood as diagnoses of social pathology. Marx's definition of capital is found in his account of the form of circulation that characterizes capital and distinguishes it from mere money – that is, from money that functions merely as a medium of exchange. There are two reasons for making this doctrine of *Capital* central to an examination of social pathology in Marx. First, because the pathologies this doctrine brings to light arise from the elementary definition of capital, they are, if Marx is correct, fundamental to capitalism and belong to it in any of its forms. Second, the pathologies revealed by capital's form of circulation are such that we will be less likely to be misled into thinking of Marx's conception of social pathology as involving dysfunctions only in a narrow sense of the term, as the example of economic crises might suggest, or dysfunctions defined only in terms of economic deprivation, as the example of structural unemployment could be taken to imply. Instead, the pathologies I focus on show that social pathology also involves spiritual dimensions, the recognition of which is essential to understanding what social, as opposed to merely biological, pathology consists in. Contrary to what one might expect from Marx the materialist, his accounts of social phenomena frequently employ the term "spiritual" (*geistig*), and not merely loosely, as in ordinary German, to mean "mental" or "intellectual" in some general sense.[8] As I will show, many of

[8] The use of *geistig* to refer to something like what Hegel means by "spiritual" can be found in the "Economic and Philosophic Manuscripts" of 1844 (MER: 73–7/MEGA: XL.472, 474, 479, 513–17, 524); in *Capital* (Cap.: 341, 375, 482–3, 523/MEGA: XXIII.246, 280, 382, 446); and in *The German Ideology* (MER: 147–200/MEGA: III. passim).

Marx's diagnoses of social pathology presuppose a normative conception of spirit (*Geist*) inherited from Hegel, according to which spiritual phenomena are those that unite, or reconcile, the demands of life with those of self-conscious freedom. Because, viewed from this perspective, social pathologies involve some deficiency in the freedom of those who carry out the material reproduction of society, the illnesses in question could also be thought of as *pathologies of social agency* that fail to meet the requirements of self-determined activity.

Defining Capital via Its Form of Circulation

The theses of Marx's to be focused on here are located in his attempt to define capital by distinguishing money that is merely money from money that is capital via the general formula for the latter's circulation (Cap.: 247–57/MEGA: XXIII.161–70). Following Aristotle's distinction between *economy* (the art of managing wealth within the household in the service of the good life) and *chrematistics* (the art of maximizing wealth) (Aristotle 2009b: I.8–9),[9] Marx begins with the claim that when money functions merely as money – exclusively as a medium for the exchange of commodities – its form of circulation can be represented by the simple formula $C - M - C$, which designates a series of exchanges in which the owner of a commodity she does not wish to consume herself sells that commodity (converts it into money) with the aim of purchasing a different commodity she does wish to consume. The point of such a series is easily explained in terms of the different uses (use values) of the commodities exchanged: the first commodity is not desired by the person who sells it, whereas the second is. In this case money functions merely as a means to achieve an end outside itself – use or consumption – which explains why the exchanges are undertaken.

The formula for the circulation of capital seems, at first glance, to be just the reverse: $M - C - M$, where the possessor of money buys a commodity for the purpose of reconverting it into money. Since money, as Marx says, is "abstract," or a bare universal – "but the changed form of commodities in which their particular use values have disappeared"[10] – no appeal to the specific use values of the commodities involved can make sense of this set of exchanges. Converting money into commodities with the

[9] Karl Polanyi remarks on the genius of Aristotle's distinction (Polanyi 2001: 55–7).

[10] This and the following paragraphs are based on Sections 1 and 2 of Chapter 4 (Cap.: 247–69/MEGA: XXIII.161–81); accounts of the same can be found in Harvey 2010: 87–107.

aim of converting them back into the same thing one started with makes sense only if there is a quantitative difference between beginning and end points, that is, only if the exchange value represented by the second M is greater than that of the first. Hence, in order for the exchanges denoted by $M - C - M$ to make sense as human actions (which they are), their formula must be $M - C - M'$, where M' is greater than M, and where the difference between the two constitutes surplus value (roughly, profit), the ultimate aim of capital's circulation. In this form of circulation, money is not a means but the end for the sake of which exchanges are undertaken: in capitalist circulation, money is an end in itself (*Selbstzweck*) (Cap.: 228/MEGA: XXIII.150), although it is more precise to say that its end is *accumulation*, increases in the quantity of exchange value with which the series began.

Marx's point that capital can be distinguished from mere money only by specifying their respective forms of circulation is equivalent to the claim that what distinguishes them is their *functions*: mere money functions only as a medium of exchange, whereas capital's function – the end served by converting money into more money – is to create surplus value. In both cases the functions ascribed to these processes correspond to conscious aims of the individuals who undertake the transactions. That is, even though the circulation of money and capital can be represented in terms of impersonal formulae making no reference to human agency – indeed, they must be so represented if we are to grasp capitalism as a system rather than as a mere collection of discrete individual acts – the phenomena those formulae describe are made up of transactions voluntarily undertaken by human agents. That human activity underlies capital's movement is precisely Marx's point when he refers to the capitalist as the "conscious bearer" of this movement, or as capital "personified," which is to say, capital "endowed with will and consciousness" (Cap.: 254/MEGA: XXIII.167–8). This point is crucial to grasping the sense in which the pathologies of capitalism are social and not merely natural. Yet the dysfunctionality Marx finds in the circulation of capital does not consist in failing to achieve the subjective aim – the generation of surplus value – that motivates it for the capitalist. (I consider below whether the same can be said of the agents who undertake the converse set of transactions, $C - M - C$.) His claim, rather, is that when the totality of social conditions that makes capital's circulation possible is revealed, other forms of dysfunctionality, still inseparable from human agency, come into view.

As the preceding claim suggests, the method Marx follows in treating the formula for the circulation of capital can be regarded as transcendental in the sense that he begins with a description of a familiar empirical phenomenon, $M - C - M'$, and proceeds to understand, and ultimately to evaluate, it by uncovering the (social) conditions of its possibility and then examining the (social) implications of those conditions as a whole. The first step toward this end is to extend the two formulae distinguished above beyond a single cycle of exchange. Since the aim of $C - M - C$ is the consumption of a specific use value (for example, to acquire bread in order to eat it), its second phase – where money functions only as a medium of exchange and is converted into the desired commodity – brings the cycle to completion in the sense that the subjective end for the sake of which its exchanges were undertaken has been achieved. As new needs and desires arise, as they inevitably do in living beings, the process is renewed, but this takes the form of simple repetition, in a new cycle formally indistinguishable from the first. Thus, $C - M - C$, when temporally extended, reveals itself to be $C - M - C$; $C - M - C$; $C - M - C$; *and so on*, where the cycle's renewal is determined by something external to it: the rhythm of needs and desires as they arise in the sphere of life. For this reason, the formula expressing the circulation of mere money is fully comprehensible from the perspective of life (or living beings) alone and hence independently of particular social conditions.[11] Exchanges mediated by money are, to be sure, more than merely biological phenomena, but biological neediness, conjoined with the inability to satisfy one's needs entirely through one's own labor, is sufficient to explain what motivates humans to enter into such exchanges.

In real exchange economies where more than mere subsistence is at issue, transactions that use money merely as a medium of exchange will also be motivated by desires that go beyond basic biological needs. Marx's assumption, however – I return to this below – is that such desires share the same rhythm of satisfaction as the needs of life: once satisfied, they cease (for a while) to drive one to seek further satisfaction. Thirst, like the desire for a good smoke, is (except in pathological cases) temporarily extinguished after a beer or cigar has been consumed. Satisfaction is followed by moments of peace in which no further desire arises, and even though tomorrow one will most likely wish again for the very same things, there is no presumption that these desires will be more intense or more demanding than those of today. Since we know that in many human societies

[11] Of course, the needs and desires that stimulate the constant renewal of the cycle are never purely biological; they are needs and desires of *social* life.

some desires do not exhibit the same finite, or punctuated, pattern of satisfaction, it is more precise to express Marx's point by saying that there is nothing about the use of money as a medium of exchange that requires or presupposes either unsatisfiable or infinitely expanding desires on the part of those who engage in such exchanges.[12] There is nothing, in other words, intrinsic to this phenomenon that implies the unbridled, "cancerous" economic growth that figures centrally in one of Marx's diagnoses of capitalism's pathologies.

The situation is different in the circulation of capital. Since what is aimed at here – money (or exchange value) – is an abstract quantity shorn of all particular qualities, this form of circulation has no natural resting points rooted in life outside it. Renewal takes place not as a repetition but as a *continuation* of the first cycle, where the original M increases even further (to M'') and where, because purely quantitative aims can in principle always be improved on through further quantitative increases, no resting point internal to the process is in sight. When extended, then, $M - C - M'$ reveals itself to be $M - C - M' - C - M'' - C - M'''\ldots$, *and so on, ad infinitum*. Here one could object that it is always open to a single capitalist to decide at any time that he has had enough and to exit the process of accumulation, in which case he would cease to function as, and hence to be, a capitalist. This is true, but Marx's interest is not in what options individual agents might have in capitalism but in how capitalism works as a system, and if all capitalists happened to decide not to reinvest in pursuit of further accumulation, capitalism would cease to exist. Moreover, we know that this is not how capitalist economies in fact function: even as individual capitalists move in and out of the process of accumulation, economic activity in general continues to be undertaken in accordance with the formula $M - C - M' - C - M'' - C - M'''\ldots$.

Similarities between the Circulation of Capital and Processes of Life

It is not clear what normative conclusions, if any, Marx thinks follows merely from this extension of the formula for the circulation of capital. There is something vaguely unsettling about a social process that exhibits the logic of what Hegel calls "bad infinity" – something's merely going on and on without an end in sight that could bestow meaning or purpose on

[12] Marx's critique here does not depend on claiming that capitalism gives rise to infinitely expanding desires for commodities that cannot be satisfied. This may be true of capitalism after Marx, but it is not his point here.

it or draw it to a satisfying close. Moreover, his reference to "the *restless* augmentation of value" (Cap.: 254/MEGA: XXIII.168; emphasis added), as well as the thought that in the circulation of capital, money becomes an end in itself, are surely signs that Marx thinks something is, at the very least, odd about capitalism. At this point in the text, however, these thoughts are not yet formed into a cohesive critique. What seems to be foremost in Marx's mind here, rather, is not a critique of capitalism but the fact that the formula for the circulation of capital raises a puzzling theoretical question: under what conditions is it possible for money to do what it must in order to be capital, namely, increase in quantity merely through a process of circulation in which, by hypothesis, commodities are purchased and sold at their full exchange value? This question, of course, is nothing but the question of how capitalism itself is possible. The puzzlement we are meant to feel in considering this requirement of capital is expressed by Marx in terms of our having stumbled upon an "occult" phenomenon we are unable to explain: "The movement in which [capital] adds surplus value is its own movement, a movement of ... *self-augmenting value* (*Selbstverwertung*). It has acquired the occult quality of being able to bring value into existence (*Wert zu setzen*) because it is value. It brings forth living offspring, or at least lays golden eggs" (Cap.: 255/MEGA: XXIII.169). Yet an occult phenomenon is not necessarily a pathological one, and even the creepy biological metaphors Marx invokes – throwing off (*werfen*) offspring and laying golden eggs – associate capital somehow with the domain of life but without implying that its life, however mysterious, is somehow diseased.

Before examining how further analysis demystifies capital's apparently occult quality of bringing new value into existence merely through its circulation, it is worth considering what lies behind the fact that, once the formula for capitalist circulation has been expanded, Marx tends to discuss it in terms borrowed from our familiarity with living phenomena: Why, and in what respects, does he take the circulation of capital to bear similarities to processes of life? Part of what makes Marx's treatment of the circulation of capital confusing is that sometimes it is capital's likenesses to (mere) life that are taken as signs of social pathology, and sometimes it is capital's differences to (mere) life that makes it pathological. Life, we should conclude, is a normatively ambiguous concept for Marx, and to think that it furnishes the sole standard in terms of which social phenomena are to be judged is to make the same mistake one falls into in thinking that Nietzsche – another philosopher who emphasizes the living character of the spiritual – looks to biological life alone for his standards of spiritual health (Neuhouser 2017a: 355–368).

The most obvious respect in which the circulation of capital is connected to life is that it is centrally involved in the material reproduction of its participants. Even if the capitalist's ultimate end is accumulation, the production of surplus value determines the economic form within which the reproduction of life takes place in capitalism. Although capitalism produces many commodities other than the necessities of life, it also produces these, and it must do so as a condition of its own survival as an economic system. Since capitalist production depends on the labor of humans, the bodies of those who perform that labor must themselves be reproduced. This basic fact is reflected, if not explicitly, in the formula for capital's circulation: the process denoted by *M – C – M' – C – M" – C – M'"* . . ., though articulated from the perspective of the capitalist, requires that other individuals (laborers) participate in the same process according to the formula *C – M – C*. The first transaction in the first formula, *M – C*, must correspond to a transaction that, from the laborer's perspective, is denoted by *C – M*: the selling of a particular commodity, labor power, in exchange for a wage, where money serves merely as a medium of exchange. Moreover, the point of this exchange for the laborer is to eventually convert that M into commodities that satisfy the needs of life, thus bringing the process *C – M – C* to a notional end.[13] This analysis makes clear that *M – C – M' – C – M" – C – M'"* . . . and *C – M – C; C – M – C; C – M – C;* . . . do not denote numerically distinct processes but one and the same process viewed from the different perspectives of its participants. That the capitalist's *M – C – M'* must correspond to a laborer's *C – M – C* makes clear that the circulation of capital is a cooperative enterprise sustained by differently situated agents whose intentions and actions must complement those of other agents if all are to achieve their own aims.

Yet the connection Marx means to draw between life and capitalist production extends beyond the mundane point that a major end served by such production is biological reproduction. In addition, he describes the circulation of capital as *structurally analogous* to the processes of life – or, more precisely, to processes of life as Hegel conceives of it. Readers unfamiliar with Hegel's account of life will have a hard time finding these themes in *Capital*, but they are unmistakably, and intentionally, there. One indication of this connection is Marx's reference to the circulation

[13] As a pupil of Hegel, Marx knows that for humans, biological needs never appear purely as such. Laborers need not only to eat but to eat in a way consistent with minimal standards of human decency. At the same time, our biological make-up places insuperable constraints on how and what we eat if we are to survive.

of capital as its "cycle of life" (*Kreislauf seines Lebens*) (Cap.: 255/MEGA: XXIII.169). In fact, his depiction of the circulation of capital points to no less than five structural similarities between it and life processes.

The first is that capital is what it is – and differs from mere money – by virtue of the specific function, or end, it serves, one that governs its countless individual transactions and unites them into a single process, much like the diverse behavior of living beings is determined and organized by the biological end of reproducing their species. In the case of capital, "exchange value itself is [its] driving incentive and determining aim" (Cap.: 250/MEGA: XXIII.164), which means that it, too, has a characteristic function defined by a single overarching aim – the accumulation of surplus value – and that it is their subordination to this single aim that makes the individual exchanges intelligible as an ongoing process.

Second, both processes of life and the circulation of capital are infinite, or endless, in a sense alluded to above by the term "bad infinity": nothing internal to the two furnishes a reason for bringing them to an end. Although both can be said to have a final end in one sense (aiming, respectively, at the reproduction of the species and the accumulation of surplus value), neither has a temporally last, or terminating, end in the sense of a natural or logical stopping point, internal to the processes themselves, at which it makes no more sense for them to go on. If this claim about the similarity between capital and life is to hold, however, we must distinguish between two aspects of life: the life-cycles of individual living beings and the pattern of reproduction of a living species. If we focus only on the former, the alleged similarity between life and capital does not obtain: an acorn achieves a kind of completion when it becomes a healthy oak, and its life as an individual does not exhibit the infinitely repeating pattern of bad infinity. However, if we regard life as the life of a species, or even as the larger system of interdependent species, then the analogy between life and capital holds: there is in principle no reason for the accumulation of surplus value or the reproduction of a living species not to renew itself indefinitely, for no birth of a new generation, no creation of a specific amount of surplus value brings the respective processes to a notional end.

This point is distinct from the claim made above that the process denoted by *M – C – M' – C – M" – C – M'"* ... is restless – void of resting points – whereas the process denoted by *C – M – C; C – M – C; C – M – C;* ... is not. First, the earlier point was supposed to indicate a respect in which processes of life *differ* from capital's circulation, and, second, it concerned not the unending character of the process in question but its internal rhythm. As its semicolons illustrate, the rhythm exhibited by $C - M - C; C - M - C; C - M - C; \ldots$ differs from that of the circulation of capital because it is *punctuated* by interruptions due to the nature of the needs or desires in the service of which the process is carried out, which allow for at least temporary moments of satisfaction where no further activity is called for. That process is unending but not restless, and the reason it avoids capital's restlessness has nothing to do with whether or not it has a stopping point but (as I discuss below) follows instead from the nonabstract, or particular, character of its end (the reproduction of life), as opposed to the abstractly universal character of capital's circulation (increasing quantities of exchange value).

A third, related similarity between life and the circulation of capital is that both are "self-determining" in the sense that their determining ends come from within themselves rather than from without: the activities through which a species reproduces itself are determined by, and serve no purpose beyond, the continuation of life itself, just as the constant renewal of capital's circulation serves no end beyond its own reproduction and augmentation. Each of the processes is, in Marx's words, a *Selbstzweck* (Cap.: 254/MEGA: XXIII.167).[14] Hegel would say that this feature makes them self-standing (*selbständige*) phenomena, implying that they are self-sufficient, self-moving, *autonomous* processes (moved by wholly internal ends) and that, as such, they approximate Aristotle's criteria for the highest form of activity, *theoria*, which, higher even than *praxis*, exhibits the same features. Although, for Hegel, life's autonomous, self-moving character gives it an elevated ontological status approximating that of absolute spirit, Marx, abandoning Hegel's metaphysical concern with gradations of being, refrains from drawing the same conclusions for the life-like process of capital's circulation.

When Marx points to the self-determining character of capitalist circulation, he sometimes calls it automatic rather than autonomous: in that process exchange value becomes an "automatic subject" (Cap.: 255/MEGA: XXIII.169).[15] The automatic character of capitalist circulation points to a fourth similarity to life: not only does its circulation take on, as it were, a life of its own but it does so – capital fuels and regulates its own

[14] This is a slightly different claim from the point, mentioned above, that in the circulation of capital, money, or exchange value, is an end in itself. Marx says both (Cap.: 234/MEGA: 23.150).

[15] "Automatic subject" may be a reference to Spinoza's conception of the human mind as an *automa spirituale*, the main idea of which is that the soul is a self-moving machine functioning in accordance with laws immanent to itself, which, because they are natural, causal laws, do not require conscious application (Marshall 2013: 2–4, 19n1).

reproduction – in the absence of a guiding will or consciousness. The sense in which this is the case for nonspiritual animal life is clear: biological species reproduce themselves by giving birth to individual living beings whose pattern of development and self-maintaining behavior are determined not consciously but by the character of the DNA that each inherits from its parents (together with environmental conditions). The circulation of capital cannot be automatic in precisely this sense because the transactions that comprise it, whether $M - C$ or $C - M$, do not take place independently of the consciousness and will of the individual workers or capitalists who undertake them. The automatic, or unconscious and unintended, nature of this process must, then, be located elsewhere.

Marx's point is that the societywide motions of capital are such that it behaves as though it were animated by a living, spontaneously acting soul. Just as it is in the nature of a living being to seek to reproduce life using internal resources, capital, too, "seeks" its own reinvestment and augmentation. Although this could not take place if there were no individuals acting in accordance with their own subjective ends, capital ceases to be capital if not reinvested and augmented. Of course, it is conceivable that the circulation of capital could simply die out, just as it is conceivable that life as a whole might come to an end, but Marx's interest lies in uncovering how capitalism functions when it does, just as the biologist is concerned with understanding how life functions when it does. If we think about capitalism as a social system rather than as a series of actions undertaken by individual capitalists, there is a sense to speaking of capital (as we often do) as blindly pursuing ends that we attribute to it by virtue of the kind of thing it is: "Capital seeks to minimize labor costs" is true not because scores of capitalists, each exercising free will, happen to choose to invest where labor costs are low but because, given what capital is (given its characteristic end), this is what rational capitalists – "the conscious bearers" of capital's motions – do. At the macro-level, then, capital flight to developing countries – a pattern that capital's development takes nearly everywhere – is not explained by referring to the psychology of individuals who "choose" to invest where labor costs are low; the reproduction of society's capital as a whole exhibits the patterns characteristic of it without being guided by an overarching will or consciousness.

That Marx calls capital an automatic *subject* points to a fifth similarity between life and the circulation of capital: both embody – or approximate – the distinctive ontological structure that Hegel ascribes to subjects. In this context Hegel uses "subject" to characterize not only beings that possess consciousness and will, but also certain unconscious and unwilled processes – most prominently, those of life – that he takes to have the same structure, or to exhibit the same form of activity, that is constitutive of conscious subjectivity. Thus, Marx's innovation consists not in thinking of life in general as a subject – this he takes from Hegel – but in finding the same subject-like qualities in the circulation of capital. The Hegelian thesis underlying Marx's claim has its origin in Fichte's account of the subject as essentially *self-positing* (Neuhouser 1990: 102–16). According to this view, a subject exhibits a characteristic mode of being, distinct from that of inert things. This thesis has two parts: The first is the claim that the subject exists only as activity (rather than as a "lifeless" substance); the second specifies the nature or structure of the activity in which the subject's existence consists. More precisely, to say that the subject is essentially self-positing is to say that it exists only insofar as it continually engages in the dual activity of "positing contradictions" internal to itself (for example, distinguishing the I from a not-I) and then "negating" those contradictions without losing its identity as a single thing.

It is not important here to understand why subjectivity in the ordinary sense of the term is supposed to have this structure. More relevant is that Hegel takes life to exhibit the same structure because a living being constitutes itself as what it is only by constantly confronting an environment it treats as distinct from itself and then assimilating to itself what is not it – by, for example, transforming what is initially other such that it can be used to serve the living being's own life purposes. A living species, too, exists as what it is – reproduces itself in its characteristic form – only through the birth and death ("negation") of infinitely many individual beings that both are and are not identical to the larger species that grounds their individual existence. (A particular fish is, essentially, its species – the species brings forth its existence and determines its essential character – but no particular fish is identical with its species as a whole.) In referring to capital as a subject, Marx has in mind its constant need to assume "contradictory" forms – first that of money, then of commodities, then again of money, and so on – in order to be what it is (in order to fulfill its defining function). A more important reason Marx takes capital to belong to the same ontological class as conscious subjects and living species is that it, too, exists only as a *sich Reproduzierendes* (a "self-reproducing something") (PhN, §352) – that is, only insofar as it ceaselessly engages in reproducing itself as the kind of thing it is. Just as a living being that ceases to maintain itself by relating to what it is not soon drops out of the domain of the living, so capital that rests and discontinues its process of self-augmentation ceases to exist as capital and becomes instead mere money, or "lifeless" value.

Perhaps there is a further respect in which the subject-character of both capital and life is relevant to Marx's purposes. There is a sense – appreciated later by Nietzsche – in which life as Hegel conceives of it exhibits an exploitative, or imperialistic, tendency to subordinate what is not itself to its own purposes: it carries out its characteristic (reproductive) activity by taking over its other, that is, by remaking and appropriating – or, in Nietzsche's language, "interpreting" – it. But the sense in which life for Hegel is driven to impress its mark upon its other goes beyond assimilative processes such as nutrition and respiration. In reproducing itself, life also remakes the world around it into a home, an environment more hospitable to its vital aims than raw nature is. Beavers build dams; bees make hives; earthworms aerate the soil. Although Marx does not note this parallel here, capital, too, not only consumes raw materials supplied by nature but also thoroughly reshapes the surrounding world, rendering that world hospitable to its own, species-defining aim of accumulating surplus value. To the beaver's dam correspond factory buildings and warehouses, rail lines and freeways, wind turbines and open-pit mines – permanent modifications of the natural world that have their *raison d'être* in the capital's single-minded drive to augment itself. In the twenty-first century it is not difficult to imagine how this similarity between capital and life is the source of a specific social pathology of capitalism: nature is remade in accordance with the imperative of capital accumulation without regard to whether the natural conditions of human life are undermined.[16]

Finally, although the processes of life and the circulation of capital are alike in approximating the ontological structure of "subjects," both also fall short of achieving full subjectivity. Both exist by reproducing themselves through activities of differentiation and appropriation, but they are not fully subjects because they do so unconsciously. The idea behind distinguishing gradations of subjecthood is that a being that consciously constitutes itself as what it is – in accordance with its conception of what it essentially is, for example – enjoys a higher degree of self-determination than a merely "automatic" subject (since the former's conception of what it is essentially is also, at least potentially, a product of its own activity). This means that even though life is in some sense autonomous and self-determining, it cannot achieve the full, self-conscious freedom available to spiritual (consciously living) beings that determine their activities in accordance with their own conception of what they essentially are. Life's

[16] Marx draws attention to this phenomenon at, e.g., Cap.: 638/MEGA: XXIII.529–30. See also Foster 1999: 371–5, 378–90.

ontological defect in relation to spirit is that it lacks self-consciousness, and for the same reason capitalist circulation, too, falls short of full subjecthood, even if there is a structural similarity between its motions and those of autonomously moving "subjects."

Normative Implications of Capital's Similarities to Life

Marx's intention to draw parallels between processes of life and the circulation of capital is unmistakable, but what, normatively speaking, do these parallels amount to? As noted above, the (mostly implicit) criticisms bound up with this comparison fall into two categories: those where capital's divergence from well-functioning life points to a social pathology; and those where it is precisely capital's similarity to life that makes it pathological. In diagnoses of the first type the assumption is that healthy (or ill) social processes mirror healthy (or ill) processes of life, whereas diagnoses of the second type rest on the idea that social *qua* spiritual phenomena are or should be fundamentally different from those of mere life and that social pathology involves something that is by its nature more than mere life behaving as though it were not. There is nothing inconsistent in holding that social well-functioning sometimes mirrors and sometimes differs from well-functioning in the domain of life, depending on the aspects of the respective processes we focus on. The fact that Marx nowhere points out this duality makes sorting out his view more complicated than it might have been, but it does not make his position contradictory. Among these two alternatives, however, there is a *dominant* line of thought in these sections of *Capital*, namely, that there is an ontological difference between human society and biological phenomena and that something goes wrong when social processes simply mirror life. The social pathologies in this category will be my main object of concern (especially in the following chapter), but before turning to them I take up an example of the other, more easily grasped class of pathologies, where ill health in the social domain mirrors ill health in life.

One respect in which the pattern of capital's circulation is pathological because of a similarity it bears to unhealthy life processes concerns the relation between reproduction and growth in each. Unlike healthy biological processes, capital's cycle of life aims not at the mere reproduction of its species (exchange value) but at its continual increase. As noted above, the unending character of capital circulation means that there is no stopping point to this process internal to its own logic, no point at which its striving could be satisfied or completed. In the case of the life of a species or the web of life comprising all species there is likewise no internal reason for reproduction to come to an end; both life and the circulation of capital exhibit bad infinity in Hegel's sense. In the case of life, the fact that it goes on and on without completion is not a sign of pathology, and, contrary to what one might first think, there is no reason to regard the same feature of capitalist circulation by itself as pathological. Instead, pathology in the latter case comes into view only when we join to its bad infinity the fact that its characteristic aim is not merely to reproduce but to grow in number, as if it had taken the "go forth and multiply" of Genesis[17] – addressed, tellingly, to humans and not to animal life in general – as its governing imperative.

To be sure, growth, too, is a biological concept, but organic growth differs structurally from the growth sought by capital: For the latter, growth is defined solely quantitatively, and for this reason it is "abstract"; it seeks only more of what it is (exchange value) without regard to the form in which that more is embodied (in specific commodities). Organic growth, in contrast, has less to do with quantitative increases than with the development of new or enhanced capacities. When organic growth does include increases in quantity – for example, when height or body mass increase – it is far from merely quantitative. For organic growth consists not in increases in the quantity of abstractly identical elements but requires instead the reproduction of intricately organized forms. Organic growth differs from the growth of capital in that it follows a tight script that allows little room for deviation because it is governed by exacting constraints deriving from the various functions it must carry out in order to live and reproduce. The important difference between qualitative and merely quantitative growth is expressed in the fact that the imperative "do what is necessary in order to reproduce the species" is immeasurably more restrictive than "do what is necessary to maximize surplus value."

Because capital seeks nothing but quantitative increases of homogenous, fungible units of exchange value, its "movement … is without bound or measure (*maßlos*)" (Cap.: 254/MEGA: XXIII.167). That is, the circulation of capital is not subject to self-imposed constraints with regard to rhythm, pace, form, or size. In the case of a living being, the functions it must carry out in order to live furnish an exacting measure for its growth; it lies in the nature of capital's function (to accumulate indistinguishable units of surplus value), however, that no such internally derived measure regulates

[17] Genesis 1:28, loosely translated.

its growth. Moreover, because the circulation of capital is governed by a quantitatively defined imperative that commands continual growth, it is indistinguishable from one that commands not simply quantitative increases but *maximization* (of surplus value). (If only quantity matters, there is no internal reason for capital not to strive for maximization rather than mere increases.) What the formula for the circulation of capital reveals is that capital's cycle of life is subject to an unceasing pressure to find ways to maximize the production of surplus value and – it is here that the pathological character of capital's circulation comes into view – to do so single-mindedly, without concern for whatever extra-economic or human costs it might entail. If it is possible to speed up the phases in which *M* is converted into *C* and back again into *M'*, capital will use all means at its disposal to do so, where the "decision" to accelerate flows simply from the kind of thing capital is. To use a concept I introduce more formally below: capitalist production eludes the *conscious control* of those whose activities make that process possible. The familiar image of modern life as racing on a hamster wheel – a furious "going nowhere" at a pace beyond all control – derives its resonance from this feature of capital's pattern of reproduction.[18] To return to the comparison with life, the logic of capital's circulation – unending, uncontrollable, ever-accelerating, blind to externalities – makes it look more like cancerous growth than the healthy reproduction of life. Capital is value that just grows and grows, without discernible bounds and without regard for its human effects; it is like the blob that ate Downingtown, Pennsylvania, in the 1950's movie of the same name.

Although the measureless growth of capital looks like a specific type of pathology in biological organisms, it would be a mistake to tie even this form of social pathology too closely to its organic counterpart. This is because there is nothing in the mere idea of uncontrolled growth in either realm that by itself indicates illness. In a biological organism such growth counts as illness only because, and when, it impedes one or more of the organism's vital functions. This is an important point for both physiology and social philosophy. As I reconstruct the concept of social pathology here, dysfunctionality is a central feature of any phenomenon deserving that label. Accordingly, finding an adequate conception of function for the social domain – and, by extension, of

[18] Rosa Hartmut turns capital's tendency to accelerate economic processes into the defining feature of modernity (Hartmut 2015).

well-functioning – constitutes the largest challenge faced by any attempt to rehabilitate the idea of social pathology. That there can be no simple identity between biological and social pathologies is a consequence of the fact that societies and living organisms are different kinds of beings, despite their possessing similarities that, in some cases, make analogies between them meaningful. This is why articulating an acceptable conception of social pathology cannot ultimately be separated from issues of social ontology, including questions concerning how to conceive of the appropriate functions of a social "organism."

That cancer in biological organisms and pathological growth in capitalist societies cannot be conceived of in strictly identical terms comes more clearly into focus if one takes seriously the suggestion above that it is the single-mindedness of capitalist growth – its proceeding without concern for the human costs it might entail – that makes it cancer-like and therefore pathological. Clearly, some evaluative presupposition, taken neither from biology nor economics narrowly construed,[19] underlies the claim that purely quantitative growth that takes place at the expense of other human goods constitutes pathology in the social domain. Diagnoses of social pathology are by their nature normative; so, too, are diagnoses of pathology in physiology. In contrast to the latter,[20] however, the norms invoked by the former rest on certain claims about the human good (or, more modestly, about the good of certain historically and socially situated human beings). To think otherwise is to fall into the trap of a scientism that fails to recognize the inescapably ethical character of the functions to be attributed to human societies. Many diagnoses of social pathology – capital's cancer-like growth, for example – do not presuppose thick or controversial conceptions of the human good: an economic system fueled by the imperative to maximize the wealth of some without regard to how it affects the good of all clearly has a susceptibility to pathology written into its genes, as it were, even when the good of all is defined in rudimentary terms.

Despite his embrace of some form of naturalism, Marx does not fall into the trap of scientism as defined above. This is made clear by the debt he acknowledges to Aristotle in analyzing the formulae for the circulation of money and capital. For in citing Aristotle's distinction between economy

[19] Such presuppositions are not foreign to the tradition of political economy, including as practiced by Smith.

[20] Many philosophers of disease maintain that even standards of health employed in medicine rest on conceptions of human flourishing that are not strictly physiological (Murphy, n.d.).

and chrematistics – corresponding to the circulation of mere money and that of capital, respectively – Marx notes that for Aristotle the former concerns itself with managing money in service of the good life. Further, in explaining the different rhythms of $C - M - C$; $C - M - C$; $C - M - C$; … and $M - C - M' - C - M'' - C - M'''$ … he cites Aristotle's claim that "the measure of possessions adequate for the good life is not unlimited," in contrast to the unlimited end governing the maximization of wealth (Cap.: 254/MEGA: XXIII.167). Indeed, this detail adds a new dimension to the claim regarding the distinct rhythms of $C - M - C$; $C - M - C$; $C - M - C$; … and $M - C - M' - C - M'' - C - M'''$ … : rather than understanding the former as a natural rhythm that follows the logic of desires arising in the sphere of life, one can take Marx's point to be that, *viewed from the perspective of what the good life requires*, the healthy acquisition of goods appears limited in quantity, constrained in pace, and attentive to the particularities of the goods it seeks, while unpunctuated and measureless pursuit of wealth for its own sake looks, at best, incomprehensible and, at worst, incompatible with human well-being.

The point that considerations regarding the good underlie even the most biological-appearing of the social pathologies Marx alludes to leads naturally into the second, more prominent class of social pathologies, where the distinction between the human good and the norms governing mere life is more salient than in the first, inasmuch as these pathologies exhibit a similarity rather than a difference to healthy life processes. The first step in examining this class of pathologies is to articulate how Marx (implicitly) conceives of the fundamental difference between processes of life and those of human societies. A first hint of this can be gained by returning to his comparison of capital's circulation to the movements of life. As noted above, Hegel regards life as exhibiting an ontological defect in relation to spirit (its lack of self-consciousness), but he does not take this defect to imply a criticism of life. Although life is less free (less completely self-determining) than self-conscious spirit, this "defect" is merely part of what it is for something to be a living rather than a spiritual being. Even if spirit is ontologically superior to mere life, it would be wrong to expect life to be more than it is or to criticize it for not being what it is not. Marx's point is that things are different in the case of capital: although the formula for capital's circulation presents its movements as the life-like motions of things (money and commodities), what underlies those motions are *social activities of human subjects*. This means that it is fitting to judge capital's motions by standards of freedom appropriate to spiritual phenomena – standards of self-conscious agency – and that their failure to be fully subjective (unconscious and unwilled) calls for critique. As I explain below, activities of self-conscious subjects that appear instead as the motions of things could be called alienated social powers (MER: 46/MEGA: I.370).[21] Such activities fall short of the normative standards appropriate to the kind of thing they are, and because of their merely life-like characteristics it is appropriate to regard alienated social powers as spiritual, or social, pathologies.

In the sections of *Capital* under examination here, Marx does not articulate the normative standards on the basis of which he judges the activities of humans in capitalism to be alienated social powers. But the standards that make sense of Marx's critique are formulated in a preliminary way at the end of Part I of *On the Jewish Question*: "Human emancipation will be complete only when the real individual human being ... has recognized (*erkannt*) and organized his own powers (*forces propres*) as social powers ... and for that reason no longer separates social power ... from himself" (MER: 46/MEGA: I.370). At the heart of this conception of unalienated social powers (or activities) is the Hegelian ideal of humans *appropriating*, or reappropriating, their own powers, in a context where those powers appear external or alien and where reappropriation turns on humans both *recognizing* (or knowing) their own powers as the social powers they in fact are and *organizing* them such that they *become social powers* in a normatively more adequate sense. There is clearly an ideal of self-determination underlying Marx's call for us to organize our powers ourselves, but his characterization of reappropriated social powers also implies that a distorting individualistic perspective on one's activities is part of what makes the powers of social members alienated and in need of reappropriation: if reappropriation requires coming to recognize our own powers as social powers, then there must be some sense in which those powers (or activities) are already social even if we do not experience them as such, and this lack of transparency with respect to what we do in the social world is part of what Marx thinks makes alienated social powers deserving of critique. (Here the ideal of self-transparency underlying his critique of commodity fetishism reappears.) One element of human emancipation, then, involves a transformation of the consciousness of social members such that they will be able to regard their activities as integral parts of a social, or cooperative, undertaking.

[21] For a different aspect of unalienated social powers – producing for others – see Kandiyali 2020: 555–587.

Contrary to what some have thought, this ideal of nonalienated social powers continues to animate Marx's thought until the end of his career. This continuity is unmistakable in his characterization in *Capital* of one of the two forms[22] of the freedom to be realized in a post-capitalist future: "Freedom [in the sphere of material production] can consist only in socialized human beings – associated producers – rationally regulating their interchange with nature, bringing it under their collective control, rather than being dominated by it as by a blind power, and achieving this ... under the conditions most appropriate to ... their human nature" (MER: 441/MEGA: XXV.828). This passage from the mid-1860's echoes the one from *On the Jewish Question* in emphasizing the conscious and socially controlled, or organized, nature of unalienated activity, as well as the need for social agents to reappropriate activities that were once alien to them (to bring them under their collective control rather than being dominated by them). What the later passage makes clearer than the earlier is that the activities to be reappropriated are themselves processes of life – productive activities, or "interchanges (*Stoffwechsel*[23]) with nature" aimed at satisfying human needs – that have been elevated so as to be appropriate to and worthy of human beings. Implicit in Marx's statement here (but also throughout the 1844 Manuscripts) is that humans are spiritual beings, that is, not only participants in life moved by biological needs but also (potentially) free subjects. Hence what makes human life activities appropriate to human nature is not only their character as free (transparent and self-determined) but also their success in satisfying the needs of life. Stretching this only a bit, one can say that reappropriated social powers are *productive of the human good* and, applying to this the ideal of transparency, that they are also recognized as such by those whose activities they are – which is just another way of saying (in terms Nietzsche would endorse) that for a human being, *affirming* what one does (without illusion) is a constitutive element of reappropriating one's own activity.

For Marx, then, the most fundamental respect in which mere life belongs to a different order of things in comparison with human society is that the latter is *self-conscious life*, where that implies that the processes through which life maintains and reproduces itself are social activities

[22] The other, allegedly more complete form of freedom is realized beyond the sphere of material production. Here labor is no longer undertaken out of biological need nor merely as a means to satisfy external ends; instead, labor – as "the development of human powers" – is valued as an end in itself (MER: 441/MEGA: XXV.828). See Kandiyali 2014: 104–123.

[23] *Stoffwechsel* is a biological concept, normally translated as "metabolism" (Schmidt 2014: 76–79; Foster 1999: 379–83).

that could in principle be known and collectively directed by those whose activities they are – that could, in other words, be free, or unalienated. Since self-conscious life is just another name for spirit, satisfaction, or health, for spiritual beings is realized for Marx in life activity that meets the criteria for free (unalienated) activity mentioned above. How, though, is this claim relevant to understanding the instances of social pathology implicit in Marx's formula for the circulation of capital? The key to answering this question, explored in the following chapter, lies in seeing how the idea of human social activity, in the form of labor, enters into his attempt to comprehend $M – C – M' – C – M'' – C – M'''$ ... by inquiring into the conditions that make such a mysterious phenomenon as the self-augmentation of exchange value possible.

CHAPTER 4

Marx: Labor in Spiritual Life and Social Pathology

This chapter continues the themes of the preceding one by examining the importance of labor in Marx's conception of spiritual life and in his diagnoses of social pathologies, especially in *Capital.* At the end of the chapter, I suggest that his conception of human society omits, or underemphasizes, important elements of the spiritual aspect of human social life that the theories of later chapters enable us to incorporate into a more adequate social ontology.

Labor and Its Normative Significance

I suggested in the previous chapter that the formula for capital's circulation masks the true character of that process inasmuch as it presents capital's movements as the motions of things (money and commodities), even though it is in fact human activity that makes those motions possible. Thus far, however, I have spoken of human activity in conjunction with the circulation of capital only in the sense that the individual transactions that make up that process – countless instances of $M - C$ and $C - M$ – are human actions undertaken in accordance with certain subjective aims: both capitalist and worker enter into those exchanges with the intention of achieving a specific end they are fully aware of. Exchange itself, then, is a human activity, a paradigm example of instrumentally rational action with clear criteria for success or failure: the agent who undertakes $M - C$ succeeds by acquiring a commodity that will enable her eventually to turn M into M', and the agent who undertakes $C - M$ succeeds inasmuch as he receives a sufficient quantity of M to purchase, in a further exchange, a different C that will satisfy the desire or need that motivated the exchanges in the first place. It is possible that a kind of social pathology, one involving a failure of purely instrumental agency, might be located simply in some systematic inability of some or all of the agents involved to achieve the specific ends that motivate their exchanges. I return to this possibility below, but the most important forms of social pathology implicit in the formula for the circulation of capital are not of this type. If the main thrust of Marx's diagnoses of pathology is to come into view, it will be necessary to find another sense in which the motions of capital represented in that formula depend on human activity. Doing so requires rejoining our analysis of the formula for the circulation of capital at the point at which we abandoned it in the previous chapter, just before comparing capital accumulation with the processes of life.

Recall that we left unsolved the central puzzle raised by *M – C – M' – C – M" – C – M'"* ..., namely, how it is possible for money in the form of capital to increase in quantity merely by its circulation. One could describe Marx's puzzle in Hegelian terms by saying that the formula for the circulation of capital as articulated thus far is not explanatorily "self-standing" in that the possibility of the phenomenon it represents cannot be grasped in terms of that formula alone. For, as Marx goes on to argue, it represents a certain movement of things without reference to the specific type of human activity that makes that movement – the (apparent) self-expansion of exchange value – possible. True to his method of "transcendental" inquiry, he takes his next task to be to uncover the social conditions that make self-expanding value possible in order to bring the nature of capitalism more completely into view.

In the following sections of *Capital*, Marx attacks this puzzle by examining more closely the specific commodity workers bring to the market to convert into money, their labor power, and distinguishing between its exchange value (which, by hypothesis, the worker receives in full in the form of wages) and the amount of exchange value the "consumption" of that commodity in the productive process – translating the capacity to labor into actual laboring activity – produces for the buyer of the commodities necessary for production, the capitalist. What we learn is that *M – C – M'* sheds its occult appearance only when it is further expanded so as to make clear that *M* can become *M'* only if a process of production *P*, involving an expenditure of human labor, intervenes between the two *M*'s. Thus, if the possibility of *M – C – M'* is to be grasped, it must be expanded into *M – C ... P ... C – M'*, where *P*, if it is to yield surplus value, must make use of a kind of activity (human labor) under specific social conditions – conditions that, as will be eventually revealed, make the systematic exploitation of human labor necessary.[1] The most important of

[1] Although exploitation will ultimately reenter my account of the pathologies of capitalism, it is not, as I make clear below, the core of those pathologies.

these conditions is the nonnatural circumstance that one class of humans appears in the marketplace as the owners of only one commodity (their labor power), while another appears as owners of exchange value in excess of what they need in order to satisfy their life needs.

The key to understanding the most important forms of social pathology depicted in this part of *Capital* lies in recognizing the nature of the specific activity that makes the self-augmentation of capital possible: human labor. This activity is different from, and spiritually more significant than, the activity involved in merely exchanging *C* for *M*. As the passages cited in the previous chapter make clear, Marx thinks of labor as *social, productive activity* that has the potential to make the life activity of human beings – the material reproduction of themselves and their society – a spiritual phenomenon subject to normative standards beyond the norms of health internal to merely biological life. Those passages also make clear what those standards require: a reappropriation of our activity that turns alienated social powers into free activity, where appropriating our own activity – making it genuinely our own – takes place along three dimensions: *knowing* (or understanding) it as it really is; collectively *controlling* or organizing it; and *affirming* it without illusion as appropriate to and worthy of the (spiritual) nature of human beings.[2] In other words, unalienated social powers must be *transparent, self-determined*, and *productive of the good* of those whose powers they are. Moreover, making social powers productive of the good, and hence genuinely affirmable, requires reorganizing, and not merely reinterpreting, our activity so as to make it social in a sense that in capitalism it is not.

How, then, does Marx conceive of human labor, and why does it play a central role in his critique of capitalism?[3] Marx begins his account of human labor in *Capital* with a set of claims about the kind of thing it is at more or less all times and places. First, labor is an activity of life – "a process [that takes place] between the human being and nature"[4] – the necessity of which derives from a basic property humans share with the rest of animal life: the need "to appropriate the stuff of nature in a form useful for their own life" (Cap.: 283/MEGA: XXIII.192).[5] Labor is not

[2] The Hegelian provenance of this conception of activity that is fully one's own will be obvious to readers familiar with "Morality" in the *Philosophy of Right*.

[3] For an engaging, updated treatment of labor and its significance, see Geuss 2021.

[4] Marx need not claim that all labor involves interacting with nature but only that at least some members of any society must interact with nature if it is to be self-sustaining.

[5] For Hegel and Marx, the fundamental activities of material and spiritual life are forms of appropriating – making one's own – what is initially foreign.

merely one thing humans do among others; it is crucial to their survival, and for this reason it is a permanent dimension of human life, even if the unprecedented development of productive forces in capitalism opens up the possibility for necessary labor to take up increasingly less of our time (MER: 440–1/MEGA: XXV. 828). At the same time, human labor differs fundamentally from the life activity of other animals because it is an interchange (*Stoffwechsel*) with nature that humans "arrange, regulate, and control through their own deeds" (Cap.: 283/MEGA: XXIII.192). Human labor is consciously undertaken activity, not merely animal behavior, and as such it is to be understood and evaluated in categories appropriate to conscious agency.

Labor is a form of subjective agency, in the first place, because, unlike the behavior of nonhuman animals, it is guided by a conscious end, the satisfaction of a particular need or desire (Cap.: 284/MEGA: XXIII.193). Digestion can function well – and even functions best – when one is unaware of doing it, whereas the idea of unconscious labor is barely intelligible. Moreover, digestion differs further from labor in that it takes place without conscious direction of the body. Although both digestion and labor involve bodily interactions with nature in the service of life, only the latter requires conscious, intentional direction of the bodily motions it consists in. This means that the physical characteristics of one's body and the natural world, as well as the particular nature of the need to be satisfied, place demanding constraints on how those motions must be undertaken.

In the 1844 Manuscripts, Marx draws attention to another aspect of the conscious (and self-conscious) character of human labor. He begins by making what can sound like the same point about labor discussed immediately above: in contrast to nonhuman animals, "the human being makes his life activity into an object of his will and consciousness; his life activity is conscious."[6] But it quickly becomes clear that he means by this something beyond the mere fact that human laborers imagine the product they intend to make before producing it: "The animal is immediately one with its life activity. It does not distinguish itself from its activity. It *is* that activity." For the human being, in contrast, life activity "is not a characteristic with which he simply merges; ... his own life is an object for him." Marx's point is that human labor is not only conscious with regard to the end to be achieved and the means for doing so. It is, beyond

[6] All quotations in this paragraph come from MER: 75/MEGA: XL.516; in the second, Marx's emphasis has been amended.

this, life activity that the human being can distinguish from himself and thereby make into an object of reflection and evaluation. Human laborers can take a stance on their activity, whereas nonhuman animals cannot. It is this aspect of the conscious nature of labor that allows for the possibility of being subjectively alienated from what one does and that underlies the normative aspiration to reappropriate one's activity by affirming it, or finding it good.

Like *poiesis* for Aristotle, labor in the service of life's needs is *productive* – it brings about something other than itself – in that it is directed at a specific end external to the activity itself, where that end derives from and is answerable to specific needs outside itself. This is not to say that labor cannot also be experienced as intrinsically rewarding but only that it is constitutive of something's being labor that part of its purpose be located in the satisfaction of a need outside itself. One consequence of the poietic character of labor is that affirming it as one's own activity involves seeing its *consequences* as good, which includes recognizing that one's labor has succeeded in satisfying the needs for the sake of which it was undertaken.

The fact that, unlike strolling in the park, labor is undertaken for specific instrumental reasons explains why it requires a degree of subjective engagement – a concentration and attention to the specificity of one's objects and environment, as well as to the needs one is trying to satisfy – not essential to amusement or relaxation. Thus, labor distinguishes itself from less consequential activities because it engages and makes demands on the subjectivity of those who perform it to an extent, and in a multitude of ways, that many other human behaviors do not. Unlike enjoying a meal or shopping for clothes, labor requires deploying specific habits and skills acquired through the discipline of previous labor. This explains why Marx speaks of labor not as one activity among others but as a human *power*, something that is not extinguished when some specific activity comes to an end but remains as an enduring resource to be called upon again and again as further needs arise. The fact that labor is the exercise of a power and that it engages the worker's personal makeup and history – specific habits and skills she has participated in acquiring – makes it plausible to think of laboring activity as (potentially) an externalization of self, or self-expression, which is itself a kind of freedom.

Let us now consider why labor is a social activity. The principal reason is that (nearly all) labor is cooperative. While most labor requires the working together of many hands – harvesting crops, building ships, educating children – this is not Marx's main point. The type of cooperation most relevant here stems from the fact that even in capitalism the totality of productive activities that enable society to reproduce itself comprises, whether recognized as such or not, a socially organized scheme of cooperation. We have already encountered this idea in two places: in the account of commodity fetishism and in noting that if social reproduction and the process of circulation are to occur, the various instances of $M - C - M'$ must correspond to the various instances of $C - M - C$, implying that the actions of social members must materially complement one another if all are to achieve their aims. The socially organized character of labor, together with its status as productive activity, means that labor is not only socially organized – divided up according to some cooperative scheme – but also social in the sense of having to realize, in some sense, a collective good. Human labor is social, or cooperative, because achieving my ends depends on my cooperators achieving theirs, and vice versa, where this is accomplished within a network of complementary acts that produce for all. Capitalist production requires cooperation in a further sense: among classes rather than merely among individuals occupying different places in the material division of labor. The need for this type of cooperation is reflected in the fact that, from the perspective of individual participants, the circulation of exchange value takes two forms, with distinct starting points, depending on the class positions of those involved: $C - M - C$ and $M - C - M'$. This complementarity goes beyond that involved in every form of the division of labor, where my specific needs must be satisfiable by the specific activities of others, and vice versa: capitalism also demands the cooperation of differently situated economic classes, defined by their respective ownership relations to the productive forces – where, of course, "cooperation" does not imply that its benefits and burdens are distributed equally or fairly.

The social, or cooperative, nature of labor emphasized above does not simply follow from its definition. As the example of Robinson Crusoe shows, it is conceptually possible for there to be labor that is not part of any socially organized scheme. Rather, the cooperative nature of labor at issue here follows from a fundamental social condition under which the great majority of human production takes place, namely, that a division of labor, simple or complex, renders each of us dependent on others for our survival and well-being. This thoroughgoing interdependence necessitates that individuals' productive activities be organized according to a cooperative scheme of some sort if there is to be even a rough correspondence between the various kinds of labor undertaken and the diverse needs it must satisfy. Borrowing Hegel's terminology, we could say that a laborer in capitalism is an *I* that is a *we* – where what one does is determined by how one's activity fits in with that of other social members – even if in this case one is likely to be unaware that one's activity is in reality a social power. Or, to put the point differently, what I do as a laborer in capitalism is constituted – although only objectively, through mechanisms of the market – as a specialized and intricately coordinated part of a cooperative project directed at satisfying not only my needs but also those of my cooperators. (We must also say, then, that a laborer in capitalism is not yet fully an *I* that is a *we* in Hegel's sense inasmuch as she does not *understand* herself and her labor as parts of a social project. Becoming fully an *I* that is a *we* requires not only knowing one's own powers as social powers but also consciously organizing them as such.)

In sum, labor is a complex form of self-conscious, intentional activity that our natural needs compel us to undertake.[7] It is of great importance to humans not merely because nature compels us to labor but also because, in the societies we are familiar with, it occupies a major portion of our waking life. Even more importantly, the complexity of most forms of labor – the intricate skills it relies on if it is to achieve its ends in conformity with its objects' constraints, as well as the transformation of the laborer it effects through the acquisition of ever more sophisticated skills – makes this form of activity well suited to be a source of subjective satisfaction and self-expression.[8] Labor is not only relevant but crucial to social philosophy for Marx because he views human societies as centrally engaged in the material reproduction of life. This reproduction is accomplished through socially organized labor such that individuals can satisfy their own life needs only by integrating their specialized labor into a complex, societywide web of laboring activity that constitutes, whether laborers know it or not, a collective project. Labor is so central to what human beings do, to what they are, and to how they conceive of themselves that it is not merely a means to reproducing the material conditions of life but also "a specific manner in which they express their life, a specific way of life that belongs to them" (MER: 150/MEGA: III. 21). Social life, then, has both material and spiritual dimensions, and since the former requires interacting with the world as a living being, both "the physical and spiritual life of humans is bound up with nature" (MER: 75/MEGA: LV: 516). It is for all

[7] Even if intellectual workers do not engage in a physical interchange with nature, their labor is still, in part, a means for satisfying their life needs.

[8] Contrary to what the young Marx may have believed, these claims do not imply that labor is the only human activity with spiritual potential, nor that it is the spiritually most elevated of human activities, nor it that it constitutes the "human essence."

these reasons that Marx regards labor both as appropriately judged by the norms of free agency and as central to a social philosophy concerned with realizing the human good.[9]

Labor in Capitalism

We are now in a position to ask how and why Marx regards labor in capitalism as destined to fall short of the standards of free activity. As already noted, Marx holds that in capitalism the cooperative character of our labor (in the main sense distinguished above) is concealed from us by the atomized way in which production is undertaken when coordinated, unconsciously, by a system of market exchange. It is this that explains why labor in capitalism is not usually recognized by those who perform it as the social activity it already is, and this lack of transparency, a consequence of a basic feature of capitalism (commodity production), constitutes a part of the spiritually pathological, "fetishistic" character of that mode of production.

Moreover, part of the legacy Marx inherits, via Feuerbach, from Hegel's conception of spiritual phenomena is the view that if human activities are unfree in the sense just articulated – unknown as the social powers they in fact are – those same activities will appear to those who carry them out as, and actually be, powers that control or dominate the very subjects whose powers they are, thereby inverting the proper relation between agents and their activities. According to this view, a necessary concomitant of the unrecognized social character of labor in capitalism is that the unconsciously coordinated productive activities of individuals manifest themselves as objective laws of the market that dictate (without there being a dictator) nearly everything about production in that society, including what will be produced, how it will be produced, and whether workers will be able to find employment at wages high enough to satisfy their families' needs. That market-determined forces control us rather than vice versa is not directly visible from the formula for the circulation of capital, but it is a consequence of a feature of that process implicitly represented by that

[9] Although the young Marx acknowledges that sexual reproduction is a major way in which humans reproduce life (MER: 156–7/MEGA: III. 29), little of his work is concerned with the family (although Engels' *Origin of the Family, Private Property, and the State* is based in part on Marx's notes). Because Marx's main aim is to understand how the capitalist economy works, he focuses on reproductive labor in that sphere rather than on the reproductive labor traditionally carried out by women in the family.

formula, namely, that exchange value, not the satisfaction of specific needs, is capitalist production's "driving incentive and determining aim" (Cap.: 250/MEGA: XXIII.164). If profitability calculated in terms of quantities of exchange value determines what and how a society produces, then "decisions" on those matters elude the control of those whose activities make up that collective project; far from determining their own activities, they are determined instead by a social power that has its ultimate source only in themselves. Insofar as it strives toward its end unmediated by the consciousness and wills of the agents who make it possible, capital is an alienated social power that serves first and foremost its own ends (those of capitalists) and only secondarily, if at all, the ends of those whose labor it depends on.

A further respect in which the activities of laborers in capitalism fail to be fully their own comes into view by focusing on the extent to which, under the conditions of capitalist production, their labor is productive of the good and hence affirmable by them. As suggested above, Marx's conception of spiritually satisfying life activity requires that such activity not only be transparent to and organized by the subjects whose activity it is, but also satisfy the ends that are the reasons for undertaking it and be seen by those subjects as doing so. In the case of labor, seeing what one has accomplished as good includes seeing that one has achieved the ends that motivated one's activity in the first place, which entails seeing that one's own life needs have been satisfied through one's labor. (Here the requirement that activity be affirmable by its agents is hard to separate from the requirement that it be controlled by them. If we cannot affirm what we have done because we have failed to achieve our ends, then there is a sense in which our activity has eluded our control: even if we organize our activity and in that sense determine it ourselves, its failure to bring satisfaction implies a loss of control over what we have done.)

In order to understand this aspect of Marx's critique of capitalism we must return to the abstract nature of capital's driving aim, its being defined exclusively in terms of homogenous units of exchange value. We saw above that whereas life achieves its vital aims only by establishing very specific relations with its surrounding world determined by the requirements of life and the circumstances of its environment, capitalist accumulation is more self-sufficient: its aim is supremely indifferent to the particular forms the commodities it produces assume, including the use values they possess. It is the abstract nature of the value capital aims at – its utter disregard for the particularities of its products – that explains Marx's statement that in capitalism "use values are produced only because and to the extent that they are the material substratum, the bearers, of exchange value" (Cap.: 293/MEGA: XXIII.201). This perfect indifference between abstract exchange value and the particular use values that are its bearers is relevant to whether laborers can recognize and affirm their activity as their own because it opens up a permanent, nearly unbridgeable gap between the two kinds of value capitalist circulation depends on: those whose labor produces commodities enter the market with aims defined in terms of use values, but their production is organized by a different end, the maximal accumulation of exchange value. Of course, capitalist circulation can achieve that end only if it produces use values of some kind – there must be effective demand for its products if surplus value is to issue from the process – but it is indifferent to what specifically it produces as long as production serves accumulation. Hence the motor driving capitalist production is perilously decoupled from the ends of those whose labor sustains it.

It may be fine for living processes to be autonomous and self-moving in the sense of being determined only by purposes internal to themselves, but something has gone badly wrong when the movements of things, which are underwritten by human activities, serve the sole end of maximizing surplus value. The internal character of capitalist circulation's aim is especially inappropriate when the activity that makes it possible is human labor. For (in every mode of production we have known thus far) labor is not generally undertaken for its own sake but for ends external to it, the satisfaction of needs laborers have independently of their laboring. No matter how unalienated it might be, labor retains an ineliminable moment of *poiesis*, aiming, in part, at ends external to it.[10] One of the dangers of the self-moving character of capital's circulation is that its criteria for success do not track the needs of those who make the process possible; when such tracking is absent, their labor fails even as *poiesis*.

We broached this topic in the previous chapter in asking whether the two complementary transactions that capital's circulation depends on, the laborer's $C - M - C$ and the capitalist's $M - C - M'$, are, even from a narrowly instrumental perspective, equally successful at satisfying the respective subjective aims motivating them. The decoupling of use value from exchange value at the heart of capitalist production opens up

[10] This claim might seem to contradict Marx's statement that in communism some labor – "beyond the sphere of material production" – will be undertaken for its own sake, where "the development of human power" counts as an end in itself. This does not imply that such labor will not also aim at satisfying ends external to it, but only that it is not determined by natural need and that it is experienced as valuable in itself (MER: 440–1/MEGA. XXV.828).

the permanent possibility that the end of accumulating surplus value is achieved while the ends that motivate workers are not, or only partially. In fact, there are three possibilities for failure, all of which point to real phenomena familiar to members of capitalist societies: First, there may be workers who need to sell their labor power in order to live but who cannot find work (the problem of unemployment); second, those who find work may receive wages too low to satisfy their needs, since, as with all commodities, the price of labor power is determined by principles of supply and demand operating beyond the control of human agents (the problem of the working poor); and, third, even if workers find a living wage on the market, the vast majority of use values actually produced may be available only to those to whom surplus value accrues (the problem of extreme inequality in access to the wealth produced in a cooperative project). Of course, the formula for capital's circulation alone implies not the necessity, but only the endemic possibility, of these outcomes. It is only when the decoupling of use value and exchange value is supplemented by doctrines developed later in *Capital* – the rising organic composition of capital and the dynamic of the industrial reserve army – that Marx can claim failures of this type to be necessary features of capitalist production.

If a laborer in capitalism is fortunate enough to find work that pays her a living wage, it may well be that her own labor succeeds in satisfying the needs that led her to labor in the first place. This means that, from her individual perspective, she may well be able to affirm the results of her productive activity as having enabled her to succeed in what she set out to do in entering the market. (There remains the question of whether her activity is transparent and self-determined, but this can be put aside for now.) Since some workers in capitalism may be fortunate in this respect, the more basic obstacle to their being able to see the results of their activity as good lies elsewhere. It comes into view only by looking at labor not from an individual perspective – with regard to what one's own activity achieves for oneself – but as a totality, as constituting a cooperative project, or social power. If one's labor satisfies one's own needs, but only on the condition that the labor of others produce the goods that do so, then it is both theoretically and morally shortsighted to assess the situation only in terms of what one's own activity yields for oneself. If capitalist production is a cooperative undertaking, it ought to be collectively affirmable, which it can be only if it succeeds in satisfying all its participants' needs. The formula for the circulation of capital explains why, viewed from the perspective of the whole, it is a matter of chance whether labor in capitalism satisfies needs in ways that can be affirmed by more than a few. For the end that determines the specific forms that production assumes – maximal accumulation – is divorced from all consideration of the collective good, even when interpreted as thinly as possible, as the satisfaction of its participants' life needs.

This is the reason Marx enjoins us not only to recognize our labor as a social power but also to organize it as such – to make it into a social power in a more robust sense than it already is in capitalism. This is the most important respect in which Marx's view of the requirements of spiritual health differs from Hegel's: it is not enough merely to understand what we already do as what it truly is in order to be free in our life activity. The ultimate goal of Marx's diagnoses of social pathology is not simply to bring to consciousness what is hidden from view in commodity production – that the labor of each is determined via market mechanisms by its relation to the labor and needs of all – but also to reorganize our labor such that it answers to standards of the collective good, including the material needs of all, thereby making that labor collectively affirmable. If we are to do this fully – uniting the ends of life with those of freedom – our concern must also be to satisfy the vital needs of all in a way that makes our laboring activity subjectively free: collectively transparent, controlled, and affirmable by us not merely in its results but also as an activity itself. Unalienated labor is not activity that has shed its nature as *poiesis* but activity in which *poiesis* is united with *praxis* (activity engaged in for its own sake), insofar as productive activity, because self-determined and self-expressive, is, at least in part, an end in itself and not carried out solely for the sake of what it produces.

Why Pathology?

It is time to return to the question of why the defects of capitalism implicit in these sections of *Capital* should be thought of as social pathologies, as opposed to normative shortcomings of some other kind. I distinguished above five instances of pathology, each of which involved a failure to satisfy at least one of the requirements of free life activity, namely: that that activity be transparent to those whose activity it is; that it be collectively organized or controlled by them; and that it be genuinely productive of their good and hence collectively affirmable. The five respects in which capitalist production was said to fail short of these criteria, and hence to imply social pathologies, were: i) Capitalism thoroughly remakes nature in accordance with its purely abstract end, threatening to undermine the very conditions of future human life; ii) capitalist growth tends to be measureless and uncontrolled with respect to rhythm, pace, form, and size (and is therefore indifferent to the human good); iii) the cooperative (social) nature of labor in capitalism is hidden from those whose labor it is; iv) laboring activity is determined not by human subjects but by non-conscious market forces; and v) production's determining end, exchange value, tracks only accidentally the good and needs of those who produce.

One feature of these pathologies that distinguishes Marx's position from most other theories of social pathology is that the pathologies in question are endemic to the form of modern society he investigates. The pathologies of capitalism he diagnoses are normal rather than anomalous phenomena, such that pathology could be said to be intrinsic to capitalist society. In other words, dysfunction – both material and spiritual – is built into, or follows from, basic features of capitalist production, namely: commodity production in accordance with the principles of market exchange; the existence of economic classes defined by their ownership relations to the means of production; and the fact that capital's organizing principle is the maximization of surplus value measured exclusively in quantitative terms. That these pathologies are intrinsic to capitalism explains why Marx is a revolutionary rather than a reformer: no cures exist for them short of substituting a new mode of production for the old. There are, however, resources in Marx's texts that might license us to formulate his position on the "cure" for capitalism in a weaker form. He claims, to be sure, to be uncovering "the natural laws" that govern capitalist production "with iron necessity" (Cap.: 91/MEGA: 23: 12). Yet at the same time he calls those laws "tendencies" and admits that extra-economic factors, such as the Factory Acts in nineteenth-century England, can act as a counterweight to these tendencies, making the empirical phenomena governed by them less harsh than the laws of capitalist production would by themselves predict.[11] One might claim, then, that what is intrinsic to capitalism is a *tendency* to pathology and that one task of the diagnostician of social pathology is to think about what political or social measures might prevent the pathological tendencies of capitalism from issuing in real pathologies.

Although this is a position followers of Marx might plausibly take (and that many have[12]), it is not Marx's. For the passage cited here makes clear that, although extra-economic measures might mollify the ills of capitalism and prolong its life, its pathological tendencies are destined to win

[11] See also Cap.: 92/MEGA: XXIII.15–16.

[12] Piketty's prescriptions to combat economic inequality can be read in this spirit (Piketty 2017: 597–745; 2020: 966–1034).

out, making the sickness and eventual death of capitalism a necessity in the long run (Cap.: 91/MEGA: XXIII.12). This insistence on the need for revolution is one respect in which social pathology on Marx's account diverges significantly from biological illness, where cures never consist in rebuilding the organism into a being of a different kind. Moreover, this aspect of his view poses a special problem for articulating the idea of dysfunction at the heart of his conception of pathology: in Marx's case what counts as a dysfunction cannot be determined by reference to how the capitalist system standardly works in the same way that a dysfunction of the heart shows up only in relation to an understanding of what the heart is supposed to do that derives from how it normally functions. Many of the pathologies examined in this chapter appear instead as perfectly functional if one takes the maximization of surplus value to be the function of capital's circulation; these pathologies can be recognized as such only by bringing in normative criteria external to the reality of capital's functioning. (This issue is more complex than it appears, for there is a sense in which Marx believes that the normative criteria presupposed by his diagnoses of pathology are indeed internal to capitalism – or, better perhaps: internal to capitalism understood as a form of life [Jaeggi 2019: 36–51]. For he takes the validity of the ideal of free social activity, embodied in unalienated labor, to depend both on the real aspirations capitalism has generated in its participants and on its having produced the material conditions of that ideal's realization in dramatically improved human productive powers.)

These features of Marx's position – that capitalism's pathologies are intrinsic and incurable and that the standards defining its dysfunctionality cannot simply be read off of how the system standardly works – only increase the urgency of asking: Why conceive of capitalism's intrinsic defects on the model of illness? As we have seen, Marx refers to human societies, both capitalist and precapitalist, as organisms, but the analogies he invokes are too weak to point directly to any determinate picture of what social illness might consist in. I have argued, however, that even the weak sense in which Marx takes capitalist society to be a "social organism of production" provides resources for conceiving of some of capitalism's defects as illnesses. The thought here is that the capitalist economy is a highly coordinated ensemble of productive activities, one function of which is to produce the material conditions of its own reproduction. (That capitalism serves this function is compatible with its driving end being the private accumulation of surplus value, since achieving the latter presupposes that the former is also achieved.) Moreover, many of the goal-directed processes through which capitalist society reproduces itself involve self-perpetuating dynamics that resemble processes through which living organisms maintain and reproduce themselves. The dynamics in question constitute something like a living system of forces, where one force affects and is affected in turn by others, with the result that their interaction acquires a life of its own, not directly dependent on the intervention of conscious agency. In a being of this sort many forms of dysfunctionality become possible, and it is this – the presence in a "living" being of a *dysfunctional dynamic* – that makes the use of the concept of social pathology appropriate. One source of such dysfunctionality in capitalist society is the circumstance that, being unguided by a conscious agent, the dynamics in question escape the control of those subject to them (who are, at the same time, the agents whose activity fuels them), producing systematic consequences not intended by anyone, not even the most advantageously situated.

What is essential to a social organism in this sense is simply the working together of differentiated individuals, who make specialized contributions to social production, which must be elaborately coordinated – if only blindly – if the goods produced are to correspond to the needs of producers. Even here, however, Marx's conception of a social organism of production is more complex than this, encompassing features that stretch the society–organism analogy and provide resources for expanding the notion of dysfunction to include spiritual pathologies. One complexity is that although much of the coordination among laborers is accomplished unconsciously in capitalism, this itself depends on there being a large amount of coordination – internal to a workplace, where workers' relations are not mediated by exchange – that, unlike coordination in biological organisms, requires conscious planning (Cap.: 477/MEGA: XXIII.377), even if not by the workers themselves. Thus, in human social life, even where unconscious coordination seems to reach its highest point, such coordination depends on conscious agency.

These points touch on the principal respect in which the social pathologies attributable to Marx fit uncomfortably with the model of illness and the central mark of that concept, dysfunctionality: the spiritual pathologies I have articulated here appear as such only if one brings into the picture external normative presuppositions about how a human society with highly developed productive capacities "ought to" function that are not derived from how capitalism in fact normally (standardly) works. In contrast, diagnoses of pathology, whether in animals or societies, involve applying immanent standards – they depend on a judgment that a living system fails to meet the standards of health immanent in its own normal functioning.

I have suggested that there may be ways of interpreting Marx that make the diagnoses of pathology I have attributed to him more immanently grounded in the reality of their objects than seems to be the case when focusing only on the formula for capital's circulation. Yet despite these possibilities, there are reasons to want to go beyond the account of the immanent dysfunctionality of the pathologies his critique of capitalism allows us to give, where these reasons stem from more than merely a desire to fit Marx as neatly as possible into the tradition of social pathology theory. These reasons are also not simply due to the fact that Marx's focus is *capitalist* social life, where the cleavage between normative social consciousness and normal functioning might seem to be distinctive of that form of economy alone. (In that case, the problem I am pointing to would be a problem of capitalism, not of Marx's view of the nature of social life more generally.) Rather, as I argue below, these reasons for dissatisfaction point to a problem at the root of Marx's social ontology,[13] at least as it is articulated in his understanding of social life and transformation generally – namely, in his theory of historical materialism.[14] Although Marx recognizes the thoroughly socialized (*vergesellschaftet*)[15] character of humans and their life activity, there is a fundamental aspect of sociality missing from – or, at best, highly undeveloped in – his conception of social life in general.[16] One might articulate this objection by saying that Marx's conception of human society is excessively materialistic, that in explaining in general what societies are and how they work it ascribes too small a role to human consciousness. The point I want to make might be formulated in this way, but if so, it must be further qualified.

Marx's framework for understanding how societies in general function and, especially, how they develop and eventually transform themselves, posits relations among four elements said to belong to any mode of production (except for very simple societies that lack class divisions): the productive forces; the relations of production (or economic structure of society); the superstructural legal and political institutions; and social consciousness, which, in class societies, is the locus of ideology (MER: 4/MEGA: XIII.8–9). Because Marx unmistakably embraces some form

[13] For an account of Marx's social ontology that emphasizes "individuals in social relations" as the fundamental elements of society, see Gould 1978: 1–39.

[14] My interpretation of historical materialism is heavily based on those of G. A. Cohen 1978 and Wood 1981: 61–110.

[15] In the *Theses on Feuerbach* (MER: 143–5/MEGA: III.535) but also throughout *Capital*.

[16] Daniel Brudney, focusing on the early Marx, emphasizes some of the spiritual aspects of social life (Brudney 1998: chs. 4–5).

of materialism, it is often thought that the consciousness of social members, including the normative acceptance of their institutions, plays no role in his understanding of social reality. This, however, is incorrect. His inclusion of social consciousness among the elements of all class-based modes of production is a clear acknowledgment that, in some form, the *meaning* social members ascribe to their social participation, including the values they take it to realize, is essential to social life. If social consciousness, including beliefs about the legitimacy of institutions, functions to shore up existing relations of production (G. A. Cohen 1978: 290–2), then Marx's materialist thesis is not that normative consciousness can be dispensed with in social life.

His view, rather, encompasses two claims, both of which deny the explanatory primacy (but not in every respect the causal efficacy[17]) of social consciousness. The first claim is that the content of a given society's social consciousness is explained by something outside itself, namely, the existing relations of production, where this explanation is functional: a certain configuration of social consciousness is dominant because it serves to shore up existing relations of production. The second claim is that what ultimately explains how societies develop and transform themselves is, again, external to consciousness – in this case, the state of development of the productive forces and the extent to which the economic structure of society continues to further their development. Neither claim implies that social life can go on without social consciousness that legitimates existing institutions in the eyes of their members. Indeed, historical materialism implies precisely the opposite: social life requires that its participants conceive of their social participation as having a "point," or as serving the good in some way. One might even say that this reveals a further sense in which social life for Marx is spiritual: generally speaking, social activity – including in capitalism – is informed by a sense on the part of agents of what makes their institutions legitimate and their own activity good. In this context one should recall Marx's statement that labor is not, in most times and places, merely a means to reproducing material life but also "a specific manner in which [social participants] express their life, a specific way of life that belongs to them" (MER: 150/MEGA: III. 21).

Since the role consciousness plays in Marx's account of capitalist functioning is frequently misunderstood, it is important to distinguish two

[17] The causal role of social consciousness is synchronic, not diachronic: internal to a given mode of production, it props up existing relations of production (by legitimizing them in the eyes of social members).

ways in which it enters that account. First, the circulation of capital depends on individuals, capitalists, and workers, acting on specific conscious aims: they exchange *M* for *C* and *C* for *M* with an explicit goal in mind, either to satisfy their needs or to accumulate surplus value. The good for the sake of which they engage in such exchanges is a particular good: the satisfaction of one's needs or an increase in one's private fortune. This form of conscious agency is essential to capitalism's functioning, but it is not what Marx has in mind in speaking of social consciousness – first, because it involves no norms beyond those of instrumental reason and, second, because it is concerned with individual actions, not with the goodness or legitimacy of the social system itself.

The second place at which consciousness enters Marx's account of how capitalism functions – as social consciousness – is more robustly normative, and it has as its object society (or its economic structure) as a whole. This form of consciousness is necessary to capitalism's functioning because workers' acceptance of the disadvantageous terms under which they participate in it requires that they take those terms and their systematic consequences to be meet certain ethical standards (to be both just and productive of the good). The beliefs Marx ascribes to the social consciousness that legitimates capitalist relations of production can be read off of his account in *Capital*, after the mystery of the source of surplus value has been resolved, of how the wage–labor relation appears to participants in the system themselves, namely, as embodying ideals of "freedom, equality, property, and Bentham" (Cap.: 280/MEGA: XXIII.189). More specifically, the social relation that makes the production of surplus value, and hence capitalism, possible appears as just because it is based on a contract that is *freely* entered into by all parties; respects the *property* rights of each; and involves the exchange of *equivalents*. It appears as good because the social system it makes possible seems to realize a version of the utilitarian ideal, where each, in following only his own interests, participates in a scheme of social production that *serves the good of all.*

The respect in which consciousness plays a diminished role in Marx's account of capitalism – and of social life more generally – is that the social consciousness it presupposes in its members is systemically inefficacious (or, as is often said, "epiphenomenal").[18] More precisely – since it *is* efficacious in that capitalism could not survive without it – social consciousness

[18] In the words of one reviewer of *Capital* whom Marx cites with approval: "Marx regards social development as a process of natural history, guided by laws … independent of the consciousness and intentions of human beings" (Cap.: 92/MEGA: XXIII.26).

in no way tracks, constitutes, or steers the course of capitalist accumulation and development, and in this sense it remains external to the actual functioning of the economy it claims to be about. In an important sense, what participants in capitalism take themselves to be doing is irrelevant to what they in fact do – external to how the system actually functions – if the latter is understood in terms of the systematic consequences of their coordinated activities and the ends those activities actually achieve. In capitalism, for Marx – as well as in other modes of production – real functioning and social consciousness fall apart. The normative beliefs of social participants *enable* capitalism to function, but they are not *internal to* its functioning, which means that any spiritual dysfunctions Marx's theory diagnoses remain external to the system's actual functioning. The problem here – for a proponent of social pathology theory – is that if the analogy between societies and biological organisms is to be tight enough to warrant applying the idea of (spiritual) dysfunction to the former, then it must be possible to ascribe (spiritual) functions to social institutions that are more directly bound up with how they actually function than is the case in the pathologies examined in this chapter. In analyzing the circulation of capital and ascribing to it a function (the accumulation of surplus value), Marx proceeds fully immanently; the problem is that this immanent function is unrelated to the standards for *healthy functioning* his diagnoses of pathology presuppose.

It is worth asking whether this aspect of Marx's view of social life derives from his focus on the (capitalist) economy rather than on institutions such as the family and the state. As Hegel recognizes, civil society functions (largely) "behind the backs" of its participants (PhR: §§181–2, 184, 186–7), which is to say: even without consciousness on their part of the good those institutions achieve or of what their well-functioning consists in. For Hegel, however, this aspect of civil society distinguishes it from social life more generally, where what Marx calls social consciousness is necessary for the healthy functioning of institutions and where, even more fundamentally, the content of that consciousness is (partially) constitutive of institutions' actual functions. I do not mean to suggest that Marx's diagnoses of the social pathologies of capitalism are themselves ill-founded but only that the conception of social life implicit in them (and in his historical materialism) is not generalizable to other social spheres or to social life generally and is therefore not rich enough to capture social pathologies more broadly, outside the domain of the capitalist economy. As I have suggested, the problem here is not merely that Marx's view falls short of being a paradigmatic example of social pathology theory but that there is an important element of social reality related to normative social consciousness that his understanding of social life in general leaves out of sight.

In subsequent chapters I attempt to make good on this claim by examining various conceptions of social reality, according to which normative social consciousness is (partially) *constitutive of*, not external to, the functioning of social institutions and by showing that these views capture an essential element of human social life that Marx's historical materialism overlooks or obscures, namely, the extent to which normative commitments (of various kinds) – and not merely need, interest, or power – are what holds society together and underlie its functioning. According to these views, determining the functions of social institutions, as well as what their members actually do when participating in them, depends on what those members *take themselves* to be accomplishing in their social activity. Since the latter includes their sense of the point of what they collectively do – of the good their cooperation serves – normative standards, including conceptions of the good, are not external but internal to the real functioning of institutions. Hence, in the case of social (but not biological) life, functioning well includes realizing the good at which participants themselves take their institutions to aim. In Marx's critique of capitalism, in contrast, it is possible to characterize how capitalism functions – as he does in laying out the implications of the formula for the circulation of capital – without taking into account the participants' understanding of what makes their cooperative project worthy of their participation. Chapter 6 begins to address this deficiency in Marx's social ontology by examining and defending Rousseau's claim that human society is "artificial."

CHAPTER 5

Plato: Human Society as Organism

This and the following chapter examine the views of two social philosophers, each of whom articulated prior to the nineteenth century (a version of) one pillar of the social ontology I take defensible theories of social pathology to rely on: Plato provides a classical defense of the society–organism analogy that, despite its ancient provenance, remains relevant today, and Rousseau articulates an innovative version of the modern view that human society is artificial, or made by us.[1] What I take from Plato is the thought that human society is to be understood functionally, as composed of parts or subsystems especially suited to serve specific functions, and that it is primarily in this respect that societies resemble biological organisms. In the case of Rousseau, the claim I emphasize concerns the artificiality of human societies, construed as a thesis not about their historical origins but about the constitutive role that normative consciousness plays in social life. Very roughly, then, the conception of human society I argue for here is a synthesis of Plato's functionalism and Rousseau's artificiality thesis, and it is not far-fetched to describe the protagonists of the following chapters, Durkheim and Hegel, as proponents of some version of precisely this synthesis.

The Functionally Organized Polis

The first systematic account of social pathology in Western philosophy – the account of a "fevered" polis – occurs in Book II of the *Republic* (Plato 1992: 368e–373e)[2] in the context of Plato's attempt to discover the nature of justice in the individual soul by first locating it within a larger, supposedly

[1] I do not mean that only moderns regard society as a human creation but that Rousseau's way of doing so is distinctively modern.

[2] Plato refers to other versions of the sick polis without calling them fevered, suggesting that there are other ways a polis can fall ill (Plato 1992: 544c, 556e, 563e, 564b).

isomorphic being, the polis. It is noteworthy that Plato's account of the polis invokes an analogy, not between society and a biological organism, but between society (or the polis) and the human soul (or psyche). I will return below to the significance of this variation on the society–organism analogy, but for now it suffices to note that, despite this variation, it remains true that his account of the polis rests on an organic analogy since he takes both psyche and polis to share essential structural features of biological organisms, as his frequent use of metaphors of health and illness when treating both the psyche and polis makes clear (Plato 1992: 372e, 373b, 407c–e, 409a, 409d, 444c–e). That Plato regards the human psyche as closer in nature to the polis than a biological organism is a sign that he understands the polis as a spiritual, and not a merely living, organism.

It is relatively easy to grasp the sense in which Plato understands the polis as a kind of organism and his reasons for doing so, and this will be the main topic of my treatment of his views. Before turning to this topic, however, it is worth considering in what respects he also regards the polis as in some sense humanmade. That he does so is suggested by the fact that his account of justice begins by asking how the polis "comes into being" (Plato 1992: 369–374e). If this is construed as a straightforwardly genetic question, then the artificiality of social life might simply be understood as the thesis that the polis is the product of human consciousness and will. If we add to this the fact that Plato then immediately discusses various human needs that social life serves, the picture that seems to emerge is one in which the polis comes into being as the intentional result of humans' attempts to satisfy needs they have independently of political membership.

That this is not the sense in which Plato upholds the artificiality of the polis, however, is signaled by the fact that he frames his inquiry into the origin of the polis in terms of "watching [its] coming into being *in thought*" (Plato 1992: 369c; emphasis added). Although this account takes the form of a genetic narrative, it should not be read as a story about how humans societies in fact originate and develop. It is, instead, a hypothetical narrative that aims to clarify two related matters: first, the essential functions of the polis, including, as we eventually see, the achievement of justice, both in the polis and in the souls of its members; and, second, the implications those functions have for how society ought to be constituted. Thus, in asking about how the polis comes to be Plato constructs not a causal but a functional explanation, where this has two related strands. The first is explanatory, aiming to explain both *the existence of the polis* and *why its parts are constituted as they are* by appealing to the functions, or human needs, they serve when brought together properly within the whole. Here Plato claims that, regardless of how a polis originates, what explains its permanence in human life is that it is well-suited to serve certain basic human ends, including those bound up with the material reproduction of society and its members. The second, related strand of Plato's functional account is normative rather than explanatory. It provides an answer – even if merely partial[3] – to the question of what makes social life good for humans, and it does so by appealing to the various ends that the properly ordered polis serves. For reasons that become clearer below, one theme that surfaces repeatedly in theories of social pathology is the inseparability of explanation and evaluation in accounts of human society.

The respect in which Plato might be taken to regard the polis as artificial, then, is weak but still worthy of note. His suggestion that creating the polis in thought reveals fundamental truths about social life implies that, even if real poleis are not originally the intentional products of human agents, determining the polis's proper functions is facilitated by thinking of it as a human creation: for methodological reasons it is fitting to view the polis *as if it were* something we create in order to realize certain goods the value of which is readily grasped by its members. Hence, the weak sense in which the polis might be regarded as artificial for Plato involves two claims: first, that we can understand why the polis is a permanent feature of human life, and why it is good that this is so, by referring to the human ends it is well-suited to realize; and, second, that a recognition of what makes social life good is generally available to its participants, just as it is to uninitiated readers of the *Republic* who have not yet acquired philosophical knowledge of the Good, and that such recognition is part of what explains the polis's continued existence (by providing its members with good reasons to reproduce it). Thus, articulating how the polis might be artificial for Plato leads directly to the functionalism of his social theory, which is also the source of his view that the polis is structured like a living organism.

The main thesis of Plato's functionalism is that human society is made up of mutually dependent, specialized parts (both individuals and groups) well-suited to carrying out specialized functions, which when appropriately coordinated satisfy the essential needs of society as a whole and the

[3] The answer is partial because the polis realizes a good – justice – that does not appear in the account of needs Plato appeals to in reconstructing the polis's coming-to-be.

individuals who compose it. His reconstruction of the polis in thought proceeds by first identifying fundamental human needs that can be effectively satisfied only within a polis and then, on that basis, identifying the essential functions of a "healthy" society. In other words – in a move that Durkheim repeats in the opening sections of *The Social Division of Labor* (DOL: 11/55) – Plato specifies the functions of a polis in terms of the vital human needs it satisfies. If to specify the functions of society is to define the human needs it serves, then a functionalist social theory must be able to determine what needs human have and which are relevant to understanding social life. In Plato's case this inquiry rests on a fixed conception of human nature and the Good. Although contemporary theories of social pathology mostly reject the ideas of a fixed human nature and an a priori knowable good, two aspects of Plato's position are worth retaining: first, that our thoroughgoing *dependence* on others for the satisfaction of our needs is fundamental to understanding social life; and, second, that *material needs* – the requirements of reproducing ourselves as living beings – play an important, if not exhaustive, role in specifying the functions of a healthy society.

In pursuing the latter point, Plato begins by identifying various material needs of humans – food, clothing, shelter – and then ascribes the different social tasks that cater to those needs to the specialized groups in society best suited to carrying them out. Specialization is crucial to Plato's vision of social life because it leads to increased efficiency in satisfying human needs. This rudimentary division of labor implies further needs – for merchants and retailers, for example – to which further functions, and hence further subgroups within society, correspond. Yet specialization alone does not guarantee increased efficiency. As Marx points out when contesting the idea that capitalist production dispenses with conscious planning (Cap.: 439–54/MEGA: XXIII.341–55), specialized activities must be appropriately coordinated if they are to result in the increased efficiency that specialization makes possible. These two points are sufficient to grasp the basic principle of Plato's functionalism in social philosophy: the survival of even the simplest of societies depends on carrying out a variety of functions, and some form of division of labor – assigning diverse functions to different specialized parts of the whole – helps to ensure that a society's vital needs will be met. This is also the main idea underlying the analogy between societies and biological organisms shared by social theories that employ the concept of social pathology: it is *coordinated specialized functions*, executed by a living being's coordinated specialized parts, that make that analogy plausible and ground diagnoses of social pathology, understood as a breakdown or disturbance in the functioning of those parts.

Since Plato's account begins with material needs, it might be thought to be grounded biologically, but if so, then in a significantly weaker sense than is the case for Comte and Spencer in the nineteenth century: the biological nature of humans – with the needs that derive from it – plays a central role for Plato in understanding the functions served by social cooperation, but this does not imply that the methods of social theory ought to track those of a science of living beings or that the structure and functions of human societies are homologous to those of living organisms. Ultimately, Plato's account of the human needs that determine the functions of a healthy society extends beyond the material needs that are most effectively satisfied in society. As Glaucon famously suggests, a scheme of cooperation that satisfied only those needs of its participants would be like a city of pigs (Plato 1992: 372d–e), rather than one in which higher, distinctively human goods can be realized. It is not necessary to explore here what Plato takes humans' nonmaterial needs to be (although his account of human psychology in Book IV suggests that the need for honor, or recognition, is among them). More important is that his functionalist account of the polis accords major significance to the satisfaction of its members' biological needs while also denying that its essential functions are limited to this. For Plato, taking into account the essential differences between the lives of pigs and those of humans is crucial to grasping the nature and point of human society.

It is tempting to read Plato's later description of the desires underlying the diseased polis's fever – desires that "overstep the limits of … need" (Plato 1992: 373d) – as implying that social pathology is the direct consequence of desires that go beyond purely natural needs. On such a reading, the desire to "recline on proper couches" or to "dine at a table" would suffice to indicate that illness had seized the polis. The extreme austerity of this position is evident, and it should prompt us to seek an alternative interpretation, a hint at which is provided by the first characterization of fever-producing desire cited above, a desire for "the endless acquisition of money" (Plato 1992: 373d).[4] It is important that what is fever-producing about this desire is not that it is directed at money (and hence goes beyond purely biological needs) but that – as in Marx's account of the circulation of capital – it is a desire for endless acquisition,

[4] Given his account of the soul in Book IV, it seems likely that Plato means to trace this desire back to the desire for honor associated with *thymos* (Plato 1992: 439e–440c). If so, he anticipates Rousseau's position, in which *amour propre* and the needs created by luxury play central roles in explaining the corrupt social state (Neuhouser 2010: 57–89)

for ever-increasing amounts of money with no foreseeable point at which satisfaction might be achieved and desire stilled. In other words, it is not a desire's distance from biological needs that makes it pathological but the dysfunctional consequences it has for society and its members. As Plato's narrative makes clear, pathology-inducing desires can arise only once they go beyond the purely natural, but this alone does not make "unnatural" desires hazardous to social health or to the well-being of those who have them. In imagining fever-producing desire as a desire for a good that is measured only quantitatively, Plato anticipates a point made more explicitly later by Rousseau, Hegel, and Marx in their critiques of desires, particularly widespread in modern societies, whose endlessness makes them impossible to satisfy.

An important feature of Plato's functionalist account of social life – which may explain why the psyche is a more appropriate analogy for the polis than the living organism – is that it is developed in the context of an explicitly ethical project to discover the nature of justice (*dikaisyne*), where this ideal, in distinction to many modern conceptions of justice, is inseparable from a conception of a good human life. In other words, the *Republic* offers an avowedly normative account of the good of social life, including its contributions to the human good in the broadest sense. For Plato, understanding what the polis is cannot be separated from understanding the essential human goods it secures. Functionalism in social theory more generally always involves a normative account of society in at least a weak sense, insofar as attributing functions of any kind to society or its parts implies ascribing ends to its constitutive processes that can be achieved with varying degrees of completeness and efficiency. That normativity of this weak sort is insufficient for a satisfying grasp of social life is made clear by the observation that not every species of well-functioning in this sense is also productive of the good: an oppressive secret police system can effectively achieve its ends, but a social philosophy without the resources to criticize such an institution is clearly deficient. (As I suggested in Chapter 2, a society that relied on an oppressive but efficient secret police would not be well-functioning as a whole. This is because ethical standards – what makes the secret police oppressive – are not deontological pronouncements of pure reason but are, instead, inseparably bound up with considerations of what is affirmable and tolerable by real human beings.) In any case, in Plato's social philosophy – as in the understanding of social health I develop here – the well-functioning of a society cannot be specified independently of a conception of the human good and an account of how it is realized in social life.

Given the tight connection between functional claims and normative standards, it should not be surprising that my description of Plato's functionalist account of the polis already provides an initial answer to the question of what social health is: a healthy polis is one in which each specialized part effectively carries out the specific function appropriate to it and in which these functions are coordinated such that all are well executed and society's essential needs satisfied. Thus, it is first and foremost (for Plato) a harmonious balance among different functions that defines social health, and in this respect the healthy society embodies the ideal underlying many Classical conceptions of physiological health, as expressed in Hippocrates's teaching that illness results from imbalance among the body's four fundamental humors. One should not lose sight of the fact, however, that this ideal is at root functionalist: specifying what harmonious balance consists in for a being depends on knowing which functions it must carry out and therefore which needs it must satisfy. Conversely, illness, whether biological or social, implies not merely imbalance but also impaired functioning.

As noted above, illness makes its appearance in Plato's genealogical account of the polis as an imbalance among society's parts that manifests itself as a fever. The diseased polis is described as fevered because it presents itself as overheated, that is, in a state of heightened agitation in which whatever internal resources it possesses for restoring its processes to a healthy tempo are limited or impaired, creating the possibility of increased agitation and an eventual spiraling out of control. In Plato's account, fever reveals itself in the polis's unconstrained tendency to expand its borders through war and the conquering of neighboring peoples, without regard to whether doing so is good for the polis (or its neighbors). Although Plato does not himself invoke this analogy, the fevered polis's behavior could be seen as akin to cancerous growth and, hence, as exhibiting a pathology beyond mere fever, perhaps a species of fever-induced cancer.

Regardless of whether this sickness is fever or cancer or both, Plato consistently refers to the polis, and not to the individuals who compose it, as ill. This is an important point for theories of social pathology because it is bound up with three closely related claims regarding holism in social theory that, on my account, any theory of social pathology is committed to. The first – the *holism of property ascription* – is that a society can be ill without its individual members being ill. When Plato describes the polis but not its individual members as fevered, he is implicitly asserting that society as a whole can have properties not ascribable to its individual members. Philosophers who are instinctively committed to rejecting holism of any type may be alarmed by such a claim, but there is nothing mysterious here, and acknowledging this point commits one only to an unobjectionably weak form of holism.[5] To take a biological example: it is appropriate to say that an organism behaves so as to restore equilibrium when one of its metabolic processes is disrupted, but it does not make sense to say that any of its parts maintains that equilibrium for the simple reason that it consists in a relation among parts that no single part can exhibit or bring about on its own; restoring such equilibrium is a cooperative venture, not one in which the heart acts on its own to restore its equilibrium, while the liver acts independently, too, and so on. Of course, the organism's tendency to restore internal equilibrium depends on no activity other than those of its parts – there is no mysterious motion of the whole over and above the motions of its parts – but restoring equilibrium is something the parts achieve only by standing in specific relations to the activity of other parts. Similarly, the polis's drive to wage imperialistic wars is a property of the polis rather than of the individuals who compose it, even if the polis's engaging in war depends on individuals acting in certain ways – some take up arms, others produce food for soldiers, others open day care centers so that women can enter the wartime economy – that, when related appropriately, constitutes a polis's going to war.

Social pathologies are holistic properties in this sense: they are ascribable to societies but not necessarily to the individuals who belong to them. A society can be ill even if its individual members are not because the dysfunction that constitutes the illness might be located, as in Plato's case, in dynamics involving relations among individuals that cannot be instantiated within a single individual: even if the unduly acquisitive desires of the polis's members are necessary for its being in a fevered state, an individual's having such a desire need not itself constitute illness, which is to say: a dysfunction of some kind internal to the individual. An individual's unduly acquisitive desire, because "unduly" so, must be bad or inappropriate in some respect, but it need not be an illness or dysfunction. It may be that Plato also regards individuals with unduly acquisitive desires as ill – as possessing a disordered, and therefore dysfunctional, psyche – but if so, then this is a peculiarity of his view deriving from the tight connection he asserts between political and psychic health that need not be, and usually is not, shared by theories of social pathology more generally. Marx's account of social pathology implies an analogous point: the lust of individual capitalists for infinite

[5] A similar version of holism is embraced by Hegel (Neuhouser 2003: 38–9).

accumulation may be necessary to capitalism's dysfunctions, but even if a defect of some kind, it is not itself an illness or dysfunction.

The second claim bound up with the assertion that the polis, not its individual members, is diseased represents holism of a different sort: a *holism of (causal) explanation.* One feature of Plato's fevered polis is that its dysfunction – its tendency to expand without limit – depends directly on the unduly acquisitive desires of its individual members. Thus, while the dysfunction in question might appear only at the level of the polis – when a number of individuals with such desires come together – it can seem as though that dysfunction were nothing more than an aggregate product of sorts, the result of bringing together individuals who have acquisitive desires independently of their social relations. In this case, the polis's dysfunctionality would appear only at the level of the whole, but it could be explained atomistically (and therefore not holistically) by summing up the forces of its elements, the desires of independently constituted individuals. This is not, however, true of Plato's fevered polis, nor does it hold for the types of social pathology I consider here. For the acquisitive desires of individuals that manifest as a dysfunction at the level of the polis are themselves due to features of society – the effects of a societywide dynamic that fosters its members' unmeasured desires – rather than properties individuals have independently of their social life. This is Plato's implicit claim, insofar as, once the argument of the *Republic* is complete and the relation between psychic and political injustice fully explored, he locates the corrective for acquisitive desires in a justly *ordered* polis where, as in the healthy psyche, its rational element rules, or provides a measure for, its appetitive part. Ultimately, then, Plato's analogy between polis and psyche extends beyond their isomorphism and its epistemological consequence, namely, that the nature of justice in the smaller unit can be read off of its appearance in the larger whole. Beyond this, he also asserts a *causal* primacy of the whole in the sense that its parts are what they are only through their relations within the whole: the desires of individuals without which the polis's illness would not exist are themselves products of the social conditions that count as diseased. An analogous point will hold for the instances of social pathology I am interested in here: the social order has explanatory primacy over its parts (human individuals) because the features of the latter relevant to understanding the illness of the whole are themselves causally dependent on how the parts are situated within that whole. As Durkheim insists, genuinely social phenomena cannot be explained by beginning with the psychology of individuals precisely because the latter takes on determinate shape only within the former: the desires that move individuals to act in ways that may result in a social pathology are themselves shaped by the social context within which they arise.

A third version of holism relevant to the claim that the polis rather than its members is the bearer of disease is *normative holism*, according to which a state of affairs can be bad for the whole even if it is not bad for at least some of its parts. The social philosophers I most rely on here, to their credit, *reject* this form of holism, holding instead that a condition cannot be bad for society if it is not bad for at least some of its individual members, even if the way in which it is bad for them may not be characterized in the same terms used to describe the defect of the whole. That is, individual members of a sick society need not themselves be sick, but if there is a pathology of the whole, it must be bad in some way for its members. Conversely, social health, if it is to count as such, must be good for individual members of society in some way. It can appear as though Plato rejects this principle when, in response to Adeimantus's objection that the polis that is being constructed in thought seems to disregard the happiness of its members, he has Socrates respond: "In founding the polis we are not looking to the … happiness of any one group … but … [to] that of the polis as a whole" (Plato 1992: 419a–420b). A careful reading of the *Republic*, however, dispels this appearance: In the first place, Plato says in the same passage that it would be no surprise if the members of the happy polis also turned out to be happier than those of a diseased polis (if happiness is correctly understood) (Plato 1992: 420b–d); in the second, the doctrine he goes on to develop concerning the intrinsic goodness of psychic justice, together with the claim that political justice is what makes psychic health on a wide scale possible, implies that a polis could *not* be happy independently of its members being so.

There is a further aspect of Plato's conception of social health that deserves mention, if only because it is a part of his view that need not (and should not) be retained by contemporary accounts of social pathology. The point in question concerns Plato's version of the claim that if social pathology is to be avoided, human desire must be subject to an external "measure" that constrains or shapes it. This idea finds expression in the emphasis Plato places on the question of which part of the healthy polis or psyche should rule, or be master of, the others. Plato's conception of health as a harmony among specialized parts is informed by a specific vision of harmonious coordination: "To produce health is to establish the parts of the body in a relation of mastering, and being mastered by, one another that is according to nature, while to produce sickness is to establish a relation of ruling, and being ruled by, one another that is contrary to nature"

(Plato 1992: 444d). Like Aristotle and Nietzsche after him, Plato conceives of health as requiring that one element of the psyche or polis – the one uniquely suited to do so – set the "measure" for the others. Although the more biologically oriented nineteenth-century theorists of social pathology – Comte and Spencer – espouse versions of Platonic hierarchy (where the brain or its social analogue plays the role of master), Durkheim emphasizes instead the ways in which coordination in society arises from below, without a supreme coordinating organ. Of course, such an organ *could* be conceived of as an impersonal, even democratically elected, agency rather than as made up of privileged individuals or groups, so that, strictly speaking, this version of organic hierarchy need not imply elitism or oligarchy, even if it has a natural affinity with those ideas. In any case, a hierarchy in which one part rules the others is an aspect of organic structure that Plato retains in his conception of a healthy polis, but it need not play a role in how contemporary theorists conceive of the coordination of parts required by social health.

Finally, two further points about Plato's account of the fevered society are worth noting. The first is that in the genealogy he relates, our recognition of illness in the polis *precedes* an account of what a healthy society is. After describing the primitive, "true" polis in which individuals' material needs are met, Socrates asks Adiemantus where justice (or health) is to be found in it, but no answer is given to that question until after the fevered polis has been constructed, the unhealthy nature of which is assumed both to be obvious and to point toward a determinate picture of social health (Plato 1992: 371e–372a, 372e, 373d–e). (Strictly speaking, justice cannot exist in the primitive polis because the simplicity of its members' needs and desires means that it needs no rational, ruling part in order to avoid illness.) This suggests a methodological point made again and again in post-Classical discussions of physiological and social illness, namely, that a diagnosis of illness need not, and typically does not, proceed by first fashioning an ideal of health and then measuring a suspected case of illness against that standard. Instead, with respect to the order of discovery, illness is generally prior to health.

An implication of this feature of Plato's accounts of illness is that it is possible to recognize instances of impaired functioning without having a worked-out conception of what healthy functioning looks like. Presumably this holds for the fever Plato attributes to the sick polis, as well as for a fever of the body: it is possible to detect that something is wrong in the fevered state without knowing that the normal (healthy) body temperature of humans is 37° C or, indeed, without knowing that there is such a thing as a normal body temperature. With respect to the polis, one can see that something is wrong in the constantly warring state without having Plato's ideal of political justice before one's eyes. This means that discerning the functions that theorists of social pathology attribute to social institutions need not precede diagnoses of dysfunction; it opens up the possibility, in other words, that empirical knowledge of real dysfunction is, with respect to the order of discovery, prior to – or, least, inseparable from – knowledge of the social "organism's healthy functions."[6]

The second point is that the *Republic*'s treatment of social health encourages us to distinguish between the two senses of "normal" relevant to conceptions of health and illness: on the one hand, what is statistically normal or most frequently occurring (or "standard"); and, on the other, what is normal in the sense of furnishing the norm for what health consists in. The *Republic* encourages us to draw this distinction because on the most natural reading of the text these two senses come apart: what is normal in the sense of "healthy" is patently not normal in the sense of "most frequently occurring." The healthy polis, in other words, is less common than the fevered one and may, indeed, never have existed.

This distance between the real and the ideal is a constant motif in the *Republic*, above all at those places where Socrates notes the shocking character, to Athenian common sense, of what he takes himself to have shown regarding the conditions of political health: that philosophers must rule; that guardians may hold no property; that citizens must be bred like livestock; and so on. The large gap between really existing poleis and Plato's vision of the healthy polis brings us back to the question of where standards of social health come from and how internal they are to how real poleis function. We can grant Plato that illness can be detected without full knowledge of health, but this response goes only so far because the picture of social health at which he ultimately arrives is far more determinate than a purely immanent methodology could provide, relying on relatively a priori conceptions of human nature and the good.

I suggested at the beginning of this chapter that one might be able to improve on this aspect of Plato's account of social health by developing the idea that society is artificial – that is, made by us in a way that enables us, at least in part, to read off the functions of social life from the consciousness of the agents who sustain it. Before examining Rousseau's version of this

[6] Durkheim embraces this principle in diagnosing social pathologies: "Studying the deviated forms [of the division of labor] will allow us better to determine … the normal state" (DLS: 353/343).

claim in the following chapter, it may be helpful to summarize the main aspects of Plato's adoption of the society–organism analogy that will follow us throughout the remainder of this book:

(1) Human societies resemble organisms inasmuch as both are functionally organized entities: they reproduce themselves via specialized and coordinated parts, each carrying out a function that contributes to satisfying the needs or realizing the ends of both the body as a whole and its parts.
(2) In human societies the ends achieved by well-functioning include but extend beyond purely biological needs, encompassing spiritual ends, such as those that spring from the pursuit of honor (or, in modern terms, from the desire for recognition).
(3) The ends realized by well-functioning societies comprise ethical ends, including the good.
(4) Regarding human societies as functional implies a tight connection between explanation (determining what a society is and how it functions) and evaluation (determining when a society fails to function well, or is ill).
(5) A defensible functionalism in social theory is committed to two versions of holism: (i) The category "ill" may apply to society as a whole without any specific part of it being ill (since dysfunction may be a property of only the whole); and (ii) the properties of individuals on which their society's being ill depends are socially caused – properties they have by virtue of particular relations to others within the whole.
(6) A defensible functionalism rejects strong normative holism, insisting that a society can be ill only if its being so is harmful to at least some of its members.
(7) A healthy society is one in which each part carries out a specific function appropriate to it and where these functions are coordinated such that all of society's essential needs are satisfied; social pathology consists in an impairment or breakdown of these functions.
(8) With respect to the order of discovery, the detection of social dysfunctions precedes a fully determinate picture of healthy social functioning.

CHAPTER 6

Rousseau: Human Society as Artificial

The previous two chapters ended with a question about how theories of social pathology can ascribe nonarbitrary standards of healthy functioning to social institutions against the backdrop of which pathological dysfunctionalities can be detected. The possibility I pursue in this chapter is that thinking of societies as artificial – or humanmade – might offer an answer to this question. The thought is that if the ends social institutions serve are inscribed into their functioning by us – the agents on whose activity that functioning depends – then perhaps we have a kind of access to those ends that is lacking when we seek the functions of living organisms from the external perspective of the biologist. Formulated differently: perhaps taking into account the spiritual nature of human societies – their being forms of life imbued with self-conscious purposes – will enable us to apprehend the ends that fuel social life and define its healthy functioning.

Convention, Consent, Authorization

If Plato gives us a start on understanding human society on the analogy of a biological organism and the various concepts associated with that analogy (function, health, holism), he is less helpful with the idea of artificiality. Rousseau's contribution to social pathology theory is precisely the opposite. Although he often uses organic metaphors when discussing the state (PE: 6/OC: 244–5; SC: I.6.x, II.6.i, III.1.xx, III.11.iii), they frequently appear alongside mechanical analogies (PE: 6/OC: 244). Yet the idea of social pathology is not foreign to Rousseau (SC: III.11.iii, IV.2.i, IV.4.xxxvi) (Honneth 2016: 195; Neuhouser 2012: 628–45): most famously, he borrows from Plato the term "feverish" to describe desires that, deriving from inflamed *amour propre*, or the drive to be recognized by others (Neuhouser 2010: 29–54) – itself a product of social ills – are "frenzied," "unbridled," and "violent" (DI: 171, 184, 199, 203/OC: 176, 189, 203, 207). Another biological motif Rousseau seems to have borrowed from Plato

(Plato 1992: 546a) is that of a normal life-cycle of a political body: "The body politic, just like the body of man, begins to die as soon as it is born and carries within itself the causes of its destruction" (SC: III.11.ii).

Despite these similarities, Rousseau's references to the state as an organism rarely invoke the idea of specialized functional parts so prominent in the *Republic*. He describes the state as a body primarily when he wants to distinguish the type of association involved in political life from mere aggregation, where independently constituted individuals come together and combine their already existing forces rather than existing as dependent "members" of a collective body with a unitary self, life, and will (SC: I.6.x). What primarily motivates his use of the society–organism analogy, then, is more a desire to avow certain forms of holism[1] than allegiance to the idea of functional differentiation. Indeed, that idea sits uncomfortably with Rousseau's conception of political society, given that, regarded as citizens, individuals are undifferentiated, possessing the same set of interests, rights, and duties as all others.

In any case, my main interest in Rousseau is not how he conceives of the organism-like character of human society but his account of the artificiality of social institutions. This thesis should be understood ontologically, as a claim about the kind of being, or reality, that human societies, or their institutions and practices,[2] possess, in distinction to the kind of being characteristic of other types of things, such as rocks or biological organisms. Rousseau's thesis that human society is artificial can be expressed most generally as the claim that society (or its institutions) has its "origin" in – or is existentially dependent on – human will and that this dependence distinguishes its mode of being from that of natural things. The obvious question raised by this thesis is: How, precisely, do social institutions depend on human will for their existence?

One might assume that the sense in which Rousseau regards society as dependent on human will follows directly from his account of the state as the product of consent, or agreement, among its members in the form of a social contract. This would help to solve the problem of how to attribute

[1] The holism Rousseau avows in these contexts holds, first, that the behavior and psychology of individual members cannot be understood unless viewed in the context of their social relations (their values and desires are shaped by those relations, not given independently); and, second, that the good of individuals is not exhausted by what is good for them *qua* private individuals, independently of social membership.

[2] For brevity's sake I will refer henceforth to "institutions" rather than "institutions and practices." While in some contexts it may be meaningful to distinguish the two, I do not do so here. This practice is consistent with that of contemporary social ontologists Vincent Descombes and John Searle.

ends or functions to social institutions because the criteria for successful functioning could then simply be read off of the expressed intentions of society's founders. For Rousseau the state would then function well if it achieved the ends for the sake of which it was instituted: to "defend and protect the person and goods of each associate … by means of which each, uniting with all, … remains as free as before" (SC: I.6.iv). Taking this route, however, would be to confuse two senses in which political society might have its source, or origin, in human will. One of these senses is normative, and this is the version of the claim implicit in the *Social Contract*: the *legitimacy* of the just republic derives from the fact that its basic framework is (in a specific sense) unanimously consented to by its members. This normative claim is silent, however, on questions regarding the causal origin of the state; that is, it says nothing about the second sense in which the state might be artificial, namely, that its *existence* depends on human will. Even if Rousseau sometimes appears to waver as to whether the contract described in the *Social Contract* is real or hypothetical, the implausibility of the former rules out interpreting the artificiality of the state in those terms: social institutions are not artificial in the sense that they come into being as the result of a real agreement among individuals at a certain point in time. Hence this social contract[3] should be understood only normatively, as setting out the criteria for legitimate political rule, according to which a state's laws and institutions genuinely obligate its citizenry if, and only if, they could be rationally consented to by all, where this means: by individuals concerned only with promoting their fundamental interests (in freedom, life, and the basic conditions of well-being).

The version of the thesis that society has its origin in human will that is of concern here involves a claim about not its legitimacy but its existence. Moreover, this claim, unlike the social contract, applies to social institutions generally, not merely to the state. As we shall see, there are various ways of interpreting the will-dependent existence of social institutions, not all of which are ontological in the sense of making a claim about the *kind of being* social institutions have. In order to find Rousseau's fullest elaboration of the artificiality thesis, we must leave behind the *Social Contract* and turn to the *Discourse on the Origin of Inequality* (or the Second Discourse). As its title indicates, the project of this text is to determine the "origin"[4]

[3] As opposed to the tyrannical contract depicted in the Second Discourse (DI: 131/OC: 131).

[4] Given that my interpretation of "origin" is not the most straightforward way of construing that term, one should remember that Rousseau's use of it comes from the Academy of Dijon's formulation of the question to which the Second Discourse is a response.

of (social) inequality, where Rousseau understands this question as having two possible answers: social inequality has its origin either in nature or in human will (which is to say, in our freedom). Because Rousseau's topic in this text is inequality, his artificiality thesis is expressed as the claim that social inequalities have their origin in human agency. His claim that such inequalities are artificial[5] entails that the institutions that make them possible or necessary are artificial as well. But artificial in what sense?

One reason Rousseau's account of the artificiality of the social is often misunderstood is that neither alternative he offers for the source of social inequality is conveyed very clearly by their defining terms, "nature" and "freedom." In attempting to uncover what is at stake for Rousseau in this dichotomy, I begin by citing the passage in which he most explicitly affirms the artificiality of social inequalities – even though the passage, in isolation, gives a misleading picture of his position. The immediate context of this passage is Rousseau's distinction, made at the beginning of the Discourse, between two types of inequality: "moral" – or what I call *social* – inequality; and "natural" or "physical" inequality (DI: 131/OC: 131). It is important to note that Rousseau's terms here do not capture well the distinction he has in mind: "moral" and "natural" must be defined more precisely if we are not to be misled by them, and equating "natural" with "physical" is highly misleading.[6] As examples of natural inequalities, Rousseau points to differences in "age, health, bodily strength, and qualities of the mind or soul," but his main point in drawing this distinction is that moral (or social) inequality, in contrast to its counterpart, "depends on a sort of convention and is established, or at least authorized, by [human] consent" (DI: 131/OC: 131). According to this statement, then, *convention*, *consent*, and *authorization* are essential elements of the (socially) artificial, although each term must be further explained if his claim is to be understood.

This characterization of the artificiality of social inequality seems to suggest the very view I dismissed above, namely, that inequalities, and the institutions that make them possible, are made by us in the sense that they are the products of real human consent and that such consent renders the

[5] *Factice* – or, equivalently, *artificiel. Factice* can mean "fake," "feigned," or "capricious," but in this case it is best construed as "artificial" or "humanmade." Strictly speaking, Rousseau does not describe social inequalities as *factice*, but he does use the term to characterize *amour propre* and "moral" love (DI: 155, 218/OC: 158, 219).

[6] "Moral" here contrasts with "physical" and the latter's synonym "natural." Rousseau uses the term to characterize the "public person," or "moral body," that issues from the social contract (SC: I.6.x) because what gives it existence is not a physical but a voluntative phenomenon: the rational consent of its members.

resulting institutions and inequalities legitimate. That this cannot be what Rousseau means to say, however, is made clear by the fact that it contradicts nearly everything the Second Discourse goes on to say about where social inequality comes from. Rather than the products of an explicit agreement among individuals, social inequalities are portrayed there as resulting from a complex chain of contingent historical events, including free acts of human beings, the systematic consequences of which are neither intended nor foreseen by those who take part in them. Moreover, Rousseau clearly does not regard the inequalities described in the Second Discourse as legitimate, and in his account in Part II of the tyrannical social contract he unambiguously rejects the view that actual consent, even if given, establishes the legitimacy of what has been consented to (DI: 131/OC: 131). Unpacking Rousseau's view regarding the artificiality of the social depends, then, on finding a way to understand the *conventional* and *consensual* nature of social inequality – as well as the *authorizing* force of such consent (which does not, however, imply true legitimacy) – in a way that coheres with the Second Discourse's account of social inequality as unintended and illegitimate.

A complicating feature of Rousseau's artificiality thesis is that it comprises three claims. Although these may not be completely independent of one another, each represents a way of denying that social inequality has its origin in nature and locates that origin instead in some aspect of human agency. Although the third of these claims is the most important for the ontological point I want to take from Rousseau, I briefly discuss the other two as well – first, because I take them also to be true, and, second, because distinguishing the three claims will help to bring the one I am most interested in more clearly into focus.

One straightforward sense in which social institutions are artificial for Rousseau is that they are the results of a complex chain of events that include free human actions. Social institutions are not put in place by nature or by God but are products of human history, which, because it depends on free human actions, could have taken a course other than the one it did. To say that something is artificial in this sense is to say that there is nothing in the basic constitution of nature, including human nature, that necessitates either that thing's existence or its being as it is. Thus, the property rules of eighteenth-century Geneva are artificial in this sense – they could have been otherwise – because their existence depends, somewhere along the line, on the intervention of free will, whereas the fact that eighteenth-century Genevans are self-interested beings, motivated by *amour de soi-même*, is not artificial but due to nature (in this case, human nature). It is important that, contrary to what readers of Rousseau often assume, "artificial" is not an evaluative term implying a criticism of whatever is said to be made by humans. Confusion can easily arise over this point because in many contexts Rousseau uses "natural," the apparent contrary of "artificial," as a term of approbation (Neuhouser 2015: 29–31, 53–5). But the fact that for Rousseau whatever is purely natural (not a product of human consciousness or will) is necessarily good (Rousseau 1979 [1762]: 37) does not imply that whatever is artificial is bad: neither conjugal love nor *amour propre* is necessarily bad, despite being artificial.

It is also important to Rousseau's position that thinking of a certain property code as having its origin in freedom does not imply that at some point a will consciously chose or created that code in the form it ultimately came to have. All that is implied is that human freedom intervened at some point in that code's causal history and that in the absence of such intervention that code would not have come about. Social inequalities, then, are the results of complex developments, some of which depend on human choices, but in such a way that, when joined with natural events that accompany them, the long-term consequences of those choices are not foreseen and therefore not intended by any human agent: social institutions are artificial without being artifacts. To take a well-known example of Rousseau's: freedom is involved when someone encloses a piece of land for the first time and says "this is mine!" (DI: 161/OC: 164), but no one intends either the shape or consequences of the system of private ownership that ultimately develops from this free act. This means that the Second Discourse makes humans causally but not morally responsible for social inequalities since the choices on which the latter depend are made without foresight of the chain of developments they initiate. A fitting epigraph for the Second Discourse would be: "Father, forgive them; for they know not what they do."[7]

In this first version of Rousseau's artificiality thesis, to reject nature as the origin of social inequality is to deny that human nature, without the contingency that the intervention of human will introduces, necessarily produces social inequality by, for example, providing us with natural incentives to seek superiority over others. Rousseau's defense of this specific claim is too complex to explicate here,[8] but two aspects of it are worthy of mention. First, denying that social inequality has its origin in

[7] Luke 23:34, King James Version.
[8] The claim is explained in Neuhouser 2015: 25–43, 55–60.

human nature implies for Rousseau that the inequalities characteristic of modern societies – inequalities we might be tempted to regard as intrinsic to the human condition – are not necessary features of any human society whatsoever, and he takes this to imply that the inequality-producing institutions we know could have been otherwise and, so, can in principle be altered by human intervention.

Second, this version of Rousseau's artificiality thesis employs "origin" in a completely straightforward sense: social inequalities, and the institutions they presuppose, have their origin in human will in the sense that their coming to be depends causally on free human actions that might have been otherwise. This claim is essential for Rousseau's ambitions to criticize and reform modern society, but it is not central to the ontological question I am pursuing here. That it is not, even for Rousseau, the main question at issue in his claim regarding the artificiality of social institutions can be seen by returning to the quotation cited earlier: social inequality "depends on a sort of convention and is established, *or at least authorized*, by consent" (DI: 131/OC: 131; emphasis added). That Rousseau shifts from talking about how inequalities are established to how they are authorized indicates that he is less concerned with the historical origins of inequality than with how, once existent, it is maintained over time. And part of his claim is that authorizing consent, however that is to be construed, plays a central role in maintaining the social institutions that make social inequalities possible. I return to this claim below.

The second version of Rousseau's artificiality thesis concerns precisely this: not how social institutions come to be but what sustains them once they are in place. It claims that free human activity is what sustains social institutions and the inequalities arising from them in the sense that their continued existence depends on our ongoing participation in them. In this context "origin" means "source"[9] in a sense similar to that in which, for example, the St. Marys and St. Joseph Rivers are the sources of the Maumee River: social institutions are our doings, and are therefore artificial, because our free activity is the standing source of their existence. While one might say, in the first case, that the St. Marys and St. Joseph Rivers fuel or feed the Maumee River, it is probably more accurate to say, in the second, that human activity is constitutive of social institutions (or social life). Unlike the first version of the artificiality thesis, which makes

[9] Rousseau uses "source" in this sense when claiming that social inequality does not have its source in human nature (DI: 124/OC: 122).

a causal claim about the temporal origins of social institutions, the second version makes an ontological point about what the being of such institutions consists in (human activity), thereby distinguishing the kind of existence they have from that of mere things (rocks and pine cones) or artifacts (houses and watches). Once mere things and artifacts come into being, they continue to exist independently of any ongoing activity that sustains them, whether their own or that of human agents. As Hegel points out after Rousseau, human societies are in this respect ontologically similar to living organisms, the continued existence of which depends on (because it consists in) their constantly engaging in the specific activities bound up with their self-maintenance. If a living organism is among the higher animals, then once its breathing or pumping of blood comes to an end, it ceases to exist (*qua* living organism) and reverts to being inert matter. Similarly, a family ceases to exist *qua* social institution when its members stop carrying out the various activities that make it a family. Biologically related individuals who merely inhabit the same living space do not constitute a family unless they also regularly engage in the activities characteristic of families. A family, one might say, is family *life*, or expressed differently, a social institution is the kind of thing that cannot exist as such independently of its own activities (the various doings of its members) that make it the kind of being it is. According to this version of the artificiality thesis, social institutions are humanmade in the sense that ongoing human activity is constitutive of what they are.

Although Rousseau endorses both the second and first versions of the artificiality thesis, it is the third that is most important to my project. This version is closely related to the second because it, too, says something about what the being of social institutions consists in and because it focuses on a specific aspect of the free activity that, according to the second, is constitutive of them, namely, the "consent" on which institutions depend. That we have not yet uncovered the whole of Rousseau's position is evident from the fact that we have not explained how convention, consent, and authorization enter into the being of social institutions. In order to do so, it will be helpful to begin with the thought that *social institutions are conventions*, for this is the best way of summarizing the third version of his artificiality thesis. For Rousseau, characterizing institutions as conventions means more than simply that they are the contingent results of free human acts. That "convention" is a technical term for Rousseau is suggested by the fact that he also uses it in the *Social Contract* when characterizing the (hypothetical) compact among individuals that grounds the legitimate republic and constitutes "the true foundation of society" and "the basis of all legitimate authority among men" (SC: I.5.ii, I.1.ii). Since it may be tempting to think that only political institutions are conventional in Rousseau's technical sense, it is important to note that he also claims that, once its children become adults, "even the family maintains itself only by convention" (SC: I.2.i). Hence, the claim that the social is conventional applies not only to the state but also to institutions more generally. Since the conventional is necessarily artificial – but not vice versa[10] – this provides further evidence that "artificial" is not inherently a term of critique for Rousseau since he regards both the (all-adult) family and the well-founded state as legitimate, obligation-creating institutions. Relatedly, even though "convention" in both French and English can suggest arbitrariness – as in the familiar thought that traffic rules are a matter of mere convention – in the sense at issue here it does not: there is for Rousseau nothing rationally arbitrary about the legitimate republic or the well-functioning family.

When Rousseau says that the republic described in the *Social Contract* is grounded in convention and that the family (of adults) maintains itself by convention, he is speaking of institutions he regards as legitimate or rational. It is important to remember, however, that his use of "convention" is broader than this. He takes language, for example, to be a convention (DI: 147/OC: 147), even though questions of legitimacy do not arise there. More important, illegitimate inequalities and the institutions that make them possible are also said to have their origin in convention. In other words, if "convention" does not imply "arbitrary," it also does not imply good, legitimate, or rational. This means that whatever "convention" means, it applies to social institutions broadly, not merely to legitimate ones, which is to say that it plays a role in Rousseau's social ontology, in his understanding of the nature of social institutions in general. Nevertheless, it will help us to get an idea of what conventions in the broadest sense are by looking first at what it means in the case of legitimate political institutions, as characterized in the *Social Contract*.

The key to grasping Rousseau's claim that the legitimate state is grounded in convention lies in noting that the term derives from the French *convenir*, which denotes the "coming together" of different things – such as the St. Marys and St. Joseph Rivers mentioned above, which come together to constitute a third river, the Maumee. In Rousseau's social philosophy, however, "convention" refers to the coming together, or agreement, of not things but human wills. It is tempting to think that a convention must

[10] The socially artificial is always conventional in Rousseau's sense; a watch, for example, is not.

be like a pact or a contract – something that comes into being when two or more wills expressly agree to something – but, as we have seen, family life, too, is conventional (when it survives the children's maturation), and Rousseau clearly does not believe that the family, in any form, comes into being through something like a contract. The agreement of human wills central to convention, then, is not conscious assent but agreement in the sense of "matching," as when we say that two copies of a document are in agreement (when they match) or that my idea of the afterlife agrees, or coincides, with yours. What makes the social contract a convention is not that it comes about through the conscious assent of negotiating parties – it does not – but rather that the principles defining it presuppose and give expression to wills that coincide, or agree, in the sense of having shared ends. Of course, not all the ends of citizens of a legitimate republic are shared with everyone else, but their wills agree with respect to the specific ends of political association, and the principles governing that association, given by the idea of the general will, concern only those shared ends. Part of the conventional nature of the legitimate republic, then, consists in the fact that its governing laws are grounded in, and promote, the convergent ends of its members. When these shared ends track fundamental human interests, the institutions that promote them count as legitimate (and hence are anything but arbitrary). (A word of warning: it will be necessary to return to the idea of the agreement of wills below because the agreement of *interests* that grounds the legitimacy of the social contract is not the same as the agreement among wills that characterizes social institutions more generally and accounts for its conventional nature.)

Another crucial aspect of the convention that grounds the legitimate republic is that it takes the form of a normative principle or rule. To say that the social contract is conventional is to say, in part, that it is the source of normative principles that govern the actions of those who belong to the state. In the case of the legitimate republic, there is one supreme principle of political association, that of the general will. The general will is often said to aim at the common good (SC: II.1.i), but its content can be spelled out more precisely by consulting the ends, noted above, that make consent to the social contract rational and bestow legitimacy on it. The convention on which legitimate political association depends, then, is a normative principle, in which the wills (or interests) of those it governs come together, or converge. It can be formulated as the imperative: "Use the collective powers of citizens to protect the life, person, and property of all such that each remains free even in being subject to the laws required for such protection!" It is the fact that a state founded on this principle safeguards the fundamental interests of each that *authorizes* such a state, and in two senses that in other contexts come apart, namely: this principle *actually makes legitimate* the state and its laws; and it makes the state and its laws *legitimate in the eyes of its members*. Finally, if we want to account for the third term of Rousseau's cited above, we might say that the principle of the general will could command the rational *consent* of every individual subject to it.

There are two respects in which this account of the conventional nature of political society falls short of answering the ontological questions I mean to raise here. The first is simply that, since it regards the social contract as merely a device for spelling out the normative criteria for legitimate political association, this way of interpreting authorization, convention, and consent fails to explain how social institutions might be made by us in any straightforward sense. The second problem is that this account does not apply to social institutions generally but only to legitimate or rational ones. If Rousseau means to offer an ontological account of the kind of being social institutions possess in general, it must apply as well to institutions he criticizes as illegitimate. This is what is at issue in the Second Discourse when he says, even of institutions that are the source of illegitimate inequalities, that they "depend on a sort of convention and [are] … authorized by consent."

It is not that convention, as Rousseau conceives of it in his normative project, has nothing at all to do with the conception of convention that figures in his more general ontological claims. One feature of both is that conventions are normative rules that tell individuals how they ought (in a broad sense) to act and that imposes sanctions on behavior that fails to comply with those rules. Immediately after setting out the convention governing legitimate political association (the general will), Rousseau discusses the necessity of the state's punitive powers in order to enforce the general will when citizens fail to comply with it (SC: I.7.viii). Importantly, this connection between convention and sanctions holds for institutions more generally: legal sanctions apply to those who violate the rules of private property or disobey laws, but sanctions of an informal kind apply as well to those who violate the more loosely defined rules that govern – to borrow examples from John Searle and Margaret Gilbert – cocktail parties and the sending out of thank-you notes (CSR: 29; Gilbert 1983: 237–9). I argue below that conventions in this broad sense – normative rules that carry sanctions for noncompliance – are essential to the third version of Rousseau's account of the artificiality of social institutions and that the rule-governed nature of social institutions opens up a new possibility for explaining their artificiality: perhaps such institutions are our creation because we, in some sense, make the rules that constitute them.

How We "Make" Social Institutions

There is one straightforward sense in which we sometimes make the rules of social life, and this phenomenon plays an important role in Rousseau's account of democratic politics. This is because the bare principle of the general will, though not contentless, is highly indeterminate. In order to move from this principle to positive laws that spell out concretely what citizens are obligated to do, the general will must be interpreted by real human beings in specific circumstances, and in democracies this interpretation – the making of positive law – is carried out by the same citizens who are subject to them. For the most part, however, the rules governing social life in general are not made by those who follow them in the sense in which, in democracies, positive laws are made by citizens. Is there, then, any sense in which the rules of social institutions more generally are made by their participants?

One answer to this question is suggested by a scene from George Cukor's 1949 film *Adam's Rib*. In this scene the attorney Amanda Bonner (Katherine Hepburn) is interviewing her client, Doris Attinger (Judy Holliday), who is accused of attempting to murder her philandering husband. In the course of the interview Bonner discovers that her client believes, to her own disadvantage, that different rules apply to the sexual behavior of men and women. This by itself is hardly surprising, but Bonner's response to her client's husband-excusing double standard is unexpected and of interest, for it suggests an answer to Rousseau's question about where social inequalities and the institutions that undergird them come from:

AB: What do you think of a man who's unfaithful to his wife?
DA: Not nice, but—
AB: What about a woman who's unfaithful to her husband?
DA: Something terrible.
AB: Aha!
DA: What?
AB: Why the difference? Why "not nice" if he does it, and "something terrible" if she does it?
DA: I don't make the rules.
AB: Sure you do. We all do.[11]

[11] Screenplay by Ruth Gordon and Garson Kanin; https://www.springfieldspringfield.co.uk/movie_script.php?movie=adams-rib

Bonner's answer to Rousseau's question concerning the origin of social inequality – in this case, the consequences of patriarchal rules that condemn women for the same behavior that is excused in men – is that it comes from us precisely because it is the product of rules that we ourselves make. Clearly, Bonner means to accuse Attinger not of inventing the rules of patriarchy but of participating in their making in some other sense that makes Attinger their coauthor. Presumably what Bonner has in mind when she says that we make those rules is the ways in which over and over in our everyday actions "we all"[12] *accept*, *apply*, and hence *reproduce* and *reinforce* the social rules we inherit – and take over as our own – from existing institutions.

If we follow this line of thought, Rousseau's thesis regarding the artificiality of social life could be formulated as the claim that social institutions in general, and not merely those of democratic lawgiving, are sustained by participants' ongoing "authorizing consent" to the conventions (or rules) that define such institutions, where "consent" refers to our acceptance[13] and application of the rules in question. The suggestion that we all make the rules of patriarchy must be understood, then, as the claim that the widespread acceptance and application of those rules is something we do that *makes* the institutions defined by those rules, and this in two senses. The more obvious of these is that accepting and applying social rules reproduces and reinforces our sense of their authority. Because of the centrality of rules to what social institutions are – more on this below – saying that we reproduce and reinforce our sense of the authority of rules implies that we thereby reproduce and reinforce the institutions defined by them. In this respect we make our institutions by *reproducing* them, and without this reproductive activity they would cease to exist.

The second sense in which accepting and applying social rules *makes* the institutions governed by them is ontologically more consequential. Here the making of social institutions must be understood less poetically than in the claim that accepting and applying their rules reproduces them. According to the second version of the claim that we make institutions because we make their rules, social members' acceptance and application of the rules in question – along with the behavior implied by that

[12] Is this Heidegger's *das Man*?

[13] This term is Searle's; I say more about it below (CSR: passim).

acceptance – is simply what it is for a social institution to exist. Social members' acting in accordance with the normative rules of institutions constitutes, or "makes up," the existence of institutions; it is what social institutions consist in. Like living beings, social institutions stop existing (as such) when their members cease to accept and follow the rules that govern the places they occupy within their institutions; social institutions are nothing but social *life*. In other words, the most important respect in which institutions are artificial is that in the absence of our authorizing consent (our acceptance of their constitutive rules) and the behavior it implies, they *cease to be*, and in this respect they depend on human doings for their existence in a way that natural phenomena do not. However we understand the consent that authorizes the rules of social life, the present claim implies that social institutions are, in contrast to natural things, "normatively constituted" phenomena (Jaeggi 2019: 86–121, 131): their mode of existence consists in human agents taking a normative stance toward those institutions' constitutive rules that, implicitly or explicitly, confers authority on them.

The well-known passage of the Second Discourse referred to above – the opening lines of Part II – illustrates how Rousseau thinks of convention, consent, and authorization:

> The first person who, having enclosed a piece of ground, took it into his head to say "this is mine!" and found people simple enough to believe him was the true founder of civil society. How many … miseries and horrors would have been spared the human race if someone had pulled up the stakes … and cried to his fellow beings: "Beware of listening to this impostor; you are lost if you forget that the fruits of the earth belong to us all and the earth to no one!" (DI: 161/OC: 164)

Rousseau's aim here is not to speculate about the historical origin of the rules of private property that allow individuals to own land but rather to illustrate the nature of that phenomenon. Describing private ownership of land as something that entered the world through human declaration at a specific point in time reflects his view that social institutions, rather than products of nature, are the results of free human activity. Moreover, the institution of private land ownership is portrayed as depending not only on the declaration of a single individual but also, and more fundamentally, on the consent of his fellow beings. The pronouncement "this is mine!" yields private property only on the condition that there are people "simple enough" to believe it, where their belief consists simply in accepting the pronouncement and the normative consequences it has for their behavior.

(If this piece of land is owned by the person who claims it, then others have no right to walk through it or use it without his permission.) The acceptance of the rules on which the private ownership of land depends, then, is collective or general – acceptance by more than just an individual or two – and this makes it conventional in the sense of being a coming together, or agreement, of human wills. To contrast this with a point made above about the normative device of the social contract: the coming together of wills in the conventions that are constitutive of social institutions more generally consists not in shared interests but in collective acceptance.[14]

Finally, the "simplicity" invoked to explain the collective acceptance of the rules of land ownership indicates that the consent that authorizes social institutions can fall well short of informed or rational consent and that for this reason such acceptance authorizes social institutions only in the weaker sense distinguished above: it implies not actual legitimacy but only legitimacy, or authority, in the eyes of those who accept the rules in question.[15] Moreover, the "simple" nature of those who accept the property claims of landholders suggests that the authorization of institutions can be weak in a further sense: accepting the validity of institution-constituting rules can include a wide range of attitudes from explicit endorsement to unreflective acceptance. And, as examples from both *Adam's Rib* and the Second Discourse demonstrate, the institutions whose rules we accept are not infrequently such that we ourselves are disadvantaged by them. Yet regardless of how robustly the concept of authorization is to be interpreted, what is most important to Rousseau's view is its implications for his artificiality thesis: even when our acceptance (consent) is unreflective or in conflict with our own good, the institutions it authorizes still count as dependent on our will, or the work of our freedom.

Rousseau's account of why even relatively passive acceptance involves freedom is obscurely formulated. He tends to describe the acceptance of institutional rules as instances of *belief*, as in the passage cited above. It is unclear what precisely the content of my belief would be when I accept another's claim to own a piece of the earth, but the reason Rousseau uses the term is that belief, for him, brings human freedom into the picture – it

[14] "Collective acceptance" is also Searle's term and is said to require collective intentionality (CSR: 89).

[15] In this respect Rousseau's "authorization" bears similarities to Max Weber's "legitimacy," which refers to the social members' belief in the legitimacy of a particular social order, not to its actual legitimacy. For Weber, too, legitimacy in this sense is crucial to the ongoing existence of social institutions, which makes them, as for Rousseau, inherently normative phenomena (Greene 2017: 295–324).

makes our acceptance of the relevant rules a kind of free consenting to them. This is because he subscribes to the Stoic view that believing of any kind requires active assent to a proposition that such-and-such is the case. Rousseau's position is probably better construed, however, by appealing to Kant rather than the Stoics. The Kantian thesis that is helpful here is that, in contrast to behavior that merely conforms ("externally") to rules, rule-*following* – the *applying* of rules – involves a kind of acceptance of those rules of which only free agents are capable: in Kant's language, applying rules, as opposed to merely being governed by them externally, involves an exercise of *spontaneity* that only agents have a capacity for. Applying rules, even unconsciously, is something we do rather than something that happens to us.[16] One consequence of this claim is that applying rules implies a sort of *responsibility* for them, even if only in the thin sense that it is within our power as practical agents to reflect on the appropriateness of the rules we follow and, in light of that, to endorse, reject, or seek to revise them. This reflexive capacity of agents, even if unexercised, gives rule-following a "could have done otherwise" character that Rousseau takes to be a necessary condition of free action. Thus, even at its thinnest, the acceptance of social institutions is something we do and for which we are responsible.

There is a long philosophical tradition that attempts to explain why institutions that disadvantage some individuals illegitimately are sometimes nevertheless regarded by them as authoritative. Of that tradition – ideology critique – Rousseau himself is a founding member, and his most prominent contribution to it implicitly depends on the eventuality just considered, namely, that the collective acceptance of rules on which social institutions depends need not extend to full-fledged reflective endorsement. The significance of this for Rousseau is that it opens up the possibility of a type of ideology critique that he and his followers – most spectacularly, Nietzsche – made famous and of which the Second Discourse is a paradigm example, namely, genealogies that, in tracing the "origin" of some phenomenon, denaturalize the social. If institutions can be sustained by an unreflective acceptance of rules – where social members are unaware of both their active role in sustaining institutions and the contingent nature of the rules they follow – then social critique can take the form of a *denaturalizing genealogies* that, by revealing the artificial character of institutions previously regarded as natural, or merely given,

[16] Searle expresses this point thus: "Nor … can we describe [social institutions] as sets of unconscious computational rules … because it is incoherent to postulate an unconscious following of rules that is inaccessible in principle to consciousness" (CSR: 5).

strip them of their apparently immutable character and place them into the "space of reasons," in which they then figure as possible objects of critique and transformation. (And this denaturalizing strategy depends on revealing the artificial character of institutions in the first sense considered, too: because they are the results of chains of contingent historical events, including the free acts of humans, they could have been different from how they in fact are.) Once the artificial character of institutions is recognized by those whose unreflective acceptance sustains them, it becomes possible for social members to assume responsibility for them in a fuller sense, by asking: Are the institutions that depend on my acceptance good, or as good as they can be?

There remains one aspect of Rousseau's picture of the will-dependent character of institutions in need of further clarification. It concerns the connection between consent and authorization or, more precisely, between the kind of acceptance of rules that is necessary if one is to apply them and a normative endorsement of those rules (and institutions) as good or legitimate. If, as suggested above, social institutions are normatively constituted – if they are grounded in conventions authorized by their participants' consent – how is the authorizing consent on which they are based to be conceived? This is a question about the type and robustness of the normativity (or authorization) that is built into social life at the ground level. It arises because in order for Rousseau's account of the artificiality of social institutions to have the generality it aspires to – to apply not only to legitimate political association but also to the illegitimate and unreflectively accepted institutions described in the Second Discourse – it was necessary above to interpret the consent that authorizes institutions very weakly, to include unreflective forms of the acceptance of rules that fall short of what we would normally call "authorization." Yet at the same time, Rousseau's account of social institutions – correctly, in my view – requires that social life be normatively constituted in a more robust sense than is present in the barest form of acceptance of institutional rules discussed above. As I argue below, the coming together of wills on which human social life depends must include some agreement regarding the *goodness*, however thin, of the institutions whose defining rules are accepted by social members. This point is of significance for theories of social pathology, for it implies that social institutions – or at least the most important among them – are predicated on some degree of shared values, which in turn helps the social theorist to define what the well-functioning of institutions consists in.

In general, Rousseau uses forms of the verb *autoriser* sparely, especially in comparison with the noun *autorité*. The verb appears in the question posed by the Academy of Dijon to which the Second Discourse is a response – is human inequality "authorized by natural law"? – and it appears again in the Discourse's closing paragraph: "It follows then that moral inequality, authorized by positive law alone, is contrary to natural right whenever it is not directly proportional to physical inequality" (DI: 188/OC: 193–4). In both cases Rousseau speaks of a state of affairs (inequality) being authorized by a law, but at other places he speaks of law (*droit*) – principles or rules of some sort – being authorized when certain conditions obtain (SC: I.9.iii). In any event, one of these cases – the authorization by positive law of inequalities contrary to natural right – shows clearly that to authorize is to accord authority or legitimacy to something that may not truly possess it.

Beyond this, the difference between a state of affairs and a rule, each a possible object of authorization, is helpful because it suggests a distinction between a social institution (or practice) and the rules that govern it that we have not yet taken into account. Rules are constitutive of institutions but not exhaustive of them; that is, institutions (or practices) have a real existence in space and time that results from their members applying, or following, those rules. When Rousseau discusses and evaluates the institution of private land ownership, he is thinking not only of the rules that govern it but also of the actions that follow from those rules, together with their systematic consequences. For Rousseau establishing the genuine legitimacy of an institution requires taking such consequences into account – his metaethical position considers consequences without being consequentialist[17] – but our according legitimacy to the *rules* of an institution might not do so. In the case of land that is privately owned by one class where others own no means of production, I may have consequence-independent reasons for taking the rules of that institution to be legitimate. That does not imply that I am correct about its legitimacy, but it is typically a consequence of my regarding rules as legitimate that I regard their consequences as legitimate, too. This is *not* the structure of Rousseau's normative position – since consequences matter in determining the genuine legitimacy of institutions – but it seems to be the picture he has in mind when he speaks of humans authorizing (taking to be legitimate) what is in fact illegitimate social inequality by their consent to (acceptance of) the conventions (rules) from which social inequalities follow nonaccidentally.

The question now is: What entitles Rousseau to claim that, in general, we authorize inequalities produced by the social institutions we participate

[17] In the sense of enjoining us to maximize some aggregate good.

in simply because, as participants, we apply, or follow, their defining rules and because following rules involves a kind of acceptance of them of which only free agents are capable? The general picture of social life that motivates Rousseau to stake out a position of this sort is one in which social institutions, even when disadvantageous to many, function *qua* institutions only on the basis of some kind of collective consent. This points to an important sense in which social institutions are moral rather than purely natural phenomena: they are maintained for the most part not by brute force but by a (tacit or explicit) consensus of some kind. When workers in capitalism perform their eight or more hours of labor, day in and day out, without sabotaging their employers' property or appropriating it for themselves, they typically do so not because they fear the state's power to enforce existing property laws but because at some level they accept, perhaps unquestioningly, the legitimacy or naturalness of the social arrangements that make it necessary for them to work for their survival while others are able to live without laboring and to enrich themselves from others' labor.[18] This point constitutes for Rousseau a general truth about human social life: *institutions that depended primarily on physical force, without any belief in their legitimacy on the part of their participants, would be highly unstable and inefficient*, not least because a large part of society's resources would have to be spent in maintaining mechanisms of coercion so oppressive that its members would perceive them as ubiquitous and inescapable. More fundamentally, *a "society" in which the threat of physical force were the sole or primary reason individuals complied with social rules* – the fact that it is difficult to imagine such a scenario is telling – *would not even count as a human society* (SC: I.5.i). That is, social life would then not have a mode of being particular to itself but would be merely a complex form of nature, where right has no place or, in what amounts to the same, where "might makes right" is the reigning law (SC: I.3.i).

The problem I am interested in here arises because the soundness of Rousseau's claim that authorizing consent is essential to social institutions in general, including the illegitimate inequalities they might engender, depends on interpreting him to mean that we (in some sense, freely) accept and apply the rules that define those institutions. While I take this claim to be true, it gives rise to a problem, inasmuch as it does not by itself imply that social institutions rest on some degree of normative consensus

[18] Asymmetric power relations between men and women, too, seldom depend only on men's superior physical power; they depend also on the belief of the participants in those relations, including many women, that patriarchy is natural or appropriate.

concerning the goodness, appropriateness, or legitimacy of institutions. What Rousseau's claim implies is that as an applier of social rules, I necessarily accept the authority of the rules I apply, albeit only in a weak sense that can fall well short of normative endorsement. When I determine my action in accordance with a rule, I need only minimally accept the authority of that rule for my action, where this means not only that I intend to do what it prescribes but also that I regard it as generating an "ought" for me such that when I discover that I have failed to do what the rule requires, I recognize my action as a mistake; I take myself to ought to have acted differently, and I recognize an obligation to correct it or to act differently in the future.

This degree of authorization is at work even in following the grammatical rules of a language, where the question of normative endorsement does not arise. It is also at play in following the rules of table etiquette: I regard it as a mistake when I notice that I have forgotten to place my napkin on my lap before eating, and I am irritated (and cannot help but roll my eyes semi-discreetly) when someone at the table chews his food too loudly. In other words, I accept the rules of table etiquette as authoritative but without necessarily believing there are good reasons for these conventions. There are also instances where I authorize rules in this limited sense even when I take them and the institutions they define to be illegitimate: I act in accordance with the rules that govern my labor contract with my employer, even though I believe that "property is theft" and that those who are rich enough to purchase the labor power of others must have come by their wealth by exploiting or stealing from others. In such a case, I live by, or accept, the rules of capitalism not because I regard the system as good or legitimate but because I need to put food on my table today and have no alternative, or because I want to avoid the disapproval of my coworkers whose company I enjoy but who do not share my moral rejection of the institution of wage labor.[19] (Note that this example depends on my belief in the illegitimacy of wage labor *not* being widely shared; if it were, "social" life would approximate merely "natural" life, where fear, need, and physical force explain why we act as we do. Moreover, there are examples of social institutions in which large numbers of their members *reject* the legitimacy of institutions whose rules govern their lives. No doubt slaves in the U. S. antebellum South widely, perhaps universally, rejected the legitimacy of slavery, following its rules, when they did, primarily to

[19] For a similar point, see Geuss 2001: 65–6.

avoid violent death. Yet slavery was widely "authorized," in both North and South, with varying degrees of enthusiasm, by U. S. citizens whose skin color exempted them from being enslaved. Even if approval was far from universal, slavery in the U. S. was an "authorized" social institution in Rousseau's sense, as are many other apparent counterexamples[20] to his claim that social reality depends on normative consensus.)

That this is a genuine philosophical issue and not merely a puzzle for interpreters of Rousseau is strongly suggested by the fact that a similar question arises in the social ontology of John Searle, a contemporary philosopher who takes pride in claiming – no doubt, correctly – not to have been influenced by the old masters of pre-twentieth century social philosophy, and certainly not by Rousseau. Examining how this question arises for Searle will help us arrive at a position on the question raised by Rousseau's account of social life concerning how robust the normatively constituted character of social institutions is.

Searle on the Normative Character of Social Reality

Questions of actual influence aside, Searle's social ontology fits squarely in the tradition of social philosophy begun by Rousseau and further developed by Hegel and Durkheim, especially with respect to the artificiality of social institutions and what makes them so. Indeed, one of the books in which Searle sets out his social ontology is called *Making the Social World*, and, like Rousseau's, its account of social institutions relies heavily on concepts of convention, agreement, and authorization. That this resemblance is more than superficial can be seen in one of Searle's fundamental theses, namely, that "collective agreement or acceptance … is a crucial element in the creation of institutional facts" and that such acceptance is "partly constitutive" of social reality (CSR: 39, 33). A further similarity to Rousseau is that Searle, too, is interested in the relation between the

[20] Contrary to what Westerners have been encouraged to believe, Eastern bloc Communism, too, enjoyed substantial (if steadily diminishing) normative authority among its citizens, without which it would not endured for seventy years. The ideals of Communism, if not its reality, were tacitly accepted by many in the Eastern bloc; moreover, Communist parties achieved significant legitimacy by virtue of the leading role they played in defeating fascism. To take another example: it is tempting to think that Blacks in the U. S. do not collectively bestow authority on the institution of the police because of its systematically racist misuse of power. No doubt there is a kernel of truth in this thought, but that matters are more complex is demonstrated by Eric Adams' win, with resounding support from Black New Yorkers, in the 2021 New York City Democratic primary for mayor. Adams ran on a promise to beef up rather than defund the city police; even among Black citizens of New York City, the police remains a social institution in Rousseau's sense.

collective acceptance of institution-defining rules and what both thinkers call their authorization.

Searle's social ontology is motivated by the desire to explain how social reality differs from natural or physical reality. Its central task is to explain what social institutions (and their corresponding institutional facts[21]) are. The crux of Searle's view is that social is distinguished from physical reality by the fact that in the former case "the attitude … we take toward the phenomenon is partly constitutive of the phenomenon" (CSR: 33). It is, then, by means of a certain attitude, or form of intentionality, that we constitute, or make, social reality, where, as for Rousseau, this is not a claim about historical origins but a constitutive claim about what it is for social institutions to exist. For Searle the form of intentionality that makes institutions what they are is the collective acceptance of their constitutive rules, which assign functions to things or persons that do not otherwise have such functions. We assign functions to things or persons by accepting rules that confer deontic statuses on them and that take the form "*X* counts as *Y* in *C*," where – this is the principal point regarding the being of social institutions – there is no difference between *X*'s counting as *Y* and *X*'s being *Y*. In the case of money – Searle's standard example of an institution – the rule might be: "Such and such bits of paper count as media of exchange within the borders of country *C*," where the paper's counting as money – which is precisely what makes it money – depends on the collective acceptance of the rules that define it as having that function. Searle speaks of the paper in question as acquiring thereby a deontic status, but because in other contexts deontic statuses might be thought to belong only to persons, it is important to add that a piece of paper's acquiring the deontic status of money confers what Searle calls deontic powers (CSR: 100–2) – rights and obligations – on the human individuals who stand in the appropriate relations to that paper: if someone possesses a piece of paper that counts as money, this accords her a right to buy certain things with it, and it imposes on me the obligation to accept it as legal tender if I have agreed to sell something to her.

Because the collective acceptance of rules plays a crucial role in Searle's account of social institutions, he, too, must face the question of what such acceptance consists in and to what extent it implies normative agreement – endorsement of the institution – among those who accept the relevant rules. In his second book on social ontology, Searle's answer

[21] E.g., that a certain piece of paper is a five-dollar bill (CSR: 2).

to this question is clear, namely, that the acceptance of rules need not imply approval of them: "Acceptance ... goes all the way from enthusiastic endorsement to grudging acknowledgment, even the acknowledgment that one is ... helpless to do anything about, or reject, the institutions in which one finds oneself" (MSW: 8). Like Rousseau, Searle needs a very weak conception of acceptance in order for his account to cover social institutions of all kinds – including human language – many of which are defined by rules that we apply only unreflectively or even with an awareness that we would reject them if we could. Once again, we seem to be led to the conclusion that social institutions are normatively constituted only in an exceedingly thin sense: they exist and function only insofar as human individuals apply – submit themselves to the authority of – their constitutive rules. Or, to put the point differently, the normative character fundamental to social life is on par with that involved in speaking one's mother tongue or buying a döner kebab with a five-euro banknote.[22]

Yet this is not the full picture of social life implied by Rousseau's or Searle's positions. In Searle's case, using money as the paradigmatic example of a social institution – together with the fact that he initially treats it in isolation from other institutions – obscures the richness of the account he is ultimately committed to. That Searle proceeds in this way is understandable: beginning abstractly, with a simple example of a single institution, brings out clearly the fundamental respect in which the being of social institutions differs from that of physical things. But this procedure also runs the risk of making the position being articulated appear simpler than it is. In order to fill out the picture of social life implicit in Searle's social ontology, I will center the rest of my discussion around the question: In what sense(s) are social institutions normatively constituted? Happily, pursuing this question will bring us back to the topic Plato introduced into the discussion of social life, the functional character of institutions.

In answering this question it will be helpful to have before us Searle's summary account of how the three "building blocks of social reality" (CSR: 1) – rules assigning deontic statuses and powers; the functional character of social institutions; and collective intentionality – are interrelated:

> Collective intentionality imposes a special status on some phenomenon, and with that status, a function ... that cannot be performed just by virtue of [the phenomenon's] physical features .... The function requires the status in

[22] For more on this topic see the discussion of "practical consciousness" in Giddens 1984: xxiii, xxx–xxxi, 7, 41–5, 290–1, 327–34.

> order that it be performed, and the status requires collective intentionality, including a continued acceptance of the status with its corresponding function.... [T]here is a functional implication carried by the description of [an] object as having a certain institutional status, as is shown by the fact that categories of assessment are appropriate under the status described that would not otherwise be appropriate. To be a husband or a citizen is already to have the possibility of being a "good" or "bad" husband or citizen. (CSR: 114)

If we examine this statement carefully, we see that normativity enters Searle's account of institutions in two ultimately related forms. The first follows simply from the rule-governed nature of social reality: institutions exist only insofar as their members collectively accept and apply their constitutive rules, but accepting and applying such rules involves subjecting one's actions to the authority of an "ought" that, through the assignment of deontic powers, specifies participants' rights and obligations in relevant contexts, thereby distinguishing wrong behavior from right. In the case of some (but not most) institutions – money, for example – these rights and obligations are clearly defined, and violations of the rules are easy to detect. Moreover, in the case of money (but not all institutions) noncompliance takes a deontological form: doing an injustice to, or violating the rights of, other participants in the institution. In other words, the "oughts" associated with an institution such as money take the form of clear directives or permissions. In this respect they are like the constitutive rules of games and unlike the regulative rules that tell participants how to play a game well and enable them to be evaluated as better or worse players of the game, all of whom must play the game "correctly" in the sense of following its constitutive rules in order to count as players at all. When I fail to follow the rules defining the use of money, I cheat someone or fail to give someone her due (perhaps even, inadvertently, myself). In any case, the mere fact that all social institutions rest on some sort of collective acceptance of their rules reveals one respect in which they are inherently normative and make a type of normative critique possible: *members* of an institution can be criticized for failing to follow its constitutive rules, as, for example, when I try to buy a subway card with a xeroxed ten-euro bill. Or, in the cases of families and states, constitutive rules make it possible for us to distinguish better husbands and citizens from worse.

The second respect in which normativity is inscribed into social institutions on Searle's account derives from their *functional* character. Recall that the rules ascribing deontic powers to members of institutions do so in the context of assigning functions to entities that as purely natural objects they lack. In other words, the content of institution-defining rules is intrinsically related to the specific social function a given institution serves: deontic statuses are assigned with some view of what the "point," or function, of money (or cocktail parties or families) is. (Again, neither Rousseau nor Searle thinks of most institutions as having been consciously created in order to serve a certain function; their causal history is typically independent of the functions they eventually come to have.) This means, however, that not only the behavior of individual members can be normatively assessed – judged as right or wrong, or better or worse, in light of the deontic powers accorded to them – but also institutions themselves, insofar as they do a better or worse job of fulfilling the function implicitly ascribed to them by their constitutive rules. Moreover, this holds for both tokens and types of institutions: the Drapers might be an unusually dysfunctional nuclear family, but the contemporary nuclear family in general might also do a poor job of fulfilling its implicit functions. Even an institution as simple as money is subject to being evaluated in this way: if its function is to facilitate exchange, then, under modern conditions, state-minted currency fulfills that function better than shells or beads.

As Searle's claim that "to be a husband or a citizen is already to have the possibility of being a 'good' or 'bad' husband or citizen" attests, the deontic powers ascribed to individuals within institutions depend on the functions ascribed to those institutions. This means that the two types of normativity distinguished above are mutually dependent. This is true for money but also for more complex institutions that assign deontic powers to their members differentially, in accordance with the particular roles individuals occupy within them. The collective acceptance of institutional rules that assign deontic powers associated with role *X* to an individual *Y* implies the possibility of judging that *Y* is a good or bad executor of *X* (a good or bad husband or citizen). At the same time, an understanding of what makes *Y* a good or bad *X* implies some conception of the function or point of the institution itself within which *Y* becomes an *X* in the first place: to understand what a good husband or citizen is is to know something about what a good family or state is.

When we think about more complex institutions such as families and states, it becomes clearer how taking money as the paradigm example of social institutions is potentially misleading. There are three related features of money that make it unrepresentative of institutions in general. First, the rules governing its use, like those of most games, are clear-cut, enabling actions to be classified neatly into one of three categories: permissible, impermissible, and obligatory. This differs from the cases of husbands and citizens, where the constitutive rules of institutions enable us to evaluate the executors of those roles gradationally: not as straightforwardly in conformity with the rules or not, but as better or worse at carrying out their roles. The gradational character of evaluation in these instances goes hand in hand with the fact that the deontic powers ascribed to the inhabitants of such roles are less precise than those governing the use of money; in these cases no merely automatic application of the rules is possible, and for this reason it is more natural to think of them as guiding principles than as rules in any strict sense.

It is important to note that when we judge husbands and citizens as better or worse, we are in most cases still judging them in accordance with the *constitutive* rules of their respective institutions, in contrast to the case of better or worse soccer players, all of whom must conform perfectly to the constitutive rules of soccer as a precondition of counting as a better or worse player. In the case of games and money, constitutive rules normally prescribe or prohibit determinately, enabling us to classify actions as right or wrong but not as better or worse. In the case of games, what determines whether a player is better or worse is not constitutive but regulative rules, and these two types of rules are strictly distinguished.[23] In the case of most social institutions, in contrast, the two types of rules often merge. It is a constitutive obligation of being a spouse that one be sensitive to one's partner's emotional needs, but one can do better or worse at this without ceasing to be a spouse. That what is at issue here is nevertheless a constitutive rule of the institution is revealed by the fact that we say of very insensitive husbands (of Doris Attinger's, for example): "He was never really a husband to his wife" or, referring to the couple, "what they have is not, or no longer is, a marriage" (Jaeggi 2019: 68–9, 90, 94, 118). That we cannot determine precisely when a family ceases to be a family in the normative sense is unimportant; the deeper point is that such judgments make sense to us, even if biologically related individuals continue to occupy the same living space and exhibit much of the external behavior we associate with family life.

The second atypical feature of money (and language) is closely related to the first: its constitutive rules can normally be followed perfectly without thinking about its institutional function, perhaps even without having any sense of that function at all. Here, too, money differs importantly from the family and the state, and even from cocktail parties. Because in these cases the conventions governing participation are too indeterminate for individuals always to know precisely which actions are required of them,

[23] For the sources of this distinction, see Rawls 1955 and CSR: 27–9.

having some grasp of the point of the institution is necessary if one is to participate in it effectively; in these cases, *applying* rules requires *interpreting* them in light of some understanding of the institution's point. Parents, for example, often find themselves not knowing exactly what their roles call upon them to do in specific circumstances. In such cases they can apply the rules governing parenthood effectively only by thinking about what the point of family life is and interpreting their situation in that light. One fails to parent at all if, for example, one regularly puts one's own good before that of one's child's, but it is often unclear what specific behaviors are implied by the rule that the child's good comes first. Hence, following that rule in the way the well-functioning of the institution requires demands practical judgment that can proceed only by consulting one's conception, sometimes inchoate and implicit, of the point of parenting and family life.

The third respect in which it is misleading to take money as the paradigm of social institutions follows closely on the second: money's function is defined thinly, without making reference to ethically tinged values, or to a conception of the good beyond the merely useful. This peculiarity of money is hinted at in Searle's claim that social institutions generally – in contrast to money – need not be functional in a narrow sense (CSR: 114). This is precisely why it is sometimes more appropriate to speak of the point of an institution than of its function. In the cases of families and states the point of the institution includes realizing or expressing values beyond mere efficiency and instrumental utility. And the same is true of many ethically less significant institutions, such as cocktail parties and bowling leagues. To grasp the point of families and cocktail parties is to grasp the specific goods each is capable of realizing, and in both cases those goods extend beyond mere efficiency, biological survival, or physical comfort. This means that much of social life is normative not only because it involves accepting and applying rules that tell us what we may and may not do, but also because accepting and applying those rules is a form of *expressing values*, and, since social institutions rest on the collective acceptance of rules, these are necessarily values shared by other members of the institution. This adds a new dimension to Rousseau's claim that social institutions are grounded on convention, the coming together or agreement of their members' wills. For now the acceptance of rules constitutive of social reality no longer involves merely accepting the authority of rules specifying what we may and may not do in specific circumstances but, beyond that, a more robust normative agreement about (aspects of) the good.

From Searle Back to Rousseau

Acknowledging the value-expressive character of (much of) social life helps us to make even more sense of Rousseau's claim that social institutions are *artificial.* For one context in which he routinely uses the term is when dealing with phenomena involving human evaluation: a prominent hallmark of artificiality for Rousseau is the intervention of human "opinions" in the constitution of a thing, and one way in which society is constituted by opinion is that shared beliefs about the good – human valuations – make it (in part) into what it is. Beliefs regarding the good are constitutive of social practices whenever social members act in accordance with some understanding of the good to be achieved by what they do together, and this implies that institutions are something we make – something that depends on our agency – for the same reasons that any acceptance of rules does: because, first, we are responsible for the values we espouse in that it lies within our power to reflect on their appropriateness and to reject or revise them; and, second, values do not usually imply concrete actions in the absence of our active interpretation of them in the light of specific circumstances.

This point helps us to see why Rousseau equates accepting an institution's constitutive rules with authorizing or endorsing it: in the case of many institutions there is a close connection between participating effectively in them and espousing, at least in part, their animating values. In the case of cocktail parties it is possible to make a good show of playing along with their conventions while silently thinking that the whole to-do is a waste of time. Here we have acceptance of rules – I recognize them as having authority and endeavor to abide by them – without authorization. One can have purely instrumental reasons for playing along with the conventions of a cocktail party – perhaps it makes my spouse happy or will lead to my next promotion – and some of us are skilled at getting the fact that our participation is grudging to pass undetected by others. It is even conceivable that institutions such as cocktail parties can endure for generations despite the fact that no one any longer appreciates the good they were once thought to realize.

This point cannot, however, be extended to social life in general. This is because there are important differences between cocktail parties (or bowling leagues) and family life (or democratic participation). The most important difference is that while cocktail parties can appear to function reasonably well without participants having strong normative commitments to the values that animate them, families belong to a set of institutions that cannot function well unless many of their members have a strong – even an identity-constituting – allegiance to the values that define them. It is no accident that when Searle elaborates his claim about the mutually implicatory relation between deontic statuses and institutional functions, his first two examples – husbands and citizens (CSR: 114) – are taken from what a long tradition of social philosophers takes to be among the basic institutions of Western modernity: the family and the state. In the case of these institutions, effective participation requires a more robust endorsement, or authorization, of the institution in question – and hence thicker normative agreement and a more substantive normative community – than do cocktail parties and bowling leagues. At most, the latter require agreement with respect to very specific aspects of the good, whereas families and Rousseauean democracies require agreement on more comprehensive conceptions of the good. It is not merely that effective participation in these institutions presupposes allegiance to a more comprehensive conception of the good than do cocktail parties; it also requires a deeper attachment to the goods pursued in family, political, and professional life than those at issue in less fundamental institutions. In general, members of families, professions, and democratic states must take their participation in such groups to aim at goods that are of special importance to human flourishing: a good parent, a good plumber or nurse, a successful citizen must recognize the aims of the family, of their professions, and of the democratic state as much weightier than those achieved in cocktail parties and bowling leagues. A parent who thought of childrearing and cocktail parties as realizing goods of approximately equal value would not be a successful parent (or "would not really be a parent"). This is because, first, the goods achieved in fundamental institutions are *more numerous and more complex* than in other institutions; second, those goods are generally taken to be (and are) *more important* than those achieved in others; and, third, the forms of cooperation required to realize those goods are *more demanding* than in less fundamental institutions, both in terms of the time and energy members must devote to them and the degree of commitment and renunciation of purely egoistic aims they demand. This is one reason participation in these institutions is aptly described as family, professional, and political *life*.

It may seem that in filling in the details of the connection between acceptance and authorization we have come a long way from Searle and his ontological concerns. To some extent this is true. There is a sense in which, once Searle has spelled out how, in the case of money, the collective acceptance of rules that assign deontic statuses allows a physical object to acquire a function it does not otherwise have, he has answered his main question of how the being of social institutions differs from that of physical objects, and the crucial role that collective acceptance plays in this account suffices to explain why, unlike those objects, social institutions have a "constructed," or artificial, existence. As I have argued, however, there is another sense in which Searle's account of money alone does not give a full picture of human social reality. And much of what he himself goes on to say about institutions is sensitive to the ways in which human social life is richer than the use of money.

If there is a link between Searle's purely ontological question and the richer picture of social life, it comes into view when, after having set out the thesis that social institutions are constructed by us, he asks a further question: "Why do people accept institutions and institutional facts?" (MSW: 107). Even if answering this question is not strictly necessary to uncovering the ontological difference between social and physical being, neither is it exactly external to that project: if social institutions are shown to depend on human agency – to be something we do, and therefore also something we could do differently – then our inquiry into what such institutions are remains incomplete as long as this "why" question about our agency goes unraised. Locating social reality in the domain of what we do, rather than in the teleologically barren domain of the physical, more or less demands that we also ask why we do it, where the response to that question will appeal to a purpose our acceptance of institutional rules serves.

As we have seen, Searle's answer to this question is that we accept institutions' rules for a variety of reasons, including reasons that a tradition he does not belong to calls "ideological" (MSW: 8, 107). He insists plausibly, however, that accepting institutional rules is typically bound up with some understanding, often inchoate and implicit, of the function or point of the institution in question, and this suggests that a sense for the function of an institution is often the reason for accepting its constitutive rules. Indeed, the answer to the "why accept?" question emphasized most by Searle – that we see that institutions increase our powers and are therefore in our collective interest (MSW: 8, 107) – suggests that understanding an institution's function often goes hand in hand with endorsing it as good or legitimate, which here, too, can be inchoate and implicit. Although Searle does not say so explicitly, the implication seems to be that without some endorsement or approval of at least some of the institutions in which we participate, social life would not be possible.

In fact, an even stronger claim is true: human social life in general is impossible in the absence of certain basic institutions whose well-functioning depends on a collective acceptance of their constitutive rules grounded in shared conceptions of the good, or in values that go beyond narrow self-interest or utility. That this is not the case for all institutions is shown by the example of money. However, when we cease to regard institutions in isolation and see them instead as existing in relation to others, especially to those that make biological life possible, a different picture of the normative agreement on which human social life depends comes into view. After all, institutions such as money and bowling leagues can exist only if humans produce the goods required for physical survival, reproduce the species by giving birth to and raising future generations, and educate their offspring such that they acquire the basic capacity to follow institutional rules. If these functions are to be carried out, humans must be united in certain fundamental institutions by normative bonds that go much deeper than collective acceptance in the weaker guises discussed above.

The claim that human social life depends on institutions that can function well only on the basis of a relatively robust agreement about the good does not follow directly from the most general account of institutions yielded by Searle's examination of single institutions in abstraction from others. It follows, rather, only when that general account is viewed in the context of certain contingent and empirical, but still fundamental, facts about human beings, such as our biological (and psychological) neediness; our radical dependence on others for the satisfaction of our needs; and the formidable complexity and demandingness of the cooperation required to satisfy those needs, including for the reproduction of biological life. A compelling claim implicit in Rousseau's social thought is that the tasks involved in satisfying basic needs, including "artificial" needs we can no longer unmake, are such that they can be carried out only if the bonds among social members rest on something deeper than narrow self-interest: if parents are to raise children (in the manner appropriate to humans), if political life requires that citizens be willing to sacrifice some of their particular interests, even to the point of risking their lives, then social members must be able to find in what they do a significance beyond mere utility that allows their activity to be experienced by them as meaningful and good. If, as Searle asserts, deontic statuses, undergirded by the collective acceptance of rules in general, are "the glue that holds human society together" (Searle 2008: 452), then, after adding certain anthropological facts to the picture, we must say that human social life as a whole is held together only on the basis of some shared allegiance to fundamental aspects of the good. On this view, acting in accordance with shared conceptions of the good is, for human beings, constitutive of basic elements of social life. If we incorporate this idea into our picture of the kind of thing human society is, our social ontology becomes thicker than those typically offered by philosophers in the Anglo-American tradition.

It may seem that by bringing in the anthropological facts of dependence and the need for sustained cooperation we have ventured beyond Rousseau's own position, directly contradicting his assumptions about our radical independence in the "original" state of nature. But this impression dissipates when one recalls that for him that original state is merely hypothetical and "might never have existed" (DI: 132/OC: 133). For one point of that hypothetical construction is to show that the radical independence of the original state of nature is possible only for beings, who lacking language, love, and reason, are unrecognizable as *human* beings. Only in Part II of the Discourse, where dependence engenders enduring, identity-constituting bonds among individuals do we encounter anything we can recognize as human beings and human society. For Rousseau dependence and enduring cooperation are anthropological facts that cannot be abstracted away from if human social reality is to come into view.

Rousseau's account of our fundamental dependence brings into the picture a specific social need I have not yet mentioned: the need we experience, whenever humans associate with one another – which is to say, whenever human beings exist – for the regard, or recognition, of our fellow beings. This need, springing from our *amour propre*, makes humans even more interdependent than mere biological neediness does because it can be satisfied only by the (favorable) "opinions" of others. Since these opinions are evaluative,[24] the need to be well regarded can be satisfied only if recognizer and recognized share certain standards of the good that ground the recognizer's valuation of the recognized. For Rousseau the requirements of biological reproduction conspire with our psychological needs to make it the case that human social life cannot get by without significant agreement on the good that underlies the collective acceptance of the conventions that govern the more fundamental social institutions. At root, then, human social life always depends on collectively accepted conventions that, because they require allegiance to a shared conception of the good, do indeed *authorize*[25] certain institutions more robustly than the thin acceptance on which some institutions taken by themselves might appear to depend.

[24] To seek recognition is to seek the good opinion of others under a certain description of what would make one worthy of their recognition; this explains why *amour propre*, too, is "artificial."

[25] "Authorize" here implies only that institutions are "*regarded* as good or legitimate."

Readers familiar with nineteenth-century social philosophy might recognize that in thinking through Rousseau's social ontology we have arrived at a close relative of the view Hegel articulates in the *Philosophy of Right*, where, in typically idiosyncratic fashion, he describes the kind of being characteristic of human society as "the *living good*" that reproduces itself through "the consciousness and will" of its members (PhR: §142). The convergence between Rousseau's and Hegel's social ontologies concerns not only the necessity for social life of some shared agreement regarding the good but also the appropriateness of evaluating the functioning of societies according to a standard of *freedom*. The point here goes beyond the claim that institutions are normatively constituted in the sense that accepting their constitutive rules implies ascribing functions to institutions, on the basis of which one can assess better and worse instances of, for example, the family and the spouses and parents that make them up. What I have in mind is a more general connection between what social institutions by their nature are and the normative standards by which it is appropriate to judge them. For in examining the kind of thing social institutions are, we come to see that their functioning well is not independent of subjective attitudes their members take to them. Because institutions function best (more smoothly, more efficiently, more reliably) when their members follow the relevant rules of their own accord – and even better when those rules correspond to their own understanding of what is meaningful and good – there is an internal connection between institutional well-functioning and a kind of freedom of their members. This species of freedom – "moral freedom" (SC: I.8.iii) – exists not when individuals conform to rules imposed on them from without but when they act in accordance with their own understanding of the good or, in the case of less fundamental institutions, in accordance with rules whose validity they accept.

Because of the complexity of Rousseau's position, it may be helpful to summarize the various aspects of his claim that social inequalities, and the institutions that make them possible, are artificial. Rousseau believes both that social institutions *come into the world* only as the (largely unintended) results of free human actions and that the ongoing *activity* of social members *is constitutive* of what those institutions are. His most important claim, however, is that "we all make the rules" that govern our participation in social life, as expressed in his statement that social inequality – and, by implication, the institutions that make it possible – "depends on a sort of convention and is … authorized, by [human] consent" (DI: 131/OC: 131). I have interpreted this statement as meaning that social institutions are constituted by normative rules whose efficacy depends on their being accepted and applied by rational agents, where repeated applications reinforce and strengthen the institutions in question. Even when collective acceptance of rules is at its thinnest (in the case of money, for example), accepting and applying them involves agency on the part of those who do so. Moreover, many institutions depend on more than a bare acceptance of their rules, namely, some awareness on the part of their members of the point, or potential good, of those institutions. In these cases, interpretation of institutional rules, rather than automatic application, is necessary, and this involves an aspect of agency beyond what is required in the simplest institutions. Further, because of human neediness and dependence, and because of the sacrifices required by the cooperation that responds to that neediness, certain fundamental institutions, especially those implicated in reproducing biological and spiritual life, cannot function, or cannot function well, unless large numbers of members endorse their institutions' ends – making talk of "authorization" appropriate – on the basis of a shared understanding of the good that those institutions enable their members to realize. Finally, because of the central role the concept of function plays in it, Rousseau's social ontology carries implications for a kind of normative assessment of institutions that bears similarities to the norms of health and illness appropriate to living organisms: First, the close connection between an institution's function and its constitutive rules makes it possible to assess institutions and their members as better or worse instantiations of the beings they are; and, second, the dependence of institutions on the collective acceptance of rules by the rational agents who are their members implies that the well-functioning of the former is inseparable from the freedom of the latter, where freedom is understood as acting in accordance with rules one takes to be authoritative or, more substantially, productive of the good.

CHAPTER 7

Durkheim's Predecessors: Comte and Spencer

The aim of the following four chapters is to articulate Durkheim's conception of social pathology and, more fundamentally, to understand why illness plays a central role in his attempts to understand modern societies and locate their deficiencies. As we know, Durkheim is not the first modern social theorist to employ the idea of social pathology, but he reflects more extensively than anyone on how pathology in the social domain is to be understood and why the concept is important. Even if the idea of social illness is pervasive in earlier social thought, only in the nineteenth century – when biology is constituted as a modern science – does it come to be the object of systematic reflection in European social theory.

Durkheim's social theory brings together and extends the ideas of Plato and Rousseau[1] examined in previous chapters: it appropriates Plato's functionalism and organicism and a version of Rousseau's account of the artificiality of human society, adapting both to the more complex social conditions of Western Europe in the late nineteenth and early twentieth centuries. Like Rousseau, Durkheim emphasizes the role moral rules play in constituting social reality. Like Plato, he thinks of human society as a functionally organized system that because of its functional character is vulnerable to illnesses of a distinctly social kind. Durkheim's social theory is indispensable to a study of social pathology not only because his treatment of these issues is extensive and relevant to contemporary social

[1] In 1892 Durkheim wrote his *thèse latine* on Montesquieu's and Rousseau's contributions to the founding of social science (Durkheim 1966). In it he refers approvingly to various of Rousseau's doctrines that he later incorporates into his own thought: the artificial character of human society (135); explanatory holism (136); the analogy between human societies and living bodies (137–9, 154, 170); the critique of social inequalities, including those arising from the inheritance of wealth (140–2); the moral character of social life (159–61); the centrality of law to social reality (174–5, 181); and the religious source of moral authority (192–3). The text also shows that Durkheim's concept of solidarity was developed in dialogue with Rousseau's account of what attaches citizens to the general will (170, 187, 192). I thank Seyla Benhabib for drawing my attention to this early text.

conditions but also because he addresses in detail how to ascribe nonarbitrary standards of healthy functioning to society and its institutions.

Because Durkheim's conception of social pathology is unthinkable without the legacy of two earlier nineteenth-century thinkers, Auguste Comte and Herbert Spencer, this chapter examines how the idea of social pathology appears in their social thought. Even if Durkheim's conception of social pathology is more interesting and scientifically respectable than his predecessors', it retains significant traces of his theoretical engagement with both. I argue that in Comte's case the influence is largely positive – Durkheim extends and refines Comte's conception of social pathology – whereas Spencer's influence is mostly negative, serving to instruct Durkheim as to what a conception of pathology in the social domain ought *not* to look like. The most important respect in which Durkheim goes beyond his predecessors is in claiming that their reliance on biological analogies disposes them to adopt impoverished positions regarding both the kind of normative evaluation sociology is able to undertake (where social health is conceived too much on the model of biological health) and the degree of agency it attributes to society's members. Both deficiencies derive from an inadequate grasp of the ontological differences between social and biological phenomena.

Auguste Comte

Comte's use of the idea of social pathology is inseparable from his aspiration to found a rigorous science of society, or "sociology," that enables us to understand the ongoing functional processes of modern societies as well as what brought them into existence. The first of these concerns – about how form and function interrelate in social life – constitutes what Comte calls *social statics*, whereas the second belongs to *social dynamics*. The principle underlying both is that a scientific understanding of human societies requires thinking of them on analogy with biological organisms and that the newly consolidating science of biology is the appropriate methodological model for a scientific sociology. Thus, when describing the first of these inquiries, Comte asserts that establishing a science of society depends on recognizing "a true correspondence between the statistical analysis of the social organism in sociology and that of the individual organism in biology" (Comte 1875 [1851]: 239–40).[2] The same principle underlies his

[2] "Statistical" here refers not to numerical data but to ongoing functional processes, as opposed to an organism's development or evolution: "statics" studies social phenomena synchronically rather than diachronically.

social dynamics, insofar as it seeks to understand the history of societies as analogous to the development of both individual living beings and their species. Here, too, biology's increasing attention, even before Darwin, to the evolution of living species, is taken to hold the key to understanding the patterns according to which human societies come into existence and develop.[3]

It is worth emphasizing Comte's distinction between social statics and social dynamics because it points to the possibility (for us) of following him in the first of these enterprises – appealing to coordinated specialized functions to understand how the parts of a society relate to one another and to the whole, as well as how, working together, they accomplish certain ends of the whole – while abandoning him in the second. Comte's immediate successors – Spencer and Durkheim – follow him in both projects, looking to living organisms as a source for understanding not only the functioning but also the history and development of human societies, or more accurately: the history and development of human society in general, understood as a single universal process. For nineteenth-century theorists of social pathology before Comte – Hegel and Marx, for example – as well as for those who come after Durkheim, the evolution of biological species plays little or no role in accounting for how societies originate and develop. Moreover, the theory of social pathology I endorse here retains some version of Comte's biologically inspired functionalism in understanding what human societies are and how they work, while distancing itself from his idea that evolutionary theories in biology are relevant to a science of society. Despite this, it is noteworthy that for all the major theorists of social pathology covered here, with the exception of Rousseau,[4] some account of a society's real history, whether necessary or contingent, is indispensable to diagnosing social pathology. What is more, most endorse some version of *universal history*, positing a noncontingent pattern of development that characterizes human society in general. Since the widespread tendency of theories of social pathology to depend on some form of social history, whether universal or local, appears to be independent of the nineteenth-century's presupposition that biology is sociology's model, it will be important to understand the thought underlying this tendency and to determine the extent to which it is compelling.

[3] For a discussion of evolutionism in social theory, see Giddens 1984: 228–33.

[4] Recall that Rousseau describes the genealogy given in the Second Discourse as "hypothetical" (DI: 132/OC: 133).

It would be wrong to assume that Comte's attempt to ground sociology on an analogy between social and biological organisms is motivated merely by his aspiration to found a science of society, together with the fact that in the nineteenth century it was the new science of biology, rather than physics or chemistry, that generated the most excitement among philosophers and scientists. In addition to these factors, biology presents itself to Comte as an apt model for sociology for the same reason that earlier thinkers compared societies to living organisms: the survival of human societies appears to require that a range of functions be carried out in social life, and doing so effectively depends on a division of labor in which diverse functions are assigned to specialized parts of the "organism" well-suited to executing those functions. Most fundamentally, then, it is the idea that already informs Plato's functionalism – that of *coordinated specialized functions*, executed by an organism's coordinated specialized parts – that underlies the analogy posited by Comte between societies and living beings.

Many aspects of Comte's social theory are readily understandable from his commitment to a biologically inspired functionalism. His sociology embraces, for example, a form of explanatory holism familiar in biology, often expressed in terms of the primacy of the whole over its parts.[5] In living organisms, wholes have primacy over their parts in the sense that understanding a particular organ – understanding both how it functions and what function it serves (or what need it satisfies) – depends on understanding its place in the functioning of the organism as a whole. In order to understand what a heart is, for example, it is necessary to say how it works together with other parts of the body to achieve a vital end of the organism as a whole. In sociology, too, explaining particular phenomena – from the division of labor to the operations of the state – requires situating them in relation to essential functions of society; more precisely, it requires explaining what functions of the whole the phenomena in question serve and how they do so. This means that diagnoses of illness will be similarly holistic: it will rarely be possible to diagnose a social pathology merely by looking at a single "organ" of the society in question. Since the functions of one part of an organism are specified in relation to the functions of others, uncovering dysfunction will require situating the workings of a defective organ in relation to those of the other organs with which it normally cooperates. Perhaps the most important respect, however, in which Comte's commitment to a biologically inspired functionalism leave its mark on his social theory (and Durkheim's as well) lies in the implication

[5] This point is emphasized using different terminology in Aron 1970: 67.

that a scientific understanding of society, like physiology and medicine, is able, via the concept of social pathology, to bring together three related tasks: explanation, evaluation, and prescription.

Comte's accounts of how the parts of society function and what functions they serve are functional explanations, but care must be taken to specify what that term means. The nature and legitimacy of functional explanations are hotly contested issues in the philosophy of biology, but in that context functional explanation is generally understood as attempting not (or, sometimes, not merely) to explain what function a certain organ serves in the organism and how it does so but, beyond that, to explain why that organ is there in the first place, and to do so functionally – that is, by appealing to the function it serves. In other words, full-fledged functional explanation includes the claim that a thing's function explains its existence – as, for example, in the claim that the heart's presence in a living organism is explained by the fact that it is able to pump blood throughout the organism, supplying the latter with the nutrients it needs in order to survive. Similarly, social theorists who embrace functional explanation in this sense claim that the existence of some *explanandum* – the state, for example – is explained by whatever essential function of society it carries out. I call explanations of this sort *existential functional explanations*, by which I mean positions that explain the existence of a thing by reference to some function it fulfills, whether in "organisms," living or social, or in nonorganic contexts (as in a clock). I argue below that existential functional explanations can be of two different types, depending on whether a thing's function is taken to explain that thing's *origin* (its coming into being) or its *persistence* (without regard to how it first came to be).

To be distinguished from existential functional explanation in both its forms is a mode of explanation relevant to organic and social life that I call *functional analysis*. Functional analysis might be considered less ambitious than the other two types of explanation because it makes no claims regarding the existence of the phenomena it analyzes (neither their origin nor their persistence). For this reason, in one narrow sense of the term, functional analysis might be said not to *explain* at all. I referred briefly to functional analysis above in describing the familiar type of explanatory holism – the primacy of the whole over its parts – that biology relies on, according to which understanding a certain organ involves grasping both how it works and what vital function of the organism its activities serve.[6]

[6] For a defense of functional analysis in biology, see Boorse 1976: 75.

Such an account "explains" what the heart is by both specifying the function it serves and showing how its particular features – its own constitution and its relations to other organs – are suited to accomplishing that purpose, but it makes no existential claims: the function attributed to the heart is not taken to imply anything about how the heart originated or even why it continues to be present in the organism at issue.

It is for this reason that weak existential functional explanation can be *distinguished* (at least conceptually) from functional analysis. That, in general, functional analysis can both take place and have a point independently of existential functional claims is demonstrated by pre-Darwinian physiology, which was perfectly able to discern the function of hearts without addressing the question of why they developed or why, once present, they persist in future generations. Even within a Darwinian framework it is possible to distinguish these two questions conceptually – "what function does *X* serve?" and "why is *X* an enduring feature of organisms that have it?" – although the commitments of that framework imply that an answer to the first is relevant to answering the second.

It is probably the case that in functionally oriented social theories a similar presumption exists (that attributing a social function to a phenomenon explains its persistence). For Hegel, for example, the various functional roles played by the nuclear family (in modern conditions) explain why (in those conditions) it reproduces itself over generations. Or, in Durkheim's case, the solidarity-producing function of (any) society's collective consciousness explains its persistence.[7] The same can be said even of Marx, who fits less comfortably into the functionalist tradition in social theory (not because he lacks the concept of social function but because social functions – in capitalism – tend to be indicative of pathology rather than healthy social life): his functional analysis of the industrial reserve army, for example, tells us what function mass unemployment serves in relation to the ends of capital accumulation and, at the same time, why it is a permanent feature of capitalism. Here, too, however – as in all these examples – the two questions are logically distinct.

If the connection between functional analysis and weak existential functional explanations is as close as the paragraphs above imply, why insist on differentiating them? One reason is that, as I argue in Chapter 8, the former is far more important to Durkheim's ambitions as a social pathologist

[7] Even in modern, "organic" societies, where collective consciousness is diminished in comparison with segmental societies, it continues to play a non-negligible role in producing solidarity. See Chapters 8–10.

than the latter: diagnosing social pathologies and thinking about how to remedy them depends on understanding what functions social phenomena serve but not on a claim about why they persist or came into being. A second, related reason is that existential functional explanations, even of the weaker sort, continue to be controversial in the philosophy of biology; if these are relatively unimportant to Durkheim's treatment of social pathologies, one can avoid making the viability of his position depend on finding a solution to this difficult philosophical issue.

There is a third, more substantial reason for distinguishing functional analysis from weak existential functional explanation: in social theory it is not clear that the connection between the two is as close as it is in biology. In the following chapter I note that for Durkheim not every feature of society serves a social function (or, equivalently, satisfies a social need): "A fact can exist without serving any purpose," and there are "more instances of this in society than in organisms" (RSM: 120/91). This means that in human societies, more so than in biological organisms, there are other – causal rather than functional – "forces" that explain why certain features of societies persist. Consider in this light Durkheim's fundamental claim, examined in Chapter 8, that the function of the division of labor in modern societies is to produce solidarity. In this case it seems odd, however, also to claim that this function of the division of labor explains its persistence. The reason for this is that the division of labor is so entrenched in modern societies that one must wonder whether its ceasing to produce solidarity would entail its disappearance. While Durkheim might claim, more weakly, that the solidarity-generating function of the division of labor explains *in part* why it persists in modern societies, its being only a (presumably, small) part of such an explanation gives us a reason to focus on functional analysis separated off from any existential functional claims that might be associated with it. The same point applies to what Durkheim calls the "sexual division of labor": its serving a social function need not play a significant role in explaining why it persists. Finally, the point is illustrated even more clearly in Durkheim's treatment of crime: that statistically normal rates of crime serve a social function (as a source of innovation and moral progress) most likely lies at a considerable distance from whatever factors – surely causal rather than functional – explain the persistence of crime in all societies we know (RSM: 98/66).

It is easy to understand why existential functional explanations of a thing's *origin* is controversial in biology (and in functionalist social theories). For, since Darwin at least, evolutionary theory in biology takes care to separate issues of origin and function: random processes of mutation explain the origin of an organism's features, independently of whatever ends those features might ultimately come to serve. Whereas it is legitimate, in both biology and social theory, to regard an organ or institution, once existent, as performing a function in the context of the living whole, it is illegitimate in both domains to assume that that function explains why the organ or institution originally came about.

One version of functional explanation of origins would be unproblematic in social theory if one posited a certain ontological difference between societies and living organisms, namely, that the former are consciously intended products of human will. As Kant points out, a kind of functional explanation is legitimate in the case of teleologically organized artifacts: both the existence and constitution of a clock can be explained by its function since a human agent designed it with the aim of achieving a specific end, the telling of time (CJ: §65). Whether human societies are like clocks in this respect is a less straightforward question than it seems because, at least since Rousseau, many social theorists, including theorists of social pathology, take social reality to differ from nature in being (in some sense) a human creation. As we saw in the previous chapter, however, Rousseau's claim is that while the being of human society depends on free actions, it is not an artifact because its various features are not consciously intended results of human will.

The issue of functional explanation is further complicated by the fact that Darwinian evolutionary theory appears to allow for existential functional explanations of the weaker sort – although this, too, is controversial – namely, those that explain not the origin of something but its *persistence* by appealing to the function it comes to serve in an organism's life: that longer necks help giraffes to survive by making the leaves of tall trees accessible to them does not explain why longer necks first came about, but it is relevant to explaining why this particular deviation survived and became a standard characteristic of future generations. In general, existential functional explanations of this type have a greater plausibility in social theory than those that purport to explain the origins of their *explananda*, and many classical social theorists – from Hegel to Marx to Comte to Durkheim – have relied on them. Nevertheless, for the reasons given above, I will bracket the question of whether such explanations have a legitimate place in social theory.

Comte's appropriation of (pre-Darwinian) biology is not very clear with respect to which modes of functional explanation or analysis it employs and what makes them legitimate. Comte employs all three modes distinguished here and is mostly unreflective about which he is using when.

What is clear is that he leans heavily on the analogy with biological organisms when thinking about how human society makes use of a specialized division of labor in order to accomplish the ends of society as a whole. Comte's account of these ends – of the essential functions of (modern) societies – also derives largely from thoughts about the vital needs of living organisms: the most prominent ends he attributes to human society are reproduction, growth, and stability. Since he regards increases in social wealth as an essential means to these ends, improvements in material productivity are also regarded as a central preoccupation of social life, as well as a sign of a society's health.

From here it is only a short step to the idea of historical progress and to Comte's conception of what such progress primarily consists in for human societies, namely, increased human control over the natural environment achieved through ever greater specialization and more perfectly adapted social "organs" and subsystems. (The importance he ascribes to *adaptability* to the environment as an indicator of social health is another sign of Comte's biologism that becomes even more important for Spencer.) This conception of social progress lies at the heart of Comte's conception of universal history, which posits an overarching telos for social development – increased control over nature in the service of social reproduction, growth, and stability – and which is given expression in his notorious "law of three stages." According to this law, every human society is destined to progress through three periods of development: the theological, the metaphysical, and the positive. The first of these is organized around military structures and values, and the second around political and legal institutions and norms, while the third is characterized by the industrial and "scientific" form of society distinctive of modernity. Comte's largely materialistic conception of historical progress, with its focus on improvements in human productive forces, acquires an ethical tinge when he adds to this the idea that in developed societies the struggle of humans against nature comes to replace the struggle of humans against one another that characterizes history prior to its positive stage. Peace, of course, might be valued because it promotes the development of the productive forces, but it seems unlikely that Comte conceived of its value exclusively in these terms.

While Comte thought of his vision of universal history as roughly mirroring evolutionary processes in the biological sphere, it is difficult for post-Darwinians to take his story seriously. Apart from the artificiality of the scheme's three stages (as well as its positing of a final stage), his idea of a unilinear development that progressively realizes a single telos – an ethically significant one at that – strays too far from what, even in the middle of the nineteenth century, would have been a compelling account of biological evolution. If there is a more plausible core to Comte's story, it consists in the idea – retained in part by Durkheim – that the evolutionary development of biological species and human societies, while not guided by an overarching telos, exhibits a common pattern, awareness of which is essential to understanding both how modern societies differ from earlier ones and what challenges they face. On this view, a central tendency of history, biological and social, is the development of increasingly complex capacities made possible by increasing specialization of increasingly differentiated "organs" or subsystems. The biological processes that lead from single-celled organisms to complex mammals seem, in any case, to share this general developmental tendency with the processes – increases in social complexity – that distinguish contemporary Western society fundamentally from the simpler societies of earlier epochs.

One strength of Comte's account is his emphasis on the need, in an internally complex organism, for the *coordination* of its diverse specialized functions. He emphasizes, in other words, that the *division* of labor is an incomplete description of what makes complexly functioning organisms possible. For a high degree of specialization without a correspondingly high degree of coordination impedes rather than promotes an organism's effective functioning. It is in thinking about this issue that Comte acknowledges a crucial difference between human societies and living organisms: whereas in the latter, the ends of the whole can be achieved without a consciousness that directs and oversees the organism's activities, in the former this coordinating function can be carried out only by a conscious agency that manages society's commercial and productive activities; assigns social roles to individuals in accordance with their abilities; and inspires and regulates social members' behavior by educating them to understand the principles that animate modern society and give it legitimacy.

It is not surprising that Comte locates the main difference between organisms and human societies in the possession of consciousness, for a similar claim is made by all the theorists of social pathology covered here. More interesting is the fact that consciousness plays for him two distinct roles in social life. The first is simply the coordination of specialized functions, which corresponds to a central nervous system in higher animals. The second is less predictable from his reliance on the analogy with living organisms, and it points to a respect in which, even for as biologistic a thinker as Comte, social reality diverges from biological life: in its second role, consciousness carries out a legitimizing function. Exactly what the content of such consciousness must be in order to accomplish the function of legitimation remains unclear in Comte, but its most important feature is that it must be universal, a consensus shared by all members of society. In earlier societies, this universal consensus was the province of religion. In the modern world, however, it is to be secured by a class of functionaries – priests, as Comte calls them – who have assimilated the doctrines of positive science (or positive "religion" [Comte 2015 [1844]: 355–444]) and convey these truths to the rest of society, thereby securing the universal consensus that gives the social organism unity and direction.

Comte's emphasis on the doctrines of positive science being conveyed to all of society's members represents a considerable deviation from the model of the biological organism, where having an overview of the functioning of the whole is itself a specialized function carried out by a single organ or subsystem: the central nervous system must have some such "grasp" of the whole, but hearts and livers need not. Even though Comte's priests, too, carry out a specialized function, that function is to secure a universal, legitimizing consensus throughout all of society. It is hard to avoid the impression that this aspect of Comte's social theory finds its impulse not in the imperatives of biological life but in an ethical demand that ultimately makes reference to the potential freedom of the individual human beings who compose the social organism. This ethical demand, characteristic of modernity, is that it must be possible for every social member to appreciate the legitimacy or the overall goodness of the society within which each plays a specialized role. This demand might be understood as having only a narrowly functional justification, for example, in the thought that a social organism carries out its functions most efficiently when each member endorses the principles governing social life and grasps their point. But since nothing in living organisms seems to correspond to this purely functional claim, it is more likely that the requirement of universal consciousness in Comte is explained by the ethical concern that the business of society as a whole be carried out in a way consistent with the free consent or endorsement of each social member. If so, then Comte's conception of the healthy social organism is informed by the same ethical principle that Rousseau, Hegel, and others regard as a condition of a legitimate or good social order: if individual social members are to be free in their social participation, they must be able to endorse – to regard as their own – the principles that organize their social life. In other words, even Comte's highly biologistic approach sneaks in some version of the Hegelian ontological distinction between spirit and mere life.

Given the functionalism of Comte's theory, it is no surprise that the central component of his vision of social health is essentially Plato's: society's health consists in a harmony or balance among the specialized parts and functions it depends on to meet its vital needs, and illness is a deficiency in the same. Thus, impaired functioning is the central element of Comte's conception of social pathology, but his attention is focused less on the proper functioning of individual organs than on their being coordinated such that, overall, the organism in question functions well. In other words, impaired functioning is taken to be due primarily to a lack of balance or coordination among the organism's functional parts.

Comte sometimes expresses his ideal of social health in terms of another idea drawn from biology: a correspondence between form and function. This raises the possibility of a different type of social pathology, a kind of structural or anatomical deformation of the social organism. The close connection Comte draws between organic structures and functions in societies is reflected in his belief that social structures can be explained by referring to the functions they enable the social organism to carry out. (Thus, expressed in the terms developed above, he attempts to give existential functional explanations of social structures.) But precisely because of this close connection between structure and function, it is probably best not to consider structural or anatomical malformation as a class of pathologies distinct from those defined by impaired functioning. In the first place, it is questionable whether, in the social case, it is possible even to identify structures independently of an understanding of the functions a structure's various components carry out. (Marx's conception of class structure, for example, depends on differentiating the functions within the overall process of social production that each class carries out.) And, second, even if it were possible to do so, it is not clear that a deviation from the normal structure should be considered a pathology unless it also impairs social functioning in some way. In any case, for Comte the main instances of structural malformation in societies are those in which functional components are inadequately organized, resulting in impaired functioning of the whole. Among such malformations the most important for Comte – and for Durkheim as well – is the absence of bonds of solidarity among both institutions and individuals, which Comte ascribes to an inadequate degree of universal consciousness, or consensus, within society.

Two further features of Comte's picture of social health are worth mentioning because they have played a major role in bringing the society–organism analogy into disrepute. Neither of these supposed aspects of social health is necessary for a theory of social pathology, and critics of such theories are correct to be skeptical of them. Both involve an overemphasis on certain ends of organic life, namely, stability and harmony. Any theory of social pathology for which specialization and the division of labor play an important role will be committed to regarding stability and harmony as aspects of social health: an overly unstable society is in danger of dissolution, and some harmony among parts and subsystems is necessary for proper functioning. But Comte elevates these ideals into the supreme virtues of healthy social life (and he is not the only functionalist social theorist to do so). This is visible in his picture of a final state of social health – the telos of social evolution, realized in history's final, positive stage – as a perfect organic totality in which a thoroughgoing harmony among parts is established; everything in the social order has its rightful place; conflicts of interests have disappeared; and every individual is aware of, and at peace with, his place within the smoothly functioning whole. On such a view, disorder and conflict always indicate illness, rather than normal development, and equilibrium, stasis, and return to the status quo are the hallmarks of social health. It follows from this, as Comte acknowledges, that medieval Catholicism was a healthy state (because of its stability and unity), while post-Revolutionary Europe, marked by change and crisis, is nonpathological only to the extent that it brings about a later stage of more perfect stability and harmony.

Comte's overvaluing of social stability and harmony is bound up with his insufficiently critical appropriation of the analogy between societies and living organisms, joined with what Nietzsche (and others) would regard as an overly harmonious picture of organic life itself. At the very least Comte's example should alert us to the danger of unreflectively deriving an account of the functions of healthy societies from those of healthy biological organisms. That theories of social pathology need not make such a mistake is demonstrated by Durkheim's more nuanced attitude to social disorder, exemplified by his partially positive assessment of crime and social conflict. A certain level of crime, Durkheim argues, is indicative of social health because it serves, *inter alia*, the functions of fostering change that results in moral progress and, when punished, of reinforcing allegiance to social norms. In looking at crime in relation to moral progress, Durkheim shows that he conceives of the functions, and hence the health, of societies in a way that, more than in the case of Comte, takes into account important differences between human social life and the well-functioning of biological organisms.

Herbert Spencer

Durkheim's appropriation of Comte's thought is shaped by his engagement with Herbert Spencer, the founder of social Darwinism and advocate of laissez-faire social policies who for a time made "survival of the fittest" the watchword of social philosophy. Although Durkheim was influenced by Spencer's ideas, especially by the latter's use of the society–organism analogy, most of the influence was negative: Spencer's biologism and attempt to apply Darwinian ideas directly to social theory served mainly for Durkheim as an example of how not to construct a science of society. As his numerous references to Spencer in *The Social Division of Labor* make clear, many of Durkheim's positions were formed by thinking about how to avoid or correct Spencer's overly hasty appropriation of Darwin and biological thought more generally. Beyond this, much of the disrepute into which the society–organism analogy and the idea of social pathology subsequently fell can be traced back to the form these ideas take in Spencer's thought.

Spencer's reliance on the concept of a biological organism is both more ambitious and more literal than Comte's or Durkheim's. It is more ambitious because it seeks to ground, not merely social science, but an account of reality in general using concepts of evolutionary development and organic structure that Spencer claims to find in biological science. In his hands evolutionary development becomes the basic category of an all-embracing, speculative metaphysics that seeks to explain the universe as a whole, including inorganic phenomena. His reliance on the concept of a biological organism is more literal than that of his contemporaries in that he applies biological concepts to the social domain much more directly than they. His work is filled with observations of the following sort: "Societies, like living bodies, begin as germs – originate from masses which are extremely minute in comparison with the masses some of them eventually reach" (Spencer 1969 [1898]: §224). The effects Spencer's biologism has on his social theory can be seen everywhere, including in the significance he attaches to adaptability in his conception of social health and in his derivation of social functions directly from the functions of biological organisms: if living beings require "sustaining," "distributing," and "circulating" systems, then healthy social organisms do so as well, and this is taken to establish a society's need for industry, roads or canals, and states (Spencer 1969 [1898]: §§241–55).

Although Spencer acknowledges certain differences between societies and biological organisms, his treatment of these is mostly superficial. In the end, survival ("of the fittest") is the main category in terms of which he conceives of the functions and health of human societies; ideas of the good that go beyond purely biological categories appear from time to time in his writings, but they are not of central concern for him. The struggle for existence is taken to be the main challenge faced by human societies and biological organisms alike (Corning 1982: 353). Although social solidarity is a topic Spencer attends to, he thinks of it primarily as serving the biological end of survival and as consisting not in ethical bonds within social life but in contract-like relations among individuals grounded in self-interest. More than for the theorists of social pathology I take most seriously, human society is for him, as Hegel would put it, less a spiritual than a "merely living" phenomenon.

Spencer's emphasis on self-interest and contract-like social relations points to an odd feature of his position: even though he is the theorist of social pathology who appears to take the analogy between societies and organisms the most seriously of all, he abandons (or misunderstands) it precisely where it is most important for social theory. For, despite his alleged organicism, Spencer's account of human society remains fundamentally atomistic and anti-holist in two ways. First, social life is principally of instrumental and utilitarian value to individuals since its primary benefit is to achieve an end, survival, that every individual has independently of social membership. (Thus, he embraces a form of normative atomism, according to which the goods realized in society are wholly reducible to the interests individuals have in abstraction from their social membership.) Second, relations of egoistic cooperation, such as contract and exchange, count for him as the basic building blocks of social life, and institutions and their functions can be understood in terms of the common benefits they provide for independently existing, egoistically motivated social "atoms." (This is a form of explanatory atomism.) Even biological organisms, Spencer goes so far as to say, are composed of units that enjoy a significant degree of existential independence from other units of the same organism (Spencer [1898] 1969: §218).

Before turning to Durkheim's views of social ontology and social pathology, it will be helpful to review Comte's position with a view to understanding what Durkheim takes from him and what he rejects. (Since

Durkheim's relation to Spencer is less complex, I will not review it here.) According to Comte:

- Human societies are like biological organisms in being functionally organized, or made up of differentiated "organs" that carry out coordinated specialized functions.
- A scientific understanding of society, like physiology and medicine, includes three related tasks: explanation, evaluation, and prescription.
- Explanations of social phenomena include functional analysis as well as existential functional explanations in both strong and weak senses.
- Illness in both societies and organisms consists primarily in dysfunction, or impairments in vital functions.
- Functional analyses of social phenomena presuppose an explanatory holism that explains a part's function by situating it in relation to the functions of the whole. This makes pathology a holistic concept since dysfunction in a part of society shows up only in seeing how the functioning of the whole or of others parts is impaired.
- Explaining social phenomena requires situating them within a universal history that ascribes a telos to human history and posits a pattern of normal development for all human societies that includes a tendency toward increasing specialization and complexity.
- The essential functions of human societies are defined by the ends of reproduction, stability, internal harmony, growth, adaptation, and (as a means to these ends) increasing productivity.
- The essential functions of human societies include some ethically tinged ends, such as peace and even a certain version of freedom.
- Societies differ from biological organisms in that members of the former are endowed with consciousness; this enables the coordination of functions to be achieved consciously, and it explains the aspiration of individuals to grasp and affirm the workings of society as a whole.

Durkheim takes over many of these Comtean doctrines, but with the following differences:

- Although Durkheim sometimes avails himself of existential functional explanations in the weaker sense, functional analysis is more important to his social science.
- Durkheim has a more complex position with respect to the role played by history in understanding social reality (including even functional analysis of it). He espouses a version of universal history

that avoids ascribing a final end to history – in both senses of "end" – but posits a pattern of normal development for all human societies, including a tendency toward increasing specialization and complexity.

- The vital needs of human societies that define their essential functions are less clearly defined and farther removed from those of biological organisms. For example, for Durkheim the end of social stability is replaced by solidarity, and the category of adaptation plays a negligible role in defining society's needs.
- The ethical content of the vital ends of human societies is more robust, more complex, and more extensively articulated.
- Related to this, Durkheim pays greater attention to the implications of the fact that societies differ from biological organisms in that their individual members are conscious beings endowed with will.
- The fact that human individuals are endowed with consciousness makes conscious coordination of specialized functions possible, but much of such coordination need not involve a conscious agency overseeing the whole.

CHAPTER 8

Durkheim: Functionalism

Durkheim's attempt to found a science of sociology begins in 1893 with the publication of *The Social Division of Labor* – or, as it has come be known in English, *The Division of Labor in Society*. (Henceforth I refer to the text as the *Division of Labor*.) Already in this first book Durkheim brings together versions of Plato's organicism (in the importance he ascribes to the division of labor) and Rousseau's artificiality thesis (in emphasizing the role moral rules play in constituting social reality). At the same time, many of the book's claims can be understood as the results of retaining insights of Comte's – including some version of the society–organism analogy, with its focus on the coordinated, specialized functions of living beings – while eliminating scientifically dubious elements of Comte's and Spencer's supposed appropriations of biology. Durkheim's debt to Comte is visible in the *Division of Labor*'s first sentence: "This book is above all an attempt to treat the facts of moral life in accordance with the method of the positive sciences" (DLS: xxv/xxxvii). "Positive sciences" is an unmistakable reference to Comte, but so too is the idea that the subject matter of sociology is *moral life*, where both of these terms are significant: social reality is (as Rousseau saw) a moral and (as Plato saw) a living phenomenon. This formulation of Durkheim's theoretical program, an apt characterization of his work throughout the entirety of his career, raises several questions: In what do the moral facts that constitute sociology's central concern consist? What makes them vital, or living, phenomena? What is it to treat these facts scientifically, from the perspective of positive science?

With regard to the last question, the *Division of Labor*'s First Preface makes clear that the concepts of social health and pathology will play a central role in the science of society it develops. As Durkheim puts the point early in that preface: "There exists a state of moral health that science alone is competent to determine" and that defines an "ideal to seek to approximate" (DLS: xxvii/xxxix). Equally clear for Durkheim at this point in his career, then, is that making social health a central concern of sociology means that it, like physiology and medicine, will include explanatory, evaluative, and prescriptive dimensions: finding the right method for understanding human society – as a functional, living entity – enables the sociologist both to distinguish healthy from unhealthy societies and to orient practice aimed at "curing" the latter.

My discussion of Durkheim in this and the following two chapters is organized around three topics: the nature of *moral facts* and their relevance to social theory;[1] what it is to *explain* or understand moral facts (and social facts more generally); and the conceptions of *social pathology* made available by his positions on the first two issues. The first topic leads directly into social ontology since it raises the question of whether, and in what sense, social reality, in distinction to the objects of the natural sciences, must be grasped as a moral phenomenon. The second involves investigating Durkheim's functionalist methodology in social theory. And the third explores the implications for a conception of social pathology of combining the first two, that is, of insisting on both the moral character of social life and its functionally organized, living nature. Because these three topics are conceptually interdependent, it is impossible to treat them completely separately. For this reason it will be necessary to some extent to address all three at once, even while recognizing them as distinguishable aspects of a single theoretical project. Since saying everything at once is not an option, however, I organize my discussion of Durkheim as follows: in this chapter I treat his functionalism with respect to both methodology and social ontology. The first part of the following chapter discusses social morality and solidarity, while its second part examines the conceptions of social pathology that Durkheim's functionalism and claims about the moral character of human society make available to social theory. Finally, in Chapter 10 I tie these themes together by reconstructing how and why Durkheim conceives of sociology as a "science of morality."

In this chapter I begin my discussion of Durkheim's functionalism by focusing on its articulation in the *Division of Labor*. The self-proclaimed task of that work is unmistakably functionalist: to discover what function the division of labor serves in modern societies. Since studying the division of labor leads Durkheim to the discovery of certain moral facts – roughly, normative rules that constrain social members' actions – the principal task of his text is to understand what function those moral facts serve. The underlying assumption that explains why these functional questions are

[1] I do not distinguish social theory, social philosophy, science of society, and sociology.

the first to be raised is that social phenomena belong to the order of living beings, or are of a "vital nature" (DLS: xxviii/xli). In this chapter I reconstruct what entitles Durkheim to begin from this assumption as well as its implications for sociology's method. Doing so will provide us with a clearer picture of what his functionalism consists in; what kind of understanding of social phenomena it promises; and how it determines what functions are to be ascribed to such phenomena. It is necessary to emphasize the reconstructive character of my undertaking. Explicating what Durkheim's functionalist project amounts to – especially determining its relation to standard forms of causal explanation – requires untangling a number of complex methodological issues that he himself sometimes fails to distinguish. In addition, Durkheim's own methodological reflections – above all in *The Rules of Sociological Method* (henceforth the *Rules*) – do not always cohere with what he actually does in *Suicide* and the *Division of Labor* when investigating specific social phenomena. To make things more complicated, there are some internal inconsistencies in his methodological reflections that must be resolved if his functionalism is to be a coherent position.

My account of Durkheim's functionalism in this chapter has four parts, corresponding to the following questions:

- What does functional explanation (and analysis) consist in?
- What justifies this approach to social phenomena?
- How does social theory identify the functions of social phenomena?
- What role does historical narrative play in attributing functions to social phenomena?

Functional Analysis and Explanation

The most consequential implication of Durkheim's assumption that human societies belong to the order of living beings is that they are taken to be *functionally organized*. If life can be thought of as purposively organized matter (Ginsborg 2006: 462), the same can be said of human societies, even if there is also an ontological difference between the two, namely, that societies achieve their purposes, in part, through conscious activity; in their case, in distinction to mere life, purposive organization depends on human subjectivity. The purposively organized character of human social life is directly connected to the aspect of Durkheim's functionalism I emphasize here: the functional analysis of societies. Just as a biologist who encounters an unfamiliar organism asks with regard to its various features what life functions they serve, so the sociologist will want to know, when considering a feature of human societies as pervasive as the division of labor,[2] what vital social function it serves. (Durkheim makes clear that function, in both biology and sociology, is to be understood in terms of the idea of a vital *need*: to inquire into something's function is to ask what vital need of the "organism" it satisfies [DLS: 11/11].[3]) This way of proceeding does not commit Durkheim to illegitimate forms of functional explanation. Instead, we should think of his fundamental assumption as a regulative principle associated with functional analysis: if a human society exhibits property *X* – or, more precisely, if *X* is a property of many societies of the same type – then the sociologist's task is to ask what function *X* serves or, equivalently, what vital social need it satisfies. Nothing in this regulative principle commits Durkheim to an a priori claim that *X* must have a function, but it predisposes him to look for one if none has yet been found. As Durkheim acknowledges, not every feature of a society serves a vital need: "A fact can exist without serving any purpose, whether because it was never adjusted to any vital end or because, after having once been useful, it has lost all usefulness but continues to exist by the sheer force of habit. Indeed, *there are even more instances of this in society than in organisms*" (RSM: 120/91; emphasis added).

Nor does this regulative principle commit Durkheim to any thesis regarding why or how something that is found to have a function originally came to be. This is clear from the very structure of the *Division of Labor*, where Book I uncovers the function of the division of labor, while Book II seeks its causes (while refraining from appealing to the function the division of labor is found to serve in Book I). Durkheim's causal account of the division of labor explains it as the effect of, *inter alia*, population growth that leads to increased population density. What is important here is not the correctness of this account but its logic: it explains the origin of the division of labor via principles of efficient causality rather than in terms of any function it comes to serve. In general, Durkheim is scrupulous in distinguishing between the functional analysis of social phenomena and

[2] Because Durkheim's first book focuses on the economy, it is easy to lose sight of the fact that the division of labor includes differentiated, enduring social roles of all types – for example, in the family (RSM: 50/3) – where differences are relatively fixed such that individuals cannot move easily from one functional role to another. In the political realm the division of labor implies specialized legislative, administrative, and judicial institutions and roles; the (relative) independence of science and scientists from political and economic imperatives is a further example of the modern division of labor.

[3] Boorse defines function thinly, as "a contribution to a goal" (Boorse 1976: 70).

causal accounts of their origins: "When undertaking to explain a social phenomenon, it is necessary to investigate the efficient cause that produces it separately from the function it fulfills" (RSM: 123/95).

Later in the *Rules* Durkheim goes a step farther: "Explanation in sociology consists exclusively in establishing relations of causality, [which] is a matter of attaching a phenomenon to its cause, or, conversely, a cause to its useful effects" (RSM: 147/124). (We will see that Durkheim contradicts this claim elsewhere.) The term "useful effects" confuses Durkheim's attempt to distinguish the two forms of admissible explanation (via efficient causality) mentioned here. But his intended meaning is straightforward: showing how the division of labor originally came about is an example of the first instance of causal explanation ("attaching a phenomenon to its cause"); the claim that the division of labor produces social solidarity (a "useful effect") exemplifies the second. Both claims invoke efficient causes; the second is merely one in which the effect happens to be socially useful. That the solidarity produced by the division of labor is socially useful – that it responds to a vital social need – is something Durkheim believes, but it does not belong to the type of explanation (connections of cause and effect) he is characterizing here. Causal explanation can determine that the division of labor produces social solidarity, but it cannot assess the usefulness of the latter or whether it is a vital social need. These issues are the province of functional analysis, which, as Durkheim consistently emphasizes, belongs to a different "order of investigation" from causal explanation (RSM: 120/91).

The thought that "explanation in sociology consists exclusively in establishing relations of causality" is familiar to contemporary social scientists, but it may seem odd to readers acquainted with Durkheim's treatments of specific social phenomena, such as crime and the division of labor, where causal explanation is clearly not the main point. That it is not reveals something important about Durkheim's project: even though the *Division of Labor* devotes three chapters to explaining (causally) the origin of the division of labor, it describes its primary task as determining "the function of the division of labor – which is to say, the social need to which it corresponds" (DLS: 6/8). The implication of this is clear: functional analysis, not explanation in terms of efficient causes,[4] is Durkheim's supreme concern. This is bound up with the "medical" aspirations of his sociology: grasping the function of the division of labor, and of the moral facts associated

[4] And not even existential functional arguments, to the extent Durkheim admits them at all.

with it, is what enables him, via the idea of dysfunction, to diagnose social pathologies and to think about how to overcome them. Durkheim's position, then, is that causally explaining how a social phenomenon originates forms part of a complete sociological understanding of it, but it is not sociology's main concern because discovering such causes does not tell us what functions a phenomenon comes to assume or whether it is a sign of health or illness.

It may seem strange to claim both that Durkheim's sociology takes medicine as its model and that it de-emphasizes causal explanation. For in most cases (but not all), if medicine is to treat the illnesses it diagnoses, it must understand their causes. Similarly, how could sociology realize its prescriptive ambitions without understanding the causes of what it aspires to cure? Untangling these issues requires distinguishing two contexts in which causal explanation can be employed, namely, in explaining how now "normal" (both widespread and healthy) social phenomena, such the division of labor, first came to be and in explaining – adducing the causes of – dysfunctional states of those phenomena, such as elevated rates of suicide. In the former case causal explanation is less important because why the division of labor came about may be irrelevant to the function it comes to assume, but causal explanation is supremely important in the latter case because without knowing the causes of abnormal rates of suicide, the sociologist will be at a loss to propose remedies for the dysfunctional state. Durkheim's most elaborated instance of the latter type of causal explanation is found in *Suicide*, where he employs his "method of comparison or indirect experimentation," or "method of concomitant variations" (RSM: 147, 151, 153/124, 129, 132), to explain why Protestant communities in Europe exhibit higher rates of (egoistic) suicide than their Catholic counterparts – namely, because "social currents"[5] that exert suicidogenic pressures on individuals are stronger in communities whose institutions produce less social integration, as Protestant churches do in comparison to Catholic ones.

Precisely determining Durkheim's position on explanation in sociology is made more difficult by the fact that elsewhere in the *Rules* he gives a broader characterization of sociological explanation that contradicts the claim cited above that explanation consists exclusively in establishing relations of efficient causality: "In order to explain a fact belonging

[5] I do not endorse this specific explanation, but it illustrates the logic of the second type of causal explanation (finding the causes of dysfunctional states).

to the order of living phenomena, it is not sufficient to demonstrate the cause on which it depends; it is also necessary, at least in the majority of cases, to discover the part it plays in establishing the general harmony [characteristic of living beings]" (RSM: 124–5/97).[6] This broader account of sociological (and biological) explanation – which includes both causally explaining a phenomenon's origin and determining the function it serves – more accurately describes what the *Division of Labor* actually does (and is consistent with my claim that the latter task is more important for Durkheim than the former). Fortunately, it is unnecessary to decide which account of explanation to ascribe to Durkheim; it is sufficient to note that he uses "explanation" in both a narrower and a wider sense. (I will avoid ambiguity by using "causal explanation" when referring to the narrower sense – "demonstrat[ing] the cause on which [something] depends" – and "functional analysis" when referring to an account of the social need a given phenomenon satisfies. For the sake of completeness, note that yet a third kind of explanation figures in his sociology: existential functional arguments in the weaker sense distinguished in the previous chapter.) The essential point – that causal explanation and determining a thing's function constitute different "orders of investigation" – is a principle Durkheim rigorously observes. It should be noted, however, that the reason the two belong to different orders of investigation is that functional analysis relies on *both* causal and functional claims, whereas its counterpart makes no claims about functions. More precisely, functional analysis ascribes a function to a social phenomenon, where this functional claim depends on a causal claim, not about that phenomenon's origin but about its "useful effects." Thus, Durkheim's functional analysis of the division of labor claims that it produces social solidarity (a causal claim), as well as that it satisfies a social need and therefore carries out a vital function. A further feature of functional analysis is that it is concerned with the causal effects standardly produced by the useful phenomenon in question and not with a causal account of its origin. A *complete* sociological explanation of the division of labor, however – explanation in the broader sense referred to above – includes both analyzing its function (including its standard

[6] Durkheim does not regard harmony as a principal goal of social "organisms." The harmony referred to here is "a correspondence between [an organism's] internal and external milieus," not the inner harmony of a perfectly organized being. He uses the term here only because he is expressing his own position in the terms used by the "standard formula," which defines health in terms of harmony (RSM: 125/97).

effects) and determining its causal origin; as indicated, these two tasks are carried out in separate books of the *Division of Labor*.

Although functional analysis depends on causal claims about the standard "useful" effects of the phenomenon to which it ascribes a function, it is logically independent of causal claims regarding that phenomenon's origin. Moreover, the two components of functional analysis – "*X* produces effect *Y*" and "*Y* satisfies a vital social need *Z*" – are logically distinct: the two claims do not mean the same thing, and it is possible for one of the claims made about a specific phenomenon to be correct and the other false. At the same time, the validity of many specific causal claims made by sociology – especially in its etiological tasks – depends on the validity of some of the functional claims it is committed to; when this is the case, causal explanation in sociology cannot be carried out independently of all attributions of purpose. Consider the case of abnormally high suicide rates, a causal explanation of which is crucial to the social pathologist's remedial ambitions. In *Suicide*, the causal explanation of elevated rates of egoistic suicide proposed by Durkheim depends on his prior conviction, argued for in the *Division of Labor*, that social solidarity, and the integration that accompanies it, is a vital social need. For it is this conviction that leads him to posit the lack of social integration as the causal factor that explains the difference in suicide rates when all other possible causes of it (that he can think of) have been excluded by the method of concomitant variations. If this is so, then the specific causal claims elaborated in *Suicide* are not epistemologically independent of the functional analysis carried out in the *Division of Labor*, and many of Durkheim's specific causal claims stand or fall with his claims regarding the social functions of the phenomena in question.

This example points to the ineliminably interpretive, or epistemologically holistic, character of Durkheim's sociology and to the necessity of judging the adequacy of his position not claim by claim but in terms of how well the whole makes sense of the totality of the phenomena it attempts to understand. (I return to this issue below.) Durkheim explicitly recognizes this aspect of sociological method (RSM: 152/130), although he would have done well to have also pointed out that this type of interpretation – forming causal hypotheses in light of one's best understanding of an organism's functional needs – is no different in principle from the methods of physiology and medicine, even if interpretation in the biological case is typically more clear-cut than in the sociological.

Another sense in which causal explanation and attributions of function, although logically distinct, cannot always be carried out separately comes to light in Durkheim's puzzling remark, immediately following the statements cited above, that it is often useful to search for the cause of a phenomenon before determining its function because "resolving the first question will often help to resolve the second" (RSM: 123–4/95). This remark is puzzling for two reasons. First, it seems not to correspond to the order of investigation actually followed in the *Division of Labor*, where seeking the causes of the division of labor (in Book II) comes after the determination of its function (in Book I). The second reason goes deeper: the suggestion that finding the cause of something can help determine its function seems to contradict Durkheim's basic claim that explanation via efficient causes is of a different logical order from attributions of function.

The first confusion can be cleared up by recalling the distinction between causally explaining the origin of a thing and determining the normal causal effects of an already existing social phenomenon. It is only in the latter case that finding the cause of a phenomenon might help to determine its function. This, however, does not yet explain how finding a thing's cause might help to determine its function, if these constitute distinct orders of investigation. The confusion here is compounded by Durkheim's allusion to a "reciprocity" between cause and effect "that has not been sufficiently recognized" by sociologists (RSM: 124/95). Fortunately, the examples Durkheim adduces clarify what he means by causal reciprocity as well as why taking note of it is important to sociological explanation (namely, because it is characteristic of phenomena in living beings such as societies and biological organisms).

Consider his elaboration of the causal reciprocity between criminal punishment and collective consciousness:[7] "The social reaction that constitutes criminal punishment is due to the intensity of the collective sentiments that the crime offends; on the other hand, such punishment has the useful function of maintaining these sentiments at the same level of intensity, for they would soon lose their force if the offenses they were subjected to went unpunished" (RSM: 124/95–6). Durkheim's claim is that there must be preexisting collective sentiments of disapprobation in order for punishment to occur, but each act of punishment has the effect of reinvigorating those original sentiments. Reciprocal causality, then, holds in cases where "the effect cannot exist without its cause but where the cause in turn needs its effect. The effect [e.g., punishment] draws its energy from the cause [e.g., collective sentiments], but the effect also restores energy to its cause;

[7] I discuss collective consciousness further below in this chapter and in Chapter 8.

consequently, [punishment] cannot disappear without its disappearance being felt [in collective sentiments]" (RSM: 124/95).

Clearly, the phenomena Durkheim means to be describing here are the feedback loops prominent in self-regulating and self-maintaining systems, which pose no mysteries of the sort that talk of reciprocal causality might be thought to involve. His claim is that in such instances the biologist and the sociologist engaging in functional analysis would do well to consider the hypothesis that the function of an effect lies in reproducing its cause, and it is only for this reason, and only in these cases, that discovering something's cause can help to reveal its function. Thus, Durkheim's maxim urging the sociologist to look for the causes of things before attributing functions to them loses its puzzling air when one realizes it applies only to living phenomena, the continued survival of which depends on feedback-dependent processes of self-maintenance; that maxim is simply one more consequence of the assumption that human societies are to be understood as living systems.

What Justifies the Regulative Principle Underlying Functional Analysis?

I have shown thus far that Durkheim has a coherent, if also complicated, position on the nature of explanation in sociology: First, functional analysis and causal explanations of origin are logically independent enterprises to be carried out in separate investigations; second, functional analysis necessarily includes causal claims about the standard effects of the phenomena deemed to be functional; third, determining the etiology of a social pathology involves causal claims, the arguments in support of which may depend on functional claims established elsewhere in the theory; and, finally, because human societies are living systems, many of its elements will have the function of reproducing, or maintaining, their own causes. None of this, however, addresses why it is reasonable for the sociologist to assume that human societies are like living beings and to adopt the regulative principle governing functional analysis. Part of the answer to this question appeals to the degree to which doing so succeeds in making sense of the diverse properties of the object of study. Apart from this, however, is there anything to be said in support of Durkheim's fundamental assumption in advance of specific empirical inquiries?

There appear to be two related considerations, both borrowed from biology, that Durkheim takes to support the regulative principle underlying functional analysis (namely, when something is found to be a widespread feature of human societies, seek to determine the vital social need it serves). One of these considerations, grounded in an analogy with natural selection, is unconvincing, while the other gives us some reason for taking seriously the assumption that human societies are living orders. Before examining these considerations, it is worth recalling that Durkheim recognizes differences between societies and biological organisms with regard to the extent to which each is functionally organized. As we saw above, he admits there are more instances of useless "organs" and processes in societies than in organisms (RSM: 120/91) and explicitly denies that the sociologist assumes in approaching his object that "each detail [of a society] has … a useful role to play" (RSM: 88/51).

If human societies were like animal species in evolving through natural selection, sociologists would have a good reason for assuming that the various features of their objects of study played functional roles in social life. There are moments when Durkheim appears to rely on some such thought, most clearly when discussing in the *Rules* the empirical markers by means of which the sociologist distinguishes normal (healthy) phenomena from pathological ones. His claim there is that the statistically normal is the best indicator of what is normal in the normative sense (healthy), and he expresses this claim in language that echoes evolutionary theory in biology:[8] the common occurrence of a particular social phenomenon across societies of the same type (or "species") is the most reliable empirical marker for its normality in the normative sense. In the case of biological organisms the principle of natural selection gives one a prima facie reason for expecting that a commonly occurring characteristic among organisms of the same species – the giraffe's long neck, for example – plays a vital function in that species' life, precisely because that function explains why the relevant characteristic was preserved over time. The problem with applying this thought to human societies, however, is that, for easily discernable reasons, it cannot be assumed that anything like natural selection is at work there: nothing in social life corresponds to Darwin's idea of reproductive success that could explain the mechanism of natural selection; the histories of biological species are immeasurably longer than those of social "species"; and Durkheim's conception of social health is too ethically robust for purely biological criteria – survival or reproductive success – to serve as markers of it.

[8] "We can be fairly certain that the conditions that are generalized throughout the species are more useful than those that have remained exceptional" (RSM: 96/63).

There is, however, a different consideration, also biological in inspiration, that provides limited support for the regulative principle guiding Durkheim's functional analysis in sociology. This consideration makes no reference to the common features of a given social "species" nor to an extended history that weeds out the maladaptive features or innovations of such a species. It derives instead from the requirements of self-maintenance for a single society. I discuss the concept of self-maintenance in more detail below, but in the present context – in searching for preliminary reasons to take seriously Durkheim's assumption that human societies are living beings – it is sufficient to rely on our ordinary understanding of the idea. The basic thought is simple: human societies are like biological organisms – and unlike stones, physical forces, or machines – in that they must maintain themselves in order to survive. Both can be understood as organized, self-maintaining systems whose continued existence depends on internally driven processes, including exchanges with the system's environment, in the absence of which, or in the case of their breakdown, societies would fail to survive. Both types of being have, in other words, vital needs, and if they are to continue to exist as living beings, they must satisfy those needs by carrying out vital functions.

Accepting this relatively weak version of the society–organism analogy is sufficient to ground the assumption that existing societies must be constituted so as to be able to carry out the functions that enable them to maintain themselves over time. The regulative principle underlying Durkheim's functionalism, however, is more robust than this. What guides functional analysis in sociology is not merely the thesis that somewhere among a society's features there must be some that are functional, but rather the stronger principle that *as a rule* (but allowing for exceptions) a society's specific features are functionally significant or, equivalently, play some role in its self-maintenance. (In selecting which of a society's features to seek a function for there may be a need to appeal to what is common to a social "species." In investigating the division of labor's function, Durkheim inquires not into a feature of French or German society but into a widely shared feature of societies of a similar type. This suggests that the idea of a social species may be indispensable to Durkheim's sociology, but it does not require that he endorse a principle of natural selection as an account of how members of a species acquire the features they share.)

This more robust regulative principle is evident in Durkheim's claim that what characterizes living beings, including human societies, is that, in general, their features "cannot endure unless [they] serve a purpose or respond to some need" (DLS: xxviii/xli). An application of this principle can be found in his response to the observation that crime is present in all known societies, where recognizing the pervasiveness of crime (even in societies of different "species") leads him directly to search for the vital function it serves: "To classify crime among the phenomena of normal sociology … is to affirm that it is an element of public health, an integral part of a healthy society" (RSM: 98/66). This is the impetus behind Durkheim's celebrated claims regarding the positive social functions of crime: a society's response to crime, punishment, serves to renew and strengthen that society's collective consciousness (DLS: 54–60/66–73); and the violation of social norms that crimes represent can, as in the case of Socrates, promote a society's moral development (RSM: 101–3/70–72).

There is a further thought that lends credence to the more robust regulative principle of functional analysis I have attributed to Durkheim: the demanding requirements of social reproduction – especially in larger, more complex societies – coupled with the finitude of resources and energy, impose an efficiency constraint on human societies:

> If the usefulness of a [social] fact is not what causes it to be, it is generally necessary that it be useful in order for it to be able to persist. For if a fact serves no purpose, this is enough to make it harmful, since then it costs without yielding anything in return. If the majority of social phenomena had this parasitic character, the organism's budget would be in deficit, and social life would be impossible. (RSM: 124/96–7)

This claim brings Durkheim's functional analyses close to the weaker version of existential functional arguments distinguished above, in which a thing's continuing presence, but not necessarily its origin, is explained by the function it is found to serve. This coheres with the fact that when discussing the historical development of societies, Durkheim commonly regards the function a social phenomenon eventually comes to have as relevant to explaining why a society developed as it did or, more precisely, why the features that serve the function in question endured. Although this does not commit Durkheim to a principle of natural selection – where socially useful traits are selected for throughout a long history involving the births and deaths of countless individual societies – it does raise the question of how to understand the mechanism by which the socially useful traits in question are preserved.

We have already noted that it is futile to appeal to the artificiality of social institutions in support of the regulative principle of functional analysis since, however one understands that artificiality, institutions are not artifacts in any sense that would allow one to infer their functions from their creators' intentions. Societies are humanmade but not in the way watches or texts are, which is to say, not in a way that guarantees that their parts work together to achieve the vital ends of the whole or the ends of their creators. Here, again, we have no choice but to appeal to the ultimate success with which the regulative principle is applied to human societies – to the degree to which it makes sense of the specific features of its object – as what in the end justifies its employment. This is not a spurious thought, but the obvious danger it brings with it – that a sufficiently clever interpreter can always find some function for a thing if he starts out convinced there is one – underscores the importance of ensuring that attributions of functions to social phenomena not be simply ad hoc. Some of what Durkheim says about the positive functions of crime has the air of arbitrariness; this is less true, however, of the arguments in support of his claims regarding the function of the division of labor, and for this reason they are the example of Durkheimian functional analysis I focus on below.

Determining the Functions of Social Phenomena

As noted above, if the ascription of functions to social phenomena is to be more than ad hoc, there must be some nonarbitrary way of both determining a society's vital needs and ascribing functions to its specific features. It is tempting to think that the latter relies on a prior account of what society's vital functions are – that, in other words, Durkheim must first establish what, in general, a society needs in order "to live" and only then look to see whether the phenomena he is concerned with respond to those needs. This is not, however, the course he takes in the *Division of Labor*, which means that we must find a more complex account of how his ascriptions of function avoid arbitrariness.

As suggested above, the best place to look for answers to questions regarding Durkheim's method is not the text explicitly devoted to this task, the *Rules*, but the specific instances of functional analysis carried out in his empirical studies, especially in the *Division of Labor*. If my interpretation of Durkheim's functionalism thus far is correct, then his thesis regarding the function of the division of labor – that it serves to produce solidarity in society – comprises three logically (if not epistemologically[9]) separable claims:

[9] In denying the epistemological separability of these claims, I mean that the investigations establishing them cannot be carried out independently of one another.

(a) a causal claim about the standard "useful" effects of the division of labor (that it produces solidarity);
(b) a claim, of functional analysis, that creating solidarity is a vital social function, or that solidarity is a vital social need; and
(c) a claim that the division of labor becomes an enduring feature of human society, and increases in scope, because it serves this vital social need (which is an existential functional explanation of the division of labor in the weaker sense distinguished in Chapter 7).

Because (c) is less important to Durkheim's position than (a) and (b), I will focus here on the latter two claims. Notice that the *kind* of claim made in (a) – that a given social phenomenon produces certain effects – presents no special difficulties for Durkheim's position since nonfunctionalist views also depend on causal claims of this sort, whereas the *kind* of claim made in (b) is unique to functionalist theories such as Durkheim's and raises special questions about the legitimacy of the very concept of vital social needs. A defense of Durkheim's ascription of a specific function to the division of labor must address, then, three questions: i) why, in general, the idea of vital social functions is essential to understanding social life; ii) why, specifically, solidarity counts as a vital social need; and iii) why, specifically, the division of labor produces solidarity.

In line with my interpretation of his position, Durkheim exhibits a clear awareness of the fact that determining the functions of social phenomena poses a major methodological problem for sociology, which is why the first chapter of the *Division of Labor* – "The Method for Determining This Function" (of the division of labor) – is devoted to precisely that. As noted above, it is tempting for a biologically inspired sociologist to proceed in this matter by attempting to derive an a priori account of vital social needs from an understanding of the vital needs of living organisms, which might include survival, reproduction, self-regulation, growth, adaptation, and so on. This is Spencer's path but not Durkheim's; their approaches differ in two respects. First, Durkheim does not begin by devising a list of the vital social functions of human society in general, or even of modern European societies, and then use that list as a schema for organizing his empirical investigations of social life. Instead, he begins with a specific social phenomenon, such as the division of labor, and then asks what functions it serves in modern social life. In determining these functions Durkheim appeals not to a predevised scheme of vital social needs but instead engages in analyses of various empirical realities – the sexual division of labor and the two types of law in modern societies, for example – on the basis of which he constructs a posteriori arguments for his claims regarding the functions of the division of labor.

A further difference to Spencer is that Durkheim is primed from the beginning to search for not merely the material but also the moral functions of the division of labor, and the claims he makes with respect to these are of major importance to his science of society.[10] For the central claim of his first book is that the division of labor satisfies not only the economic and "civilizational" needs of society (by increasing its productive forces and developing human capacities more generally) but also a further vital need of complex, "organized societies" (DLS: 242/289),[11] namely, the need for social solidarity – which, as Durkheim insists, is a moral phenomenon. If human society is to be conceived of as an organism, it is nonetheless an organism *sui generis*, one in which moral relations have an essential place and constitute a vital social need.

Durkheim's arguments establishing the functions of social phenomena are broadly empirical, but they are not empiricist in the strict sense of that term. By this I mean that his specific sociological investigations are guided by a general conception of the kind of thing human society is, and even of its vital needs, that is not arrived at through induction from experience. This general conception of society's vital needs, while not empty, remains highly indeterminate in the absence of empirical inquiries that establish in what precise ways specific human societies satisfy their vital needs. It is for this reason that I describe Durkheim as employing regulative principles in his theorizing, for example, in the thesis that human societies are in important respects analogous to biological organisms or that they have both material and moral vital needs. Such principles orient Durkheim's empirical investigations, and in that sense they shape what he finds in them, but they also acquire further legitimacy, as well as more determinate content, when they prove to be empirically fruitful, that is, when they open the investigators' eyes to phenomena and to connections between them they would otherwise miss. In this regard Durkheim's method for determining the functions of social phenomena is implicated in a version of the same

[10] It is not that Spencer completely neglects society's moral character – one of his sociological texts is called *The Principles of Ethics* – but Durkheim has a much more nuanced and substantial account of morality and its importance for social life.

[11] Durkheim uses this term to refer to organically structured societies, characterized by an extensive division of labor and organic solidarity, in contrast to the less complex, segmentally structured societies that exhibit mechanical solidarity. This is a narrower sense of "organized" than that employed in the claim that all societies are to some extent functionally organized. I use "organized societies" throughout in the narrower sense.

circular holism that characterizes other interpretive enterprises – of texts, for example, or of works of art – and that is involved even in forming and testing hypotheses in the natural sciences. This is a description of scientific method that Durkheim himself might have rejected, but it is the best characterization of what he actually does, both in the *Division of Labor* and in other works, and of what gives his specific claims plausibility. What ultimately makes the thesis that the division of labor functions to promote social solidarity in organized societies compelling is not discrete confirmatory empirical facts but the whole ensemble of social phenomena, connected to solidarity in diverse ways by his theory, that are thereby brought into view and made intelligible.

Durkheim's failure to appreciate the circular character of his method – and the consequent fact that whatever warrant his functional claims possess depends less on isolable facts or single arguments than on the plausibility and coherence of the whole picture he develops – sometimes leads him to offer bad arguments that he takes to establish his theses conclusively, independently of how the content of his science is confirmed more holistically. An example of this is the argument he gives at the beginning of the *Division of Labor* purporting to prove that the division of labor must have a moral function (before establishing what that function is). Recall that Durkheim's fundamental regulative principle – that human societies are living beings – entitles him both to assume provisionally that something as widespread as the division of labor serves some function in society and to proceed to look for that function via empirical inquiry. His concern in the argument now under discussion is more specific, namely, to show that the function in question is moral.

The argument Durkheim offers is as simple as it is speculative and unconvincing: if the division of labor served only a material (or "civilizational") function, he claims, we would be unable to find a reason for its existence. This is because, from the purely material perspective, the division of labor provides no net benefits to human society: as Rousseau might have said, it serves no needs *beyond those it itself creates*. As Durkheim argues, it is only because the division of labor increases human fatigue that we need, as compensation, the goods of civilization that the division of labor makes possible. Hence, if the significance of the division of labor were limited to the satisfaction of material needs, "its only function would be to mitigate the effects it itself produces, to dress the wounds it [itself] makes" (DLS: 15/17), and if it responded only to needs it itself created, it would have no genuine raison d'être. Because the division of labor is so widespread in human societies, however, it must respond to a social need of more universal importance. But if this need is not material, it can be only moral; hence, the division of labor must carry out some moral function in human societies.

The defects of this argument are too numerous to go into here. It is worth noting, though, the alarmingly strong character of the existential functional claim it relies on: a social phenomenon could not be as widespread as the division of labor is unless it responded to a need that it itself did not create – or, alternatively, unless it served a function that, unlike mere civilization but like morality, had an "intrinsic and absolute value" (DLS: 15/17). For various reasons the weaker functionalist position I have attributed to Durkheim could not endorse conclusions of this sort, and, fortunately, the merit of his claims regarding the moral function of the division of labor does not rest on the merits of this one argument. If I am correct in reconstructing Durkheim's method as employing regulative principles that implicate him in a hermeneutic circle – a moving back and forth between a general conception of society and of its basic needs, on the one hand, and analyses of specific social phenomena that reveal (or suggest) their functional significance, on the other – then he has no need of, and should not look for, an argument proving that the division of labor has a moral function independently of his specific arguments for what that moral function is. Contrary to what Durkheim appears to assume in this argument, his claim that the division of labor has a moral function cannot be established independently of the particular arguments that seek to reveal what its moral function is.

Having established to his own satisfaction that the division of labor must play some moral function in human society, Durkheim proceeds to develop an extended argument establishing what that moral function is: "The true function" of the division of labor "is to create a sentiment of solidarity among two or more persons" (DLS: 17/19). More precisely, he argues for what, at this point in the inquiry, he calls the *hypothesis* that this function is to create social solidarity. The argument that follows is more complex and more holistic than the simplistic argument criticized above: it appeals to various considerations, including semi-empirical speculations about the sexual division of labor,[12] as well as an intricate account of the relation between two types of law – penal (or repressive) and civil (or restitutive) – and their significance for social solidarity.

[12] Durkheim recognizes the patriarchal division of labor as more a cultural artifact than a fact of nature (DLS: 18–21/20–4).

Ultimately, then, Durkheim's central argument in the *Division of Labor* extends over many pages in several chapters and gains whatever force it has in the same ways any good interpretation does: by weaving together mutually reinforcing arguments, interpretive suggestions, and analogies so as to form a compelling picture of how the division of labor produces social solidarity and why the latter is a vital social need. Even the weightiest moves in this extended argument – distinguishing two types of law; associating each with a different way of binding society together; taking the number of laws in a society to reflect degrees of solidarity within it – are exercises in interpretation.[13] Straightforwardly empirical facts are marshaled in support of Durkheim's claim – for example, data about the changing proportions of penal to civil laws – but the plausibility of its central idea, that certain facts about law and its development can serve as "a visible symbol" (DLS: 24/28) of truths concerning social solidarity, comes more from its overall interpretive power than from directly verifiable empirical evidence.

This brings us back to the methodological problem noted above: whether Durkheim's functionalism allows him to make more than merely ad hoc claims regarding the functions of social phenomena. Since he rejects a foundationalist strategy that sets out an account of vital social needs before all empirical inquiry, he must find another means of justifying his functionalist claims. As I have argued, his strategy for doing so depends implicitly on criteria similar to those that distinguish good from bad – compelling from arbitrary – interpretations of any sort; his arguments are implicated in a hermeneutic circle that moves back and forth between claims about the nature and needs of society (or of modern European society) in general and functional claims about particular features of the society, in the course of which our understanding of both of these poles becomes more determinate. Is it possible, though, to say more about how such a strategy works in, for example, the principal argument of the *Division of Labor*, in which a specific moral function – the generation of solidarity – is ascribed to the division of labor?

Let us first consider the general thought that the concept of vital social functions is central to understanding social life – that, in other words, human societies, like biological organisms, have vital needs of some kind.

[13] This can be seen in Durkheim's arguments for replacing the standard distinction between public and private law with the sociologically more useful one (based on the kind of sanction attached to each) between penal and civil law (DLS: 28–9/32–4).

As noted above, Durkheim does not take the organism–society analogy to be so tight that he derives a conception of vital social needs and functions directly from those of living organisms. At the same time, it would be incorrect to conclude that the analogy plays no role in shaping what Durkheim looks for when he sets out to determine the function of the division of labor. He says, for example, that, like biology, a science of society can tell us "what is necessary for life" and that knowledge of the requirements of life translates directly into practical rules because we can suppose, in both cases, that the "organism" in question "wants to live" (DLS: xxvii/xl). The same idea figures in his appeal to "the essential conditions of collective existence" (RSM: 141/118) as providing guidance for recognizing the vital needs, and hence the vital functions, of the societies he studies. What Durkheim means here is that there are certain general conditions that must be met if human societies are to exist at all (or to maintain themselves) and that these conditions orient our thinking about what the vital ends of societies are. It is reasonable to assume that central to these conditions are *material reproduction* – the production of the human bodies and the material goods needed to sustain social life over time – as well as, especially in organized societies, ways of *coordinating* the specialized activities on which those societies rely. As we will see below, other vital needs – of a moral character – enter the picture when we take into account the spiritual (conscious and free) nature of the human individuals that make up society.

Even when focusing on the moral needs of society Durkheim sometimes appeals to considerations of what a society needs in order to live in order to establish that it has such needs. He says, for example, that what makes anomie a social pathology is that societies need "cohesion and regularity" – anomie's opposite – "in order to live" (DLS: xxxv/vi). And in explaining why morality has a vital function he claims that, because it reins in the natural egoism of humans, "morality is the indispensable minimum, the strictly necessary, the daily bread without which societies cannot live" (DLS: 13/14). At other times Durkheim moves away from biological analogies when specifying society's vital functions, as when he declares that anomie counts as "a morbid phenomenon" because it "runs contrary to the very goal of every society, which is to suppress or at least moderate the war among men by subordinating the physical law of the stronger to a higher law" (DLS: xxxiii/iii).

Clearly, in such passages Durkheim bases his understanding of vital social needs on something more than a conception of what a society needs in order merely to survive. Social survival might well be compatible with a degree of conflict among its individual members, and his mention of "a higher law" clearly refers to a moral law of some kind. That a society needs cohesion and regularity in order to survive is something that, expressed in this abstract form, it shares with biological organisms, but from the beginning Durkheim's inquiry into what this implies for human societies is guided by a general idea of how the latter, in distinction to organisms, achieve these ends, namely, via moral rules that regulate its parts, bringing them to cohere in ways not possible in biological organisms. This claim reflects an ontological view that ascribes a different order of being to human societies from that which characterizes mere life: only in the former are moral phenomena possible, and societies make use of that possibility in securing the cohesion and regularity they need to survive. Any living being needs cohesion and regularity, but self-maintenance in societies looks very different from its counterpart in elephants, which in turn looks very different from self-maintenance in amoebae. Again, there is no single argument or fact that conclusively supports Durkheim's choice to treat social life as moral in nature. It, too, is justified gradually and only insofar as it helps to make sense of social phenomena that would otherwise remain invisible or unintelligible. (I return to this aspect of Durkheim's view in Chapter 9.)

Perhaps the category that best captures Durkheim's general criterion in thinking about the vital needs of society is *self-maintenance*,[14] defined as a society's ability to endure over time by reproducing itself in ways *consistent with the kind of being it is*. The latter qualification is necessary in the case of both biological and social life because mere survival is, for both, too thin a description of what self-maintenance "aims at." Mere survival, social as well as biological, is consistent with stagnation, depletion, malformation, shrinkage, and severe dysfunction. It is for this reason that Durkheim concludes in the *Rules* that, even in the case of living organisms, no informative account of health or sickness – and therefore no informative account of vital functions – can appeal merely to very general ends, such as survival, without also referring to the specific survival-enhancing functions "normal" for an organism of a particular kind, given the form of life typical of its species. It would not be wrong to say that for Durkheim survival figures among a human society's vital needs, but that claim is seriously under-informative in the absence of further details about what society must do in order to maintain itself in a healthy manner, as the kind of being it is, including its moral characteristics.

[14] Durkheim uses *se maintenir* at DLS: liv, 214/xxxiii, 255. Radcliffe-Brown discusses the same topic using the language of how "social systems perpetuate themselves" (Radcliffe-Brown 1957: 84).

The passages cited above have two further implications. The first concerns the category of vital needs, and hence vital functions, in both sociology and biology, but the point is easier to see if we focus on the latter domain. In the case of living organisms, "vital need" can refer to ultimate needs, such as survival and reproduction, but it often refers to more specific, intermediate needs, such as the giraffe's need to digest leaves or to circulate blood (DLS: 11/11). Intermediate needs serve the organism's ultimate needs, but neither biology nor sociology gets very far in understanding the functions of its objects by having only their ultimate needs in view. Instead, an adequate understanding of what a living being's vital needs are presupposes familiarity with the specific being in question; merely knowing that a giraffe must survive and reproduce – and that it must be internally organized so as to perform those functions – provides little information about how specifically it does so or in what sickness and health for it consists. In the case of societies, too, it is necessary to know a great deal in particular about "the conditions of existence to which they are now subject" (DLS: 14/14) in order to know, for example, that the solidarity generated by moral rules constitutes a vital need. Claiming, then, that a certain biological organ or social institution carries out a vital social function is far from claiming that this is the only way an organism or society could satisfy its ultimate vital needs. Vital functions are more particular than this: digesting leaves and circulating blood are two ways giraffes maintain themselves and reproduce their species, but many other organisms do not meet their vital needs in these ways. Part of the specificity of organisms' vital functions is due to the fact that each specialized function must be coordinated with other specialized functions of a given life form. Specific vital functions are successful only insofar as they work together with other specific functions in the same organism. That the division of labor is able to serve a vital function in modern society is inextricably bound up with the specific ways in which other vital functions in that same society are carried out.

The second implication of these passages is that Durkheim's sociology is a moral science, and not only in the weaker sense that it takes as its object moral phenomena (phenomena that cannot be understood without acknowledging the role that morality plays in their functioning) but also in the stronger sense that it judges its phenomena from a perspective that is itself in part moral. Durkheim's statement that preventing war among social members is a higher social end than those associated with merely physical laws is not due to a slip of his pen, an instance where his own moral views illicitly find their way into his scientific pronouncements.

Durkheim's analysis of the moral function of the division of labor is replete with observations concerning the latter's connections to justice, freedom, fairness, and autonomy, and he does not use these terms as shorthand for what modern society *takes* justice, freedom, fairness, and autonomy to be. Durkheim's theory aspires to be normative not only in distinguishing proper from improper functioning but also in distinguishing ethically better from ethically worse ways of organizing social life. The moral, or ethical, ambitions of Durkheim's sociology – he uses these terms interchangeably – are so far out of line with contemporary conceptions of what a social science can or should deliver that it is difficult for us to recognize that this was Durkheim's aim and that it is not an accidental feature of his theory but one deeply rooted in his approach to social phenomena.

If it is legitimate to seek to understand human societies using the concept of vital social needs, what undergirds sociology's specific ascriptions of functions to social phenomena? Readers of the chapter of the *Division of Labor* devoted to laying out the method for determining the function of the division of labor may be disappointed in what they find there. For it does not present a precise canon articulating a unitary set of rules to be followed in any sociological investigation whatsoever. Later, in the *Rules*, Durkheim attempts to do something like this, but it is no accident that this is the least compelling of his major works. Someone who knew only the *Rules* among Durkheim's texts could be forgiven for wondering why he is regarded as a founding figure of modern sociology. In contrast, his more narrowly focused empirical studies of crime, anomie, and the division of labor continue to strike contemporary readers as rich and enlightening, as presenting compelling and still relevant pictures of social pathology despite respects in which they are outdated, overly ambitious, or unconvincing. This contrast is due to the fact that Durkheim's method in the best of his works is less formalizable, more holistic, and more interpretive than the founder of a new science in the late nineteenth century could be expected to admit.

Finally, there is a further methodological tool Durkheim employs in developing his conception of vital social needs, which in turn is relevant to his involute procedures for ascribing functions to social phenomena. It is a tool he borrows from medicine that echoes Plato's procedure for discovering what health in the polis is and that frequently plays an important role in theories of social pathology more generally. Recall that in both medicine and social theory, recognizing illness often precedes a determinate picture of what health in the relevant "organism" consists in. Just as Socrates is unable to say what health in the polis looks like before he has discovered and reflected on its feverish state, so Durkheim appeals to what he takes to be obviously dysfunctional conditions of contemporary anomie in arriving at his picture of the vital needs of, and hence the functions that must be carried out in, organized societies. In the initial paragraph of his first extended treatment of anomic forms of the division of labor Durkheim sets out this methodological principle clearly:

> Studying the deviated forms [of the division of labor] will allow us better to determine the conditions of existence of the normal state. When we recognize the circumstances in which the division of labor ceases to bring forth solidarity, we shall better understand what is necessary for it to have that effect. Pathology, here as elsewhere, is physiology's valuable aid. (DLS: 353/343)

Even if this statement presupposes that the central function of the division of labor is to produce solidarity – a thesis argued for in the text's preceding three hundred pages – the point is relevant to determining the vital needs of organized societies in advance of knowing that solidarity is one such need. This is because the phenomena Durkheim goes on to analyze – class conflict, economic crises, the enduring dissatisfaction of workers – make clear that something in the contemporary world cries out for improved cohesion and integration of society's parts.

There is no simple answer, then, to the question of how Durkheim determines the function of the division of labor or whether his specific claim in that regard is well grounded. The objection that his answer is merely arbitrary or ad hoc cannot, however, be sustained. For Durkheim offers a variety of arguments – indeed, a variety of *types* of arguments – possessing varying degrees of plausibility that, taken together, present a powerful case not only for his claim about the function of the division of labor but also for the general picture of human society that emerges from his arguments in support of that specific claim. This general picture includes the ideas that human society is a living being; that it depends on moral functions to maintain itself; that it depends on a collective consciousness of some kind in order to function; and that cohesion and regularity are among its vital needs. A survey of some of the claims that figure in Durkheim's complex argument may help to illustrate this point:

- Human societies, like biological organisms, require some means of coordinating the activities of their parts – they need cohesion and regularity – if they are to maintain themselves and function effectively. (This point appeals to an aspect of the society–organism analogy that has considerable plausibility.)

- Human societies, unlike biological organisms, depend on rules of a broadly moral character to achieve cohesion and regularity. (The discussion of laws and mores – of both their nature and their effects – helps to establish this; that discussion appeals to an implicit, commonsense understanding of social life and to empirical evidence regarding, for example, ancient and modern legal systems.)
- The division of labor is capable of creating solidarity and making social life possible. (This claim is based on a discussion of one familiar form of the division of labor – the sexual – that illustrates how specialization and the mutual dependence resulting from it can strengthen social bonds.)
- Modern societies differ from less complex "mechanically," or "segmentally," structured societies both in form and in the nature of the solidarity that holds them together. (A variety of empirical and historical claims supports this thesis, including an extensive comparison of modern and premodern legal systems, as well as an interpretive claim that facts about law, an empirically accessible phenomenon, serves as a marker for both degrees and types of social solidarity. These considerations are also supported by cogent arguments against opposing conceptions of the function of, for example, penal law [DLS: 32–3/37–8].)
- Increases in the division of labor within a society are generally accompanied by a corresponding increase in the number and complexity of civil laws, as well as by a growing preponderance of civil law in relation to penal law. (These are straightforwardly empirical claims backed up by research into the development of law in past and present societies.)

Historical Narrative and Functional Analysis

This is the place to return to a question postponed above: To what extent does Durkheim's functionalism depend on an historical account of social development? The specific version of this question I want to consider is whether a compelling functional analysis of a human society depends on knowing that society's course of development, where the history in question might be universal or more local in nature. In the biological case these two projects are largely independent: an account of the functions of an organism's parts can typically be given without recourse to its development history – physiology, after all, did not have to wait for Darwin in order to grasp the functions of hearts and livers (Boorse 1976: 74). But, again, the considerations that make this true for biology do not obviously apply in the case of societies. One reason that determining the functions of biological organs is relatively independent of both ontogeny and evolutionary theory is that the biologist approaches his object with a fairly determinate picture of the functions he can expect to find in it: it is reasonably clear prior to any actual examination of a certain organism what needs it must fulfill merely by virtue of being a living thing – survival, reproduction, self-regulation, and so on. This is not the case with human societies or, if so, then to a significantly lesser extent than with biological organisms. One way of formulating the question as to whether, expressed in Comte's terminology, social statics can be carried out independently of social dynamics is to ask whether a society's vital needs can be recognized without an account of the historical processes through which that society developed into what it now is. Or, formulated differently: Is it possible to attribute a function to some feature of society without understanding to what specific problem in its past that feature arose as a response?

There are two types of historical narrative in Durkheim's texts, one a version of universal history, the other more local in scope. His (mostly implicit) version of universal history posits a very general pattern of development that all human societies tend to undergo, although contingent factors may prevent certain societies from doing so.[15] The underlying idea is that, just as there is a normal pattern of development for the members of a biological species, human societies, too, pass through specific stages, most relevantly, from homogenous, segmental societies to organically structured ("organized") societies characterized by high degrees of specialization and co-ordination – that is, by an extensive division of labor. This type of universal history should not be confused with that offered by Hegel, according to which human history is unitary not because every society or historical epoch goes through the same pattern of development but because, appropriately interpreted, the history of humanity is a single, continuous developmental process in which each epoch plays a specialized

[15] The claim that for Durkheim a single pattern of development holds for all societies might seem odd, not least because of his apparently categorical distinction between segmentally and organically structured societies. However (and in the same passages that underscore that distinction), he asserts that later societies (of Christian Europe) pass through the same stages as earlier ones (Greek and Rome) and that when transhistorical comparisons are made, one must compare them "at the same stage of their life" (DLS: 121/146). The view underlying Durkheim's universal history seems to be that the same pressures that led to a more thorough division of labor in modern Europe – increasing size and density – tend to be at work in societies generally, and when such pressures arise, they tend to produce the same pattern of development in all.

role and where later epochs build on the accomplishments of earlier ones. (Comte's and Spencer's universal histories are versions of this type as well.) Universal histories of this sort tend to be triumphalist, and they are often robustly teleological, whether because, like classical theodicies, they posit a final end to history in which the human potential for good or freedom is fully realized or because, going one step farther, they also regard that end as explaining why history took the course it did.

Even if traces of a Comtean triumphalist version of universal history can be found in Durkheim, the more modest version of universal history – positing a general, nonteleological pattern of development that all human societies are prone to undergo – is more characteristic of his thought. There can be no doubt that he attributes to human society in general tendencies toward, for example: increasing complexity and size; increasing individualization; increasing social solidarity of the organic kind; differentiation and growing independence among a society's political, economic, and scientific functions; and a decreasing role for religion in creating social cohesion. He refers to such universal tendencies as laws of social development and even, in Spencerian fashion, claims to find a similar law of development – from the simple to the complex and differentiated – in the animal world, in the evolution of species (DLS: 139/167). It is also true that Durkheim *affirms* these tendencies and that his universal history is essentially a tale of progress. But three features of his view distinguish it from more robust universal histories: the progressive character of the general pattern of development – that it produces something good – plays no role in explaining why the process occurs; he posits no end to human history in which the good is completely realized (DLS: 279/332); and because there is no such end in which history culminates, whatever progress it exhibits does not have a theodicean significance, justifying a violent or evil past.

The important question is what role, if any, this universal history plays in Durkheim's functional analyses of society and his diagnoses of social pathologies. If my claim is correct that his conception of a well-functioning society is ethically tinged throughout, then it would be implausible to deny that his account of the goods that tend to be realized in history inform to some extent the perspective from which sociology undertakes its functional analyses and diagnoses of pathology. The core idea, shared in some form with both Marx and Habermas, seems to be that we need some idea of where history in general is destined to lead us in order to recognize the specific goods that inform the conceptions of well-functioning and pathology to be employed by a nonutopian critical social theory.

Nevertheless, because the idea of universal history, even in the relatively weak forms it takes in Durkheim, is deeply controversial, the question I want to consider is what his social theory looks like if we abstract from his account of the general tendencies of human history.

Three points are relevant here. First, discounting Durkheim's universal history does not weaken the normative foundations of his position. It is often assumed that an account of universal history can justify the normative standards on which the evaluation and critique of societies rely, replacing more a priori strategies employed by other thinkers. But merely knowing where history is headed does not establish the goodness of its endpoint. As Kant, the young Marx, and Weber are well aware, the goodness of such an endpoint must be established independently of the claim that history tends to move in a certain direction. Versions of universal history that come closer to justifying the normative standards that result from historical development – as in the histories of Hegel, the mature Marx, and Habermas – view the history they outline as rationally, not merely causally, necessary, where this typically requires interpreting history as a kind of learning process, in which the standards for what counts as progress, or learning, are formal and ahistorical, and therefore not derived from the historical account itself. That Durkheim takes later societies to be "more advanced" than earlier ones (DLS: 329/391) might motivate one to reconstruct his universal history as a rational learning process of some kind, but Durkheim does not provide these resources himself.

The second point is that the more local histories Durkheim invokes in his empirical studies yield a form of normative justification that his universal history is incapable of delivering. I return to this issue below in discussing the history of corporations in Western Europe and the transition from segmental to organized societies, but the main point is that these local histories reveal a specific problem encountered by a society in meeting its vital needs and then shows how, in that context, a certain development serves to solve that problem. Progress here can be understood as such without appeal to a universally valid pattern of development; it counts as progress not because it approaches some standard of the good or some supposed endpoint of human development but because it solves a specific problem in accordance with normative standards already implicit in the recognition of that situation as problematic.[16]

[16] A similar view is articulated in Jaeggi 2019: 221–7.

The third point is that Durkheim's specific functional analyses and diagnoses of social pathology – including his account of the function of the division of labor and the corresponding pathologies to which contemporary societies are vulnerable – do not obviously rely on his account of universal history. A review of the five claims I isolated above in reconstructing Durkheim's argument establishing the function of the division of labor reveals the essential independence of his functional analysis from his universal history. Or, to put the point differently: what is central to Durkheim's sociology is the category of function, and his arguments for which functions specific phenomena carry out and when pathologies are present do not require him to situate those phenomena within an account of how human society in general is destined to develop. Interpreted in this way, Durkheim is more a follower of Darwin than of Hegel on the topic of universal history. For Darwin posits no universal pattern of biological development and makes no appeal to an endpoint of such development, but this does not render his theory incapable of diagnosing problems of dysfunctionality and of taking certain responses to them as advances, at least in the limited sense of being solutions to those problems.

De-emphasizing the importance of universal history militates against conceiving of social pathology on the model of a misdevelopment, appealed to in both biology and Freudian psychoanalysis. In both cases, positing a normal pattern of development, the phases of which all healthy members of a species pass through, makes it possible to speak of stunted or abnormal development – or, in Freud's case, of regression. For Durkheim, as I reconstruct his position, the only admissible conception of a misdevelopment consists not in a society's failure to conform to the pattern of development normal for its "species" but in a failure to find solutions to the specific instances of dysfunctionality it confronts.

Indeed, one can find in Durkheim's texts local historical narratives that are more important than his universal history to his arguments determining the function of the division of labor. One such narrative is his long discussion of the history of corporations in Europe, which points to specific historical changes, the challenges they posed to existing societies, and how corporations responded – or, later on, ceased to respond – to those challenges effectively (DLS: xxxvi–lvii/viii–xxvii). Such local histories are consistent with Durkheim's universal history, but their theoretical relevance in determining the function a given phenomenon serves in some society does not depend on their being inserted into a universal history. His local histories invoke only the weaker brand of existential functional explanations distinguished above, arguing, for example, that corporations became a fixture of Western European society at a particular time because under those conditions they responded to a vital social need. In the case of corporations, this vital social need was social cohesion, but cohesion never appears in his treatments of specific social phenomena as a purely generic need but always as a need for a certain kind of cohesion determined by specific external and internal conditions of the society in question (DLS: 179/211). In this respect Durkheim's attributions of functions to social phenomena depend on empirical knowledge of specific societies and their circumstances, and for this reason they generally avoid the musty air of metaphysical speculation characteristic of Spencer's musings on similar topics.

An illustration of how a historical narrative, detached from a universal history, enters into Durkheim's functional analysis can be reconstructed from his account of not the function but the causes of (the origin of) the organized form of society that came to dominate Western Europe in the late Middle Ages. Throughout the five chapters of the *Division of Labor* that set out this account, Durkheim constantly reminds his readers that the function the division of labor comes to play in organized societies cannot be appealed to as a cause that first produces it. It might seem strange, then, that the causal account offered in Book II could have any relevance to the argument, centered in Book I, concerning the function of the same phenomenon. Yet this is precisely the case, I will argue.[17] In order to grasp the logic of this ensemble of claims, I consider only its main moves, with an eye not to defending those claims but to reconstructing, and revealing the power of, his functional analysis of the division of labor.

Since, as Durkheim recognizes, no society lacks a division of labor in some form, his account in Book II does not aim to explain what causes a rudimentary form of the division of labor to develop in some putative presocial state that lacks one. In other words, his aim is to discover the causes of the division of labor not in general but only in the more extensive form of it characteristic of organized societies. It is probably more precise to say that a genuine division of labor – understood in the narrower sense in which Durkheim generally uses the term – does not yet exist in segmental societies. The latter rely on individuals carrying out different functions within social life, but these different functions do not yet imply specialized individuals, where differences in characters, capacities, and ways of life

[17] As noted above, Durkheim says in a later text that discovering something's cause can help in discovering its function (RSM: 123–4/95); see also DLS: 269/319.

have become relatively fixed, making it difficult for individuals to assume different functional roles (DLS: 79, 206–7/93, 246–7). Organized societies, in other words, are characterized not only by differentiated functions but also by specialized "organs" that cannot easily perform the function of another. In this sense the guild system of medieval Europe represents a genuine division of labor. In any case, Durkheim's causal account of the origin of the division of labor in Book II presupposes the existence of the more rudimentary of the two types of societies distinguished in Book I for the purpose of contrasting mechanical and organic forms of social solidarity; its aim is to explain what causes a society with the former structure to develop into one with the latter.

Durkheim's causal explanation appeals ultimately to "changes in the social environment" (DLS: 200/237) that are themselves not explained (DLS: 288n8/330n) but that "happen to" the society in question, independently of both human intention and the potential benefits such changes might bring. Two such changes essential to his account are population growth and increases in social density (along the two dimensions distinguished below). Although obviously related, neither necessarily leads to the other, and, according to Durkheim, both must be present if a more extensive form of the division of labor is to arise that is incompatible with a segmental social structure. On his account, these two changes weaken the boundaries demarcating social segments and fill up the empty spaces between them, as a result of which the segments lose their individuality – not particularity (because the segments are essentially alike) but their status as more or less independently functioning units – and so "coalesce" in a way that "renders the social material susceptible to entering into new combinations" (DLS: 200/237–8). This increase in society's "material" density – in "the real [spatial] distance between individuals" – is accompanied by an increase in "moral" density, an increase in the possibilities for interaction and influence among social members.

At this point Durkheim's explanation takes an unexpected turn, motivated by the following consideration: if the causal account of the extension of the division of labor stopped here, it could claim at most to have shown that increases in population and social density create the conditions under which a more extensive division of labor becomes possible. Durkheim's aims, however, are more ambitious: "We claim, not that the growth and condensation [increasing density] of societies *permit*, but that they *necessitate*, a greater division of labor" (DLS: 205/244; Durkheim's emphasis). In order to make the stronger claim, Durkheim, citing Darwin, appeals to "the struggle for life" (DLS: 208/248). The Darwinian principle that Durkheim takes to apply to social life in this context is that increasing biological differentiation among species moderates the struggle to survive that would otherwise be produced by homogenous growth in population and density. This is because differentiation brings with it specific differences in vital needs – butterflies do not need the same things as bees – and this puts species in less direct competition in their pursuit of survival. Again, similarly to Darwin, Durkheim offers a causal explanation of social differentiation that does not invoke the function it comes to serve (DLS: 206/245–6), but since we are interested more in the form than the content of his argument, the details of this explanation need not concern us. (If Durkheim were simply lifting a principle from biology and assuming its validity for societies, he would be guilty of Spencer's error: taking himself to have discovered universal laws of evolution that apply to all domains of reality. At most, Durkheim can regard the similarity in principles as merely analogical, in which case the principle's plausibility in the social domain must be established independently of its status in biology.[18] Durkheim recognizes this and provides an argument of the appropriate kind [DLS: 209–13/249–54], but here, too, I refrain from assessing its tenability.)

As noted, Durkheim's argument that increases in population and social density necessitate the development of the division of labor rests on a claim about "the struggle for life." The claim is that under those conditions, together with the weakening of segmental social structure and the rise of differentiation caused by it, a division of labor (in the robust sense) becomes necessary because without it the struggle for survival would be intensified, resulting in the death of social members, who face "no alternative other than to disappear or be transformed" (DLS: 211/251). "New conditions of [social] existence," in other words, require that individuals become specialized "in order to live" (DLS: 217/259). Or, equivalently, specialization occurs because "a disruption of equilibrium within the social mass creates conflicts that can be resolved only by a more developed division of labor" (DLS: 212/253).

Especially in this last formulation, Durkheim's claim appears to be a functional explanation of precisely the sort that his causal explanation is supposed to avoid: while the differentiation of individuals – their

[18] Durkheim explicitly acknowledges that "reasoning by [biological] analogy" is illegitimate in sociology (DLS: 277/329).

acquisition of features and capacities that distinguish them from others – has its own independent causes, the division of labor proper, where differences among individuals have calcified into fixed functional roles, comes about because it serves a vital need of society, its members' survival. Yet – although this is far from explicit in his text – Durkheim is aware that his argument cannot be functional, and he believes he avoids this pitfall because the need that figures in his explanation is a bare need of life: survival. When Durkheim concludes that "the division of labor is … a result of the struggle for life" (DLS: 213/253), he means to emphasize the biological nature of the need that explains the division of labor, and this is important to him for avoiding an illicit functional explanation of that phenomenon because of the status he accords to the human "instinct of self-preservation" (RSM: 121/92). This instinct counts for Durkheim as a *psychological* feature of *human nature*, which he takes to mean that it is a motivational force operating "mechanically," internally to each individual, for which no social explanation is necessary, precisely because it is not a social phenomenon or, in his terms, a social fact. ("Mechanically" here means both "atomistically" – there is a cause internal to each individual sufficient to explain the behavior at issue[19] – and "not requiring consciousness, in the manner of instinct"; the term should not be confused with the "mechanical" solidarity ascribed to segmental societies.) In other words, Durkheim thinks his argument avoids an illicit functional explanation because the division of labor is merely the product of an aggregation of forces internal to each individual, independent of any social fact, impelling them to do what their survival requires of them. This implies that individuals behave, whether consciously or not, so as to bring about some end (their survival), but the explanation remains causal rather than functional because the totality of behavior that produces the division of labor can be accounted for by aggregating the independently motivated motions of individuals. Formulated differently, because the mere survival (of individuals) is the need to which the division of labor responds, Durkheim takes there to be an adequate explanation of why it comes about without referring to some collective benefit it results in beyond the aggregation of benefits to individuals produced by their individually motivated behavior. Because the instinct for survival is a psychological feature of human nature – a force internal to individuals operating at all times and

[19] Atomistic explanations account for complex phenomena as the vector sum of independently acting forces, where the whole is nothing more than the sum of its independent parts.

places – and because it suffices to explain the division of labor under the conditions described, Durkheim takes his account of the latter's origin to be causal (or mechanical) rather than functional.[20]

My reason for reconstructing this account is not to assess its success as an argument but to illustrate how a historical narrative – in this case, a causal narrative – detached from a universal history can be relevant to establishing the function of a social phenomenon. In the case at hand this relevance comes into view only when considerations regarding the "collective" or, equivalently, "common" consciousness – "the ensemble of beliefs and sentiments common to … the members of a given society" (DLS: 38–9/46) – are introduced. Here again Durkheim's train of thought takes an interesting turn. Rather than argue that increasing differentiation weakens the collective consciousness, he claims that such weakening is an additional condition in the absence of which differentiation itself could not occur: individuals first need some normative distance from the strict and all-pervasive demands of the collective consciousness of segmental societies in order to develop new traits and capacities. For this reason the weakening of the collective consciousness, along with its increasing abstractness and indeterminacy, is regarded as a secondary cause of the division of labor, which itself is brought about not by the differentiation of individuals but by the same causes that produce the increased struggle for survival invoked above: presumably population growth and increased social density, together with the erosion of inter-segmental social boundaries that accompanies them, make it more difficult for individuals to identify with the much larger social whole in which they have ceased to play a crucial part (DLS: 229–38/272–83).

The weakening of the collective consciousness in segmental societies figures in the overall argument of the *Division of Labor* in two respects. It plays, as we just saw, a causal role in explaining why the division of labor comes about, but it also contributes to establishing the function of the division of labor in organized societies. The key to grasping the latter point lies in seeing not only that the erosion of collective consciousness brought about by increases in population and social density has its own further causal consequences but also that this development poses a functional problem for segmental societies: since it was the collective consciousness that guaranteed the cohesion and solidarity of society in its more

[20] Durkheim addresses these issues at RSM: 121–23/92–5, but they are implicit already in DLS: 279–80/332.

homogenous form, the weakening of that consciousness means that in the newly developing society something needs to take its place if cohesion and solidarity are to be reestablished. In other words, an event in Durkheim's causal narrative can have implications for his functional analysis of society subsequent to that event just in case he is already operating, as he clearly is, with a functional understanding of the earlier society – that is, with an understanding of the vital social needs the collective consciousness serves in that society. If one functionally crucial element of social life disappears or is weakened, a problem is created that must be solved if the society in question is to maintain itself, and this provides guidance for functional analysis in the form of the question, "What, if anything, in the new social configuration can play the role previously played by the now defunct or weakened phenomenon?"

To what extent, though, does a move of this sort help establish the function served by the division of labor in more heterogeneous societies? One obvious limitation of such a move is that recognizing the erosion of collective consciousness as a problem to be solved depends on having already attributed a function to that consciousness. This means that if grasping a social development as a problem is to aid us in discovering the function of some subsequent phenomenon, the later attribution of function depends on the earlier, and we seem to have merely pushed our epistemological quandary back a step: What entitles us to the functional claim regarding the earlier phenomenon, or to the even more basic claim that cohesion and solidarity are vital social needs? That there is a threat of infinite regress here cannot be denied, but just as not all circles are vicious, so some appearances of infinite regress are more virtuous than others. In the case at hand the situation is so: to the extent that the problem for social cohesion and solidarity occasioned by the weakening of the collective consciousness plays a role in Durkheim's claim that the division of labor, once it has taken root, carries out those same functions, the validity of that functional claim depends on that of the preceding functional claim, but the grounding relation between the two is not foundationalist or unidirectional. Again, we have a situation in which the plausibility of the argument's parts depends on the plausibility of the whole and the parts face the test of empirical validity not as isolated claims but together, as a more holistic "web of belief." This means that one cannot assess the validity of Durkheim's claim about the division of labor's eventual function without also considering how persuasive his claims are about the collective consciousness and the role it plays in securing cohesion and solidarity in segmental societies (claims which I take to be among the most powerful in the *Division of Labor*).

The second limitation of the argumentative strategy I have attributed to Durkheim's historical narrative is equally obvious: the mere fact that the new society is confronted with a problem – it needs to find a new way to guarantee cohesion and solidarity – does not imply that it is the division of labor that, once established, plays this role. This claim must be argued for on independent grounds, and this is what Durkheim means to have done in Book I via his mutually reinforcing claims regarding the sexual division of labor, the two types of law and social solidarity, and so on. After having both examined his causal story regarding the origin of the division of labor and grasped the functional problem that one event in that story poses for the society undergoing transformation, we are equipped with a better sense of what we should attempt to find in the division of labor – what effects relevant to cohesion and solidarity it might have – if his claim about its function in the later society is to be convincing.

I end my discussion of Durkheim's functionalism by noting a variety of claims made by him that deserve to be retained in a contemporary account of what human societies are and in what ways they are vulnerable to falling ill:

- Human societies are functionally organized: they are made up of differentiated subsystems that carry out coordinated, specialized functions.
- A scientific understanding of society, like physiology and medicine, includes three related tasks: explanation, evaluation, and prescription.
- Functional analysis, not causal explanation or existential functional explanation, is the primary concern of a theory of social pathology. The regulative principle guiding the functional analysis of society has an affinity with a weak form of existential functional argument: if *X* is an enduring feature of a society (or of societies of a certain type), it is likely to have endured because it carries out a vital social function. Causal claims are necessary for theories of social pathology in two respects: ascribing a function to *X* implies that *X* has certain useful effects; and responding to social pathologies (usually) requires understanding the unhealthy conditions' causes.
- Functional analysis can be aided by knowing the specific history of the society being studied but need not rely on a conception of universal history.

CHAPTER 9

Durkheim: Solidarity, Moral Facts, and Social Pathology

Durkheim's claim that the principal social function of the division of labor is *moral* – that it secures *cohesion* and *solidarity* in "organized societies" (DLS: 23/26) – must now be examined in detail with an eye to understanding how he employs these three key terms. Doing so is essential to grasping Durkheim's understanding of social pathology, as well as his provocative idea, reconstructed in the following chapter, that sociology is a "science of morality" (DLS: xxv/xxxvii) in which diagnoses of social health and illness are at once diagnoses of moral health and illness (DLS: xxvii, xxxix/xxxix, xii). These views of Durkheim's are relevant to social ontology, insofar as he takes society to belong to an order of being different from that of the biological – for which not only "social" but also "moral" is an appropriate name. Human society is not merely a living being but one whose life is intrinsically moral in character. This point is expressed in Durkheim's statement that "morality is the indispensable minimum, the strictly necessary, the daily bread without which societies cannot live" (DLS: 13/14). In order to understand what holds societies together and enables them to live as they do, we must take into account moral facts that inform society and the relations among social members.

This chapter has two main parts. The first reconstructs Durkheim's account of social cohesion and solidarity, and their relation to "moral facts." The second treats Durkheim's general conception of social pathology and his treatment of some specific instances of it, most prominently anomie and closely related pathologies.

Moral Facts and Solidarity

Before proceeding, two confusing points about Durkheim's use of "fact" – as it occurs in both "moral fact" and "social fact" – must be noted.[1] The first

[1] For more on social facts, see Karsenti 2006a: 11–25.

is that "fact" can refer both to a phenomenon and to what explains such a phenomenon. For example, societywide rates of suicide count as social facts, as do the social causes that explain them; it is a social fact that in Europe Protestant societies have higher suicide rates than Catholic societies (S: 25, 83/46, 119), and the social conditions that causally explain the difference in these rates are social facts as well (S: 273–4/348–9, 461–2).

Second, "moral fact" and "social fact" are not identical concepts, even though I have used them interchangeably thus far. The latter is more inclusive than the former: all moral facts are social facts, but not all social facts are moral. What distinguishes all social facts – the distinctive object of sociology – from nonsocial facts is that the phenomena or explanations they consist in cannot be understood atomistically, as the aggregate results of the motions and forces of independently acting individuals. We encountered this distinction in the previous chapter in explaining why Durkheim takes his causal account of the origin of the division of labor not to appeal to social facts (or, as I expressed the point there, to functional explanations). That account appealed to the aggregative effects of the human instinct for self-preservation, where the force impelling individuals to act so as to promote their survival was taken to have a biological source, internal to each individual, and operative independently of social structures and relations. In other words, the instinct for self-preservation, together with its aggregative effects, is taken by Durkheim to require no social explanation and for that reason not to count not a social fact.

The domain of facts that are social but not moral is nonatomistic in the sense defined above and does not involve individuals following social rules to which notions of obligation and social sanctions are attached. Nonmoral social facts (insofar as they are explanations) make their effects felt by causally affecting individuals' behavior and emotions (without imposing obligations on them), but those individuals are so affected only because of some relation they have to other affected individuals. The most prominent representative of this class of social facts is what Durkheim calls "social currents,"[2] exemplified by "great movements of enthusiasm, indignation, or pity" that in large assemblies can cause individuals to experience such feelings without them "having their origin in any particular consciousness" (RSM: 52–3/6). Durkheim also appeals to currents of this sort when

[2] The other type of non-moral social facts is "facts of structure," or "anatomical" or "morphological" social facts, e.g., "the number and nature of the elementary parts of which society is composed; the manner in which they are arranged; ... the distribution of population ...; the number and nature of the means of communication; ... etc." (RSM: 57/12).

explaining the mechanism by means of which deficiencies in social integration (anomie) cause individuals to commit suicide at a higher rate in Protestant, as compared to Catholic, societies. In this case the social current constitutes "a collective force, with a determinate quantity of energy, that impels men to kill themselves" (S: 263–4/336).[3] What precisely these causally efficacious social currents are remains a mystery in Durkheim's texts; fortunately, such nonmoral social facts play a much smaller role in his sociology than moral facts (Karsenti 2012: 21–36), and my reconstruction of his position ignores them.

What, then, defines the domain of the moral (or ethical) such that it distinguishes human social forms of life from nonhuman forms? The *Division of Labor* offers the following definition: "The domain of the ethical … comprises all rules of action imposed on behavior as imperatives to which a sanction is attached" (DLS: 15/16). Hence the domain of moral facts – or, more precisely, the part of it that *explains* moral phenomena – consists in rules that constrain social members' behavior (RSM: 2/4), but the power by means of which they do so is normative rather than causal. They are rules that subjects capable of rational agency follow, rather than rules that govern their behavior as the law of gravity determines the motions of the planets, and in this respect they are akin to conventions for Rousseau. Such rules are moral in a further sense (also employed by Rousseau when calling the artificial "public person" that results from the social contract a "moral body" [SC: I.6.x]). Here "moral" refers to the mental or spiritual, in contrast to the physical, which explains why Durkheim says of moral facts, as well as of social institutions, that they are "ways of acting, thinking, and feeling" (RSM: 2/4). (Notice that moral facts are described both as *rules* that constrain behavior and as ways of acting. These slightly different descriptions reflect the ambiguity regarding the term "fact" noted above.) Moral rules may be codified in law, in which case the sanctions attached to them are legal punishments, but they may also be moral principles or, even less formally, social mores (*moeurs*), in which case the sanctions attached to them are disapproval or ostracism. They are *social* facts (in form) because the practices they govern cannot be explained atomistically, as the product of independently acting individuals, but they are also social (in content) in that they regulate relations among social members (DLS: xxviii/xli) and

[3] Durkheim seems to be confused here: as he says in the same passage, anomie is a moral phenomenon, but as such it should not produce effects in the mechanical manner of social currents. Suicides due to anomie have their source not in mechanical currents but in the weakening of the normative authority of social (moral) rules.

do so by constraining their egoism (DLS: 13/13–14), thereby diminishing the possibility of conflict, which in the absence of such rules would pose a constant threat to human coexistence (DLS: xxxii–xxxiii/iii).

Beyond this, moral rules regulate human action not in the name of mere utility, and hence with the conditioned authority of Kant's hypothetical imperatives, but as morally obligatory demands taken by those subject to them to possess a *sacred* authority or, as religion comes to play a lesser role in social life, a supra-utilitarian, "unconditioned" authority grounded in some secularized version of the sacred, such as Kant's categorical imperative. Since in both more and less religious societies the true source of this higher authority is the awe individuals feel for society itself, moral rules are social also with respect to the source of their motivating power. In making this claim Durkheim introduces a distinction important to his theory between how moral life is experienced by participants and what, from sociology's perspective, explains that experience: the authority of religious or moral imperatives is not typically experienced by those who follow them as deriving from a reverence for society.

Finally, Durkheim's conception of the moral reminds one of Kant's (and Hegel's) in a further respect: although he emphasizes the constraining and disciplining aspects of moral rules, it is important that they also enrich and elevate human life in ways unavailable to merely animal beings. Morality is "a source of life *sui generis* from which emanates a warmth that inflames and re-animates our hearts, opening them to a sympathy that melts our egoism" (DLS: lii/xxx).[4] Borrowing terms coined later by Max Weber, one could say that for Durkheim morality is what makes "value-rational" action possible, as opposed to merely instrumentally rational action, as well as what enables us to regard our lives and world as having a meaning beyond that of pleasure and utility.

As noted above, Durkheim defines moral facts as rules, but he uses the term to refer not merely to the rules but also to the practices (or institutions) they govern and the phenomena they make possible. Thus, the division of labor counts as a moral fact even though it is not, strictly speaking, a rule. It is a phenomenon or practice in which the activities of individuals are governed by moral rules in Durkheim's sense: "When I carry out my task as brother, husband, or citizen; when I execute the commitments I have contracted into" (RSM: 50/3), I am participating in the division of labor by virtue of following the moral rules that define the

[4] I am indebted to Conor Cullen for this point.

rights and duties of brothers, husbands, citizens, and contracting parties. (Note, again, how broadly Durkheim conceives of the division of labor; it includes not only specialized economic functions but also differentiated social roles of all types.) Such rules count as social facts because both what they require of me and the sanctions that follow on my noncompliance are "external" to me, which is to say, defined by a social practice that I as a single individual do not determine. Durkheim also regards "the beliefs and practices of religious life" as moral facts, as well as "the system of signs I use to express my thoughts, the money system I use to pay my debts, instruments of credit I use in my commercial relations, the practices of my profession, etc., etc.," all of which "function independently of the use I [as a single individual] make of them" (RSM: 51/4). Clearly, the phenomena that qualify for Durkheim as moral facts could also be described as practices or institutions.

Not surprisingly, Durkheim attributes a social function to morality, insofar as it satisfies two social needs: *reducing conflict* and *regulating cooperation* (or coordinating specialized activities). However, both concepts need further qualification. First, morality does not seek to eliminate, or even to reduce, all forms of conflict but only those with socially destructive consequences. Insofar as conflict is intrinsic to forms of competition that enrich social life, including economic competition (DLS: 302, 313/357, 371), social space must be created within which such conflict can be played out but also regulated such that its effects do not threaten society's health. Moreover, extrapolating from Durkheim's views on the unhealthiness of a condition in which crime or suicide were completely eliminated, it seems likely that he would say the same thing about even some forms of conflict that are potentially destructive: reducing such conflicts is an important function of morality, but eliminating them altogether would undermine a society's vitality and capacity for change.

Second, the coordinating function of moral rules is not to be understood merely as one of regulating cooperation in a narrow sense. If cooperation is understood broadly to mean "sharing a common task" (DLS: 79/93), then regulating cooperation could well be said to be a principal function of morality. Sometimes, however – especially when discussing Spencer (DLS: 219/261–2) – Durkheim uses "cooperation" more narrowly, referring to the coming together of independent individuals in order to work together for mutual benefit. In these instances Durkheim is at pains to avoid Spencer's view of society as cooperative in this narrow sense, as a great scheme of voluntary exchange that makes it possible for individuals to establish relations for the purpose of satisfying needs they have independently of their association. Durkheim's holism leads him to emphasize the socially conditioned nature of the needs, capacities, and natures of the individuals whose activities are regulated by moral rules. As he notes when discussing the sexual division of labor (DLS: 18–21/20–4), social processes of differentiation precede and condition the specialized needs that a division of labor satisfies. Rather than responding to already existent needs and taking advantage of already specialized capacities, the division of labor first makes such needs and capacities possible. Moreover, it forms individuals such that their identities are partially constituted by the positions they occupy within the division of labor. (Here, too, the sexual division of labor is illustrative.) Interdependence plays a major role in Durkheim's vision of social life, but it is not the dependence among butchers, brewers, and bakers he emphasizes – where already constituted individuals with complementary needs temporarily come together for mutual benefit – but the deeper, permanent need socialized individuals have of one another, which extends so far that outside their social relations they experience themselves as incomplete (DLS: 17–18, 22/19, 25). Moral rules regulate the working together of society's parts – which Durkheim typically understands as "organs" (institutions or groups) rather than individuals – but not their cooperation if this is understood on the model of the capitalist market.

If morality's principal social function is to reduce conflict and regulate cooperation as described above, it is not surprising that moral rules play a key role in creating social solidarity. Because this claim is central to Durkheim's sociology, it is worth considering in detail how he conceives of solidarity and its relation to the division of labor. The first thing to note is that in Durkheim's understanding of the vital needs of society, solidarity essentially takes the place that stability occupied for Comte and Spencer. Solidarity is for Durkheim the principal source of social stability, but it is more specific than mere stability, or more precisely: stability remains an important concern for him, but the vital social need associated with it is best described as *stability achieved through solidarity*. Mere stability, in other words, is not an unconditioned value in Durkheim's social theory; rather, social health requires that stability be achieved in a certain manner, one important element of which is the "consent" (to be described below) of social members (DLS: xxxii/iii). Moreover, it is possible for there to be too much stability in a society, just as it is possible for there to be too little crime or suicide. Stability carried too far becomes ossification and paralysis, which in both biological and social organisms is incompatible with the vitality and malleability characteristic of health.

As we have seen, Durkheim's most prominent functional claim concerning the division of labor is that it is the principal source of social solidarity in organized societies (DLS: 23/26): it "produces a moral effect, and its true function is to create a sentiment of solidarity among two or more persons" (DLS: 17/19). It is important not to be misled by the latter claim. First, social solidarity includes sentiment but is not exhausted by it. If solidarity is a moral phenomenon, it must include not merely sentiment but individuals taking themselves to be bound by rules whose authority, in their eyes, goes beyond social utility or enlightened self-interest. Sentiment plays a necessary role in letting such rules govern one's actions, but taking oneself to be obligated to do so is also essential. Second, social solidarity is not adequately described exclusively in terms of relations, whether sentimental or not, among individual social members; in addition, solidarity involves some relation between individuals and society as a whole.

Regardless of how these points are to be understood, Durkheim's claim raises a more fundamental question: Why does the division of labor produce solidarity, as opposed to being the condition that makes solidarity necessary? The key to understanding this claim lies in his assertion that the division of labor "makes societies possible that without [it] would not exist" (DLS: 21/24). Durkheim's point is that the division of labor – and the entrenched specialization it implies – increases the interdependence of individuals (or, equivalently, decreases their self-sufficiency) to such a point that their relations must be constant and enduring. In such a situation individuals are, and perceive themselves to be, incomplete on their own and profoundly dependent on others. This, together with their constant interaction – explained by the principle, exemplified by sexual difference, that "opposites attract" – explains how sentiments of solidarity based on need for others develop. Durkheim sometimes expresses this point by saying that sociability (or "fellow feeling"), the desire of dependent individuals to seek out contact with those on whom they depend, is a result of the division of labor. Although in segmental societies, where solidarity is mechanical rather than organic and is grounded in similarities among its members, individuals are far from self-sufficient. Durkheim's point is that the specialization intrinsic to a thoroughgoing division of labor greatly increases dependence and is for that reason a principal source of the sentiments on which solidarity depends.

It is less clear how sentiments of sociability and frequent interactions turn into rule-governed social relations, but it seems plausible that increases in population, social density, and specialization make it necessary for social collaboration to become more regular than merely spontaneous interactions allow for.[5] Somewhat clearer is that in order for rules, whether laws or mores, to serve this function, they must attach themselves to something already existent in segmental societies, namely, a conception of the sacred that can imbue the new rules with obligating authority.

My main concern, however, is not to reconstruct the development of social solidarity in organized societies but to articulate what Durkheim takes it to consist in. Most succinctly, solidarity is what holds societies together and enables them to live and maintain themselves, especially insofar as doing so requires the coordination of specialized functions. In human societies, solidarity depends on social members regulating their own conduct in accordance with moral rules that coordinate their activities. In premodern, segmental societies solidarity-generating rules apply equally to all individuals and are accompanied by strong sentiments shared by all, which constitute the society's collective consciousness: "Thou shalt not kill" – or lie or eat human flesh or have sex with a sibling – regulates the behavior of all social members. In modern societies, in contrast, the division of labor brings with it increased dependence among individuals and greater social complexity. The more specialized activities become, the more difficult the task of coordinating them, and, *pace* Comte, this task soon outstrips the capacities of a single consciousness. In this case the division of labor can produce solidarity only when the individuals who occupy its various positions follow rules that apply only to their specialized roles, the content of which coordinates their actions so as to serve the self-maintenance of society.

Apart from sentiments of sociability and a recognition of the obligatory character of the moral rules that coordinate action, several other elements need to be brought into the picture. First, as already noted, social solidarity exists only when the interactions governed by moral rules are frequent and regular and the relations among individuals enduring, such that it makes sense to speak of social bonds or attachments, in contrast to temporary cooperation based on punctually made agreements. It is these enduring social conditions Durkheim has in mind when he says that the division of labor makes societies possible that would not be otherwise. One aspect of such bonds is that individuals' relations to others penetrate to their self-conceptions. They do so, however, differently from how social bonds develop in segmental societies. In organized societies social bonds

[5] This claim involves some version of an existential functional explanation insofar as it explains the origin (or perhaps only the endurance?) of such rules in terms of a social need they serve.

develop not because individuals resemble, but because they complement, one another:

> The image of the person who completes us becomes inseparable in ourselves from our self-image, not only because the two are frequently associated but, even more, because the one image is the natural complement of the other. For this reason the other's image becomes an integral and permanent part of our consciousness such that we cannot get by without it. (DLS: 22/25)

Finally, social solidarity involves not only bonds among individuals but also the attachments of each social member to the society as a whole (or to its subgroups). Social solidarity is present only when social members "live a common life" (DLS: xliii/xvii) or pursue a "common project" (DLS: 23/27), which includes some degree of experiencing that life or project as one they undertake together. Such individuals are subjectively attached to the group in the context of which they engage in a common project with others; they care about that group's interests and take account of them in determining their own actions. In segmental societies individuals are attached directly to their society as a whole via the collective consciousness shared by all, but in organized societies, where collective consciousness is weaker, individuals' attachment to their societies must take a more complex form. In effect, social solidarity in organized societies results from the *different ways* individuals are attached to various subgroups of their society; that is, in the modern world social solidarity depends on an ensemble of attachments, each different in form, to a variety of social groups, most importantly: one's family, one's profession, and one's nation (DLS: 27/31). Such attachments create organic, rather than mechanical, solidarity because society's cohesion depends on the formation of specialized, semi-autonomous "organs," or subsystems – the proximate objects of individuals' attachments – that are organized so as to serve the vital functions of the whole.[6] (It is no coincidence that these subgroups correspond to the social spheres Hegel regards as central to modern social life and the sources of our attachment to society as a whole: the family, civil society, and the national state. Likewise, the idea that each social sphere is the site of a distinctive form of social relation, with its own form of attachment

[6] Habermas makes a similar point: "Whereas primitive [sic] societies are integrated via a *basic normative consensus*, the integration of developed societies comes about via the *systemic interconnection of functionally specified domains of action*" (Habermas 1987: 115).

to the group in question, is fundamental to Hegel's social philosophy [Neuhouser 2003: 135–7].[7])

There is one more aspect of social solidarity in need of mention, and it, too, involves a Hegelian thought: social solidarity requires not only that social members accept the normative authority of the rules that coordinate their activities but also that their working together satisfies, to a significant degree, each individual's particular interests. In an organized society, solidarity results only if collective social ends are achieved in ways not overly burdensome to the collaborating individuals, for example, by demanding that they regularly forgo satisfaction of their own particular interests. This requirement of social solidarity is especially prominent in organized societies because one effect of the modern division of labor is that social members become more highly individualized than in less complex societies, one aspect of which is that they distinguish their particular interests from collective ends and care deeply about them. For this reason solidarity is likely to be weak in an organized society where too much distance separates particular and "universal" ends. At the same time, having no interests of one's own distinct from those of the whole cannot be satisfying to modern social members whose aspirations include being an "individual." In discussing modern social pathology (anomie) and its primary remedy (professional corporations) Durkheim expresses this idea as follows: "A nation can maintain itself only if there comes between the state and its particular members an entire series of secondary groups close enough to individuals to draw them strongly into those groups' sphere of action and to lead them in this way into the general torrent of social life" (DLS: liv/xxxiii). Or, formulated in Hegelian language, a society's universal will must be mediated by the particular wills of its members. If we take this principle seriously, Durkheim's moral rules begin to look less Kantian than they did initially. Solidarity for him requires that individuals recognize the rules coordinating their activities as morally obligatory, but the activities they are thereby obligated to undertake are not mere duties, indifferent to the good of the individual bound by them but, at the same time, ways of achieving what the social members take their own particular good to consist in.

What emerges from these considerations is that social solidarity for Durkheim depends on "consensus," although care must be taken in interpreting this concept. Whereas Comte uses this term to refer to conscious acceptance of the authority of social rules, it has a wider sense for

[7] Honneth (2021) and Gangas (2009) explore commonalities between the two thinkers. Knapp (1985) treats questions of actual influence.

Durkheim. For him "consensus" in general means simply "going in a common direction," regardless of whether the common direction is achieved with or without consciousness. (Think of the use of *senso unico* in Italian to designate one-way traffic.) In its most general meaning, then, consensus does not necessarily involve the conscious assent to moral rules I have emphasized thus far. Indeed, consensus for Durkheim is typically understood as "spontaneous" consensus – as an unforced, unplanned working together of an organism's parts that is not governed by consciously followed rules.[8] What Durkheim points out with the idea of spontaneous consensus is the variety of ways individuals and social "organs" adjust themselves more or less unconsciously ("spontaneously") to the motions of other proximate individuals and "organs" engaged in a cooperative enterprise. Examples of spontaneous consensus include, presumably, rowers who without conscious thought attune their movements to those of their fellow rowers (in order to achieve a shared aim) (Hume 1896: 490) and cooperation in a free market economy (where no shared aim exists, and collectively beneficial outcomes are effected behind participants' backs by institutional mechanisms).[9] Durkheim's account of solidarity in organized societies is more complex than it might appear precisely because it incorporates *both* spontaneous consensus and consensus effected by moral rules. Although he believes that solidarity in all human societies relies on a significant degree of consensus of the former type, the aspect of consensus I emphasize (as does he) is what – perhaps following Rousseau – he calls "consent" (DLS: xxxii/iii), which involves accepting the authority of the rules of social life. That there is a moral element to this acceptance is clear throughout Durkheim's corpus: "It is the awakening of consciousness that broke the condition of equilibrium in which the animal slumbered; hence only consciousness can furnish the means for re-establishing it ... When appetites are not constrained automatically by physiological mechanisms, they can be held back only by a limit they recognize as just, ... [by] an authority they respect" (S: 209/275).[10]

[8] Durkheim sometimes uses "spontaneous" to describe conscious assent (S: 209/275), but this usage is rare.

[9] Habermas connects Durkheim's spontaneous consensus with his own idea of "a norm-free regulative mechanism," or "a systematic integration of society uncoupled from the value orientations of individual actors" (Habermas 1987: 115–16). On this suggestion, spontaneous consensus is defined not by the presence or absence of consciousness but by its not being guided by values or moral rules.

[10] Here, too, Rousseauean echoes are unmistakable (SC: I.8.i, I.8.iii); see also S: 212/279 and DLS: xxxii–xxxiii/iii.

Durkheim's emphasis on the need for individuals to recognize both the authority of the social rules that obligate them and the convergence among collective and particular ends that mark a healthy society is part of a larger claim that solidarity depends on a consensus that, in distinction to spontaneous consensus in a biological organism, "goes through" the consciousness of social members. The spontaneous consensus characteristic of biological organisms produces stability, but social solidarity is not equivalent to any means whatsoever of producing a stable organism (DLS: 312–13/370). For the feelings and beliefs of individuals that make solidarity possible for humans also make their social participation *free* in the sense of being consensual in a completely ordinary sense of that term: what social members do is governed by rules they themselves take to be authoritative and productive of the good. This aspect of Durkheim's view points to a further sense in which healthy social life is a moral phenomenon: not only does its functioning depend on rules taken by its members to possess normative authority, it is also the case that society's cohesion as a functioning entity – what allows it to "hang together" – is achieved, in part, through the normatively imbued consent of its members, which is itself a kind of freedom. This explains why the solutions to anomie endorsed by Durkheim must extend beyond purely legal measures, the efficacy of which depends ultimately on the threat of force, and must instead penetrate individuals' consciousness in a way compatible with their freedom.

A cynical reader of Durkheim might conclude that normative consensus is important for him only because, in the case of "organisms" composed of human individuals, it is more effective at producing social stability than legal measures. But this overlooks the general tenor of Durkheim's sociological project and the extent to which, for him, a science of society is necessarily a normative, even an ethical, enterprise. Durkheim, as is well known, endorses a positivist conception of social science, but his positivism is closer to Comte's than to later twentieth-century versions of it. Social science for him is a "spiritualized" version of medicine in which the idea of social health takes into account ethical concerns such as freedom and the good. Durkheim's commitment to science requires that his ethical position be naturalist in some sense, but that in no way implies ethical neutrality.

As this discussion makes clear, the connection between solidarity and cohesion is also very tight: consensual solidarity is what enables an organized society to function coherently as a whole. Perhaps the best way of explaining this connection is via another concept central to Durkheim's sociology, *integration*. This concept has a dual grammar for Durkheim. It can be used to characterize either individuals (in their relations to society) or society itself: in an integrated (cohesive) society, individuals are well integrated into the social groups to which they belong. The first of these uses is evident in Durkheim's claim that the division of labor ensures social cohesion because it "integrates the social body, [or] assures its unity" (DLS: 23/26–7). Social cohesion, then, is a social unity achieved by, among other things, the integration of individuals into society. What the latter consists in is best explained by Durkheim's description of its opposite: in a disintegrated society, "the individual is disengaged from social life; … his own ends take precedence over common ends; … his personality … tends to place itself outside the collective personality; … he recognizes no rules of conduct other than those based on his private interests" (S: 167/223). Since Durkheim equates this condition with egoism, a failure of integration – a form of anomie – could also be described as a condition in which individual self-interest has gained the upper hand over collective ends. At the same time, it is important not to equate the integration of individuals with a subjective disposition such as egoism or altruism. Integrated individuals are indeed less egoistic than their disintegrated counterparts, but social integration requires, beyond a mere disposition, a practical allegiance to socially recognized *rules* – to objective, or thinglike, "facts" external to the consciousness of particular individuals that coordinate their activities. As will become clearer in my discussion of anomie, it is impossible to overemphasize the importance for Durkheim of the rule-governed nature of the moral phenomena under discussion here. It is (in part) rules, and the social sanctions accompanying them, that account for the "external," thinglike character of social reality and make it possible for individuals' attachments to others to serve the universal function of regulating society's work. To speak with Hegel once again, it is such rules that make it possible to attribute to Durkheim a vision of social reality as objective spirit (the topic of Chapter 11), or as mind that acquires a real existence in the world.

That Durkheim associates social cohesion with functional unity and integration is no surprise, but there is a further aspect of cohesion that is less obvious: its relation to society's "energy" or "vitality" (DLS: 63/76). There are two reasons one can expect a cohesive society to enjoy an abundance of vital energy. First, a cohesive society is one whose parts work together (more or less) harmoniously, and a harmonious organism is generally more "energy efficient" than one whose parts are often at odds with one another. (Like stability, however, social harmony is not an unconditioned good for Durkheim since too much of it is also a threat to vitality, especially to a society's capacity to develop in new ways and to react to external challenges.) Second, in a cohesive, integrated society individuals bring more energy to their social activities than in a disintegrated one because more of themselves is present in what they do; they typically recognize their social activity as meaningful and good, even if some of their particular desires and ends may not be satisfied within it. Integrated individuals, to speak with Hegel, are free from subjective alienation, which makes clear why it is sometimes difficult to distinguish anomie, or deficiencies in integration, from alienation; though not identical, the two are closely related.

Social Pathology

Defining the Concept

The central idea behind Durkheim's conception of social pathology is dysfunctionality. At its core, social pathology consists for him in impaired functioning – more precisely, in a disturbance in a vital social function of society or, equivalently, a dysfunction within a living social system that impedes the satisfaction of one of its vital needs. Moving beyond this idea in order to define social pathology more precisely, however, quickly leads to complications. These complications are evident in the problematic third chapter of the *Rules*, the announced task of which is to set out the "rules for distinguishing the normal from the pathological" (RSM: 85/47). One reason this text is unsatisfying is that it moves back and forth between two questions not sufficiently held apart: "What does social illness consist in?" (or "How is it to be defined?") and "By what empirical marks does the sociologist recognize social illness?" To the first question, "a happy development of vital forces" is one sound but relatively uninformative answer considered by Durkheim, whereas suffering exemplifies an initially plausible but ultimately inadequate answer to the second (RSM: 87/50).

A second reason this chapter of the *Rules* is unsatisfying is that setting out the necessary and sufficient conditions of social pathology proves to be an elusive goal. Durkheim's problems here result from a failure to give specific content to the two concepts at the core of social pathology: *impaired functioning* and society's *vital needs*. Given his insistence on tying the possibility of a science of society to the viability of the concept of social pathology (RSM: 85–6/47–8), this failure might appear fatal to Durkheim's sociological project. Before drawing this conclusion, however, one should note that even today attempts to define physiological illness run into problems similar to those that plague Durkheim's efforts in the

Rules. Just as it would be imprudent to conclude that our inability to give a precise definition of physiological illness undermines physiology or its application in medicine, so Durkheim's inability to articulate the necessary and sufficient conditions of social pathology is not itself a sign that his project is a failure. As both doctors and patients can confirm, imperfectly precise concepts can be useful and tolerably rigorous; as Aristotle noted, "precision is not to be sought for alike in all discussions, any more than in all the products of the crafts" (Aristotle 2009a: I:3).

After considering, and rejecting, a number of possible definitions of pathology in terms of dysfunction – for example, "whatever disturbs ... the organism's perfect adaptation to its environment" or "anything that diminishes [the organism's] chances for survival" (RSM: 87–8/50–1) – Durkheim concludes that a satisfactory definition of pathology cannot be given by referring only to vital needs as generally described as survival and adaptation. In some species, for example, it is normal (healthy) for the reproductive act to result in death or for pregnancy to increase the chances of the mother's death. Moreover, old age entails imperfect adaptation to the environment, but it is not therefore an illness. The first of these qualifications makes clear that an informative account of physiological pathology must be species-relative, referring to what is normal for the kind of organism in question (RSM: 88, 92/51, 56). The second implies that illness in biological organisms is relative to developmental phases: what counts as normal in the old may be pathological in the young. Durkheim draws conclusions for social pathology from both of these points, but the first is more important for our purposes. It is essentially the very point made above in relation to healthy self-maintenance in the social domain, namely, that it must be understood not as mere survival but as a society's maintaining itself as the kind of being it is. (Sociologists sometimes describe this as a society's maintaining its "structural continuity" [Radcliffe-Brown 1957: 25].) In other words, any account of what illness consists in, whether for organisms or societies, will be relative to the form of life characteristic of a given species (Canguilhem 1978: 75, 84, 99, 162).

As Durkheim recognizes, this poses a significant challenge for a theory of social pathology. Whereas the biologist can approach her objects with a fairly clear idea of their species-specific characteristics – of how, for example, a giraffe's form of life differs, and implies a distinct picture of functional normalcy, from that of a muskrat's – in the case of societies it is unclear what the species-specific form of life relevant to diagnosing social pathology might be. It is clearly Durkheim's intent to make his diagnoses of social pathology relative to various nonbiological forms of human social life – "species," or "types" of societies (RSM: 108/76) – but in carrying out this task, at least in the *Rules*, he takes two wrong turns, both attributable to his following too closely the analogy between biological and social forms of life.

The first wrong turn is Durkheim's assumption that human societies can be classified into species using a biological model – a "genealogical tree of social types" (RSM: 115/85) – and that such a classification is sufficiently informative to ground an account of the pathologies of social life. In classifying societies in this context Durkheim appeals to "social morphology" (RSM: 111/81), relying on the thought, borrowed from biology, that there is a close connection in organisms between form and function. (Later sociologists suggest, plausibly, that social form cannot be apprehended independently of functional analysis.[11]) His most prominent distinction in social type is that between segmental and organized societies, but the *Rules* expands this classification scheme to include other simple (and premodern) social forms: hordes, clans, polysegmental simple societies, and doubly composed polysegmental societies, which are further divided into cities and tribes (RSM: 113–15/82–5). To his credit, Durkheim recognizes that this scheme is still too broad to be of much help in determining functions and potentials for dysfunction in complex modern societies; an informative account of the form of a given society whose pathologies are to be diagnosed requires "an entire course of long and specific inquiry" (RSM: 114/84). At the same time, in this text Durkheim clings stubbornly to the idea that examining the form of society in a narrow sense – determining "the number of composite units [of a society] and the way in which they are combined" (RSM: 111/80) – is sufficient to discern the functions relevant to diagnosing social pathology in specific societies.

Durkheim has a reason for holding to this restricted principle of classification: he wants his classification of social species to serve also as a *genealogical* tree of social types, and he believes that the development of more complex out of less complex societies is best explained in terms of the combination and recombination of already existing units. This attempt to conjoin a classification scheme of social species with an account of their evolution is another respect in which he means to emulate biology, and it points to one of his reasons for regarding a universal history of human society as a central concern of sociology. This is one place Durkheim's

[11] In human societies "the social structure … can only be *observed* in its functioning" (Radcliffe-Brown 1935: 396; emphasis in original); see also Turner and Maryanski 1979: 41.

social theory falls prey to an illegitimate biologism (and it gives us a further reason for distancing a contemporary theory of social pathology from the project of universal history). It is revealing that when Durkheim turns from methodological discussions to the actual subject matter of sociology – in *Suicide*, for example – he uses a more flexible system for classifying societies than his discussion in *Rules* implies. At points he appeals to cultural factors, such as differences in religion – not merely to morphological traits – to distinguish societies, including the ways in which various pathology-vulnerable functions are carried out within them.[12]

The second wrong turn Durkheim takes in thinking about how diagnoses of social pathology must be relativized to a society's form of life is his taking what is *statistically* normal to stand in as a marker for what is *normatively* normal – healthy, as opposed to ill.[13] Recall Durkheim's recognition that a general definition of social pathology as impaired functioning, together with a general description of a society's vital needs, is not sufficiently informative to underpin specific diagnoses of social pathology because each type of society satisfies its vital needs in different ways, and therefore healthy functioning depends in each case on what is characteristic for that specific type. In responding to this theoretical problem Durkheim is guided by the thought that comparative sociology is able to tell us what a species' characteristic functioning looks like because it can survey a number of examples of that species and determine what is statistically normal, or "average," for it,[14] just as the biologist might learn that skin pigmentation in humans is the characteristic way that species protects itself from harmful radiation because most humans are born with pigmented skin (and because pigmentation has the effect of decreasing the harm caused by radiation) (RSM: 92/56).

In social theory, however, there are two difficulties with using the statistically normal to define, and to count as a sign of, what is healthy for a

[12] This is implicit in his treatment of egoistic suicide in Protestant and Catholic societies. He nowhere suggests that higher rates of suicide in the former constitute a pathology, implying that, for these purposes, Protestant societies should be compared only with one another. This seems to imply that he takes Protestant and Catholic societies to belong to different species. See also his assertion that because of cultural differences, suicide rates in "Greek Catholic" societies cannot be compared to those in Protestant or Roman Catholic societies (S: 105/149).

[13] See the interesting discussion of "abnormal" and "anomaly" in Canguilhem 1978: 73–9, 83, 88.

[14] This explains why the third chapter of the *Rules* discusses both questions distinguished above: "What is social illness?" and "What are its empirical marks?" In defining pathology here Durkheim refers to (species-relative) characteristic functioning, but in order to give content to this idea he appeals to statistical normalcy, which, in addition to further specifying what pathology is, provides a marker for detecting it.

particular species. First, sociology has at its disposal a much smaller number of specimens to survey than biology does, so that averages, if available at all, are significantly less informative in the former case. The second difficulty is that the theory of natural selection in biology gives us a reason for associating the statistically normal with healthy functioning: the features that come to characterize a species do so because they enable its members to reproduce. There is no reason, beyond an uncritical commitment to biologism, for supposing that a similar principle holds for human societies. Even if the specimens of societies were numerous enough to make average characteristics statistically significant, we would have no reason to suppose that those characteristics were for that reason functionally significant. More importantly, if a principle of selection were at work, the characteristic selected for – fitness to survive or to reproduce – would be too normatively thin to ground an account of social pathology of the sort Durkheim is looking for. It is worth noting here again, however, that Durkheim's reflections on his method do not capture how he actually proceeds when analyzing concrete social phenomena. His discussion of abnormally high rates of suicide, for example, is replete with diagnoses of social pathology, but nowhere are those diagnoses relative to a classification of social species. What is more, they do not depend on considerations of what is statistically normal. On the contrary, they occur alongside the claim that anomie is "currently in a chronic state" in the world of commerce and industry: "the state of crisis of anomie there is constant and, so to speak, normal" (S: 215–17/283–5); "in our modern societies anomie is … a regular and specific factor in suicides" (S: 219/288). In this important part of Durkheim's corpus, diagnoses of social pathology are made simply on the basis of detecting impairments in a society's vital functions – in this case, in social integration: social disruptions, an unrestrained free market, increased rates of divorce, and the decline of religious authority weaken the moral rules that coordinate social activities and generate solidarity, and this deregulation, or absence of rules (*dérèglement*) (S: 214/281), is, as we will see below, what the pathology of anomie consists in. (*Dérèglement* can mean simply "malfunction" or "disorder," but its root, *règle*, draws special attention to Durkheim's point that anomie is a dysfunction due to the absence or inefficacy of rules.)

Defining Anomie

In order to bring the more fruitful aspects of Durkheim's conception of social pathology into view, it is necessary to turn away from his meta-reflections on the topic and reconstruct his position out of his actual treatments of specific pathology-relevant phenomena, such as crime, suicide, economic crises, divorce, and so on. Although anomie is by far the most prominent of the social pathologies Durkheim considers, it is not his only example. Another pathology he mentions is a society's inability to adapt effectively to its changing environment (RSM: 87/50). Moreover, he describes certain social conditions as hyperactivity (DLS: 213/254), hypertrophy (DLS: 213/255), and even cancerous growth, in which sizeable numbers of individuals fulfill no social function or harm society and, so, "live at the expense of the organism" (DLS: 291/3). Another pathology is said to be found in those "rare and pathological" conditions in which, as a result of rapid social transformations, a society's mores are at odds with its laws because the latter have failed to catch up with changes in the former (DLS: 25–6, 34/29–30, 40). Since both mores and laws count as moral rules for Durkheim, the mismatch between them is closely related to (moral) anomie, even though this involves not the absence or inefficacy of rules but contradictions between different sets of them.

Because impaired functioning is central to Durkheim's conception of social pathology, one should expect all pathologies, anomic or not, to involve dysfunctionality of some kind. As Durkheim recognizes, it is this that allows for the possibility of other, nonanomic forms of social pathology (DLS: 292/344) as long as they represent impairments in the functions essential to a society's self-maintenance. Since some requirements of self-maintenance are biological, there is a class of possible social pathologies, not thematized by Durkheim, whose dysfunctional character consists in a (socially caused) failure to meet the material requirements of social reproduction. Other possibilities include pathologies whose dysfunctions are not biological, for example, forms of socialization that fail to produce new generations motivated to reproduce their social institutions (too few adults motivated to start their own families or to enter the labor force), or a society that organizes itself only with violence and discord among its parts, or with considerable waste of its members' energy. In addition, a society can become ossified, losing the plasticity it needs to adjust to the changing demands of its environment (the maladaptation mentioned above) or of its own growth (DLS: 309n25/361n).

Anomie is Durkheim's preferred example of social pathology, then, not because there are no other kinds but because he takes it to be the most pervasive pathology of contemporary organized societies (S: 215–17, 219/283–5, 288). Given Durkheim's general conception of illness, we can expect anomie to be a way in which a society systematically fails to satisfy its vital needs. Anomie clearly fits this description. It obtains when *rules* that can *coordinate* activities and *integrate* social subsystems (and individuals) are absent or inefficacious, such that social cohesion – but also, as we will see below, the well-being of social members – is thereby diminished. Thus, anomie is a pathology in which the parts of a society – whether its "organs" or its individuals – are inadequately coordinated because of the absence or inefficacy of rules that could otherwise do so. In general, then, an anomic society lacks a kind of internal cohesion, and in this respect the pathology it represents conforms to the classical conception of physiological illness as consisting in or caused by an imbalance among an organism's parts or functions (Canguilhem 1978: 42, 165).

As suggested, anomie is a broadly defined category for Durkheim, even if all its instances are marked by the absence or inefficacy of rules that coordinate or integrate a society's parts. That rule-related dysfunction is central to his conception of social pathology is demonstrated by the fact that even the nonanomic pathology he discusses in most detail – the "forced (*contrainte*) division of labor" (DLS: 310/367) – consists in a deficiency in the rules that coordinate and integrate society's parts. What distinguishes this pathology from anomie is that, rather than involving the absence or inefficacy of rules, it is a condition in which rules are excessively forceful – in this case, overly rigid or constraining. In other words, this form of social pathology – what I will call *hypernomie* – is so intimately related to anomie that the two can be regarded as belonging to a single family. Later, when discussing the instances of anomie most important to Durkheim – where the absent or inefficacious rules are moral in character – I return to hypernomie because, though it is not officially a form of anomie, understanding what it is and why it is pathological sheds light on his underlying conceptions of social pathology and of human society in general, including how the latter both is and is not like a biological organism.

Because Durkheim classifies a wide variety of phenomena under the rubric of anomie, it will be helpful to establish a classification scheme that organizes them into a manageable number of categories and distinguishes them from nonanomic pathologies. Doing so is more difficult than one would expect, and it is complicated by the confusing way Durkheim classifies the main social pathologies discussed in the *Division of Labor*. The first edition of that text divides the pathological forms of the division of labor into three categories, corresponding to the three chapters of Book III. Only the first of these categories, discussed in the first chapter, bears the label "anomic." The second chapter treats the form of pathology – the forced division of labor – that I call hypernomie. The much shorter third chapter treats a further, allegedly nonanomic but still pathological form of the division of labor (DLS: 323/383), which, because it is difficult to figure out exactly what phenomenon Durkheim means to distinguish, I ignore here.[15] Finally, in the preface to the *Division of Labor*'s second edition, Durkheim discusses the specifically moral form of the anomic division of labor that, shortly after the publication of the first edition, he came to regard as the most urgent social pathology of those associated with the division of labor.

Let us bracket hypernomie for now and focus on the so-called anomic pathologies. If we recall that anomie's defining characteristic is the absence or inefficacy of rules that coordinate the activities or integrate the parts of society, then two main classes of anomie can be distinguished, according to the type of coordinating or integrating rule at issue. The two types are, first, moral rules – "rules of conduct" (DLS: 303/359) – which obligate social members morally or legally and "go through," or are mediated by, individuals' consciousness of those rules and, second, nonmoral rules, which coordinate social activities independently of consciousness and without imposing obligations on those whose activities are coordinated. Rules of the first kind must be followed, or applied, by social members in order to be operative, whereas rules of the second kind can function, and usually function best, behind the backs of those whose activities they coordinate; in the latter case, one might speak of "regulation without rules" (in order to designate regularities that come about without rules being applied by conscious subjects). Because many cases of anomie consist precisely in the absence of rules, however, it is probably more accurate to distinguish the two classes according to whether the *remedy*, as Durkheim envisions it, lies in finding a moral or a nonmoral rule to solve the problem of *dérèglement* – or, in other words, whether healthy functioning requires conscious or unconscious regulation. Because the overactive rules characteristic of hypernomie are moral in nature, it is a close relative of anomie only in the latter's moral guise; in fact, hypernomie and moral forms of anomie are more similar to each other than either is to nonmoral anomie.

[15] The pathologies of the third chapter are compared to circulation problems in organisms, where "functions are distributed such that they do not offer sufficient material for the activity of individuals," resulting in faulty coordination among society's parts (DLS: 323/383). But why the pathologies of the first chapter involve the absence of rules while these do not is far from obvious. Perhaps the pathologies of the third chapter involve disturbances in a society's spontaneous consensus, where coordination is not rule-governed. I am indebted to Eva von Redecker for discussion of this point.

In summary, all forms of anomie consist in *dérèglement*, a condition in which rules coordinating social members' activities are absent or inefficacious – or, more precisely, a condition in which the lack of coordination can be remedied by the presence or increased effectiveness of such rules. Since the rules in question can be either moral or nonmoral (where the former includes laws), anomie, too, assumes both moral and nonmoral forms. Hypernomie, because it consists in excessively constraining moral rules – one sees here again that considerations of *freedom* are relevant to some diagnoses of social pathology – is not strictly a form of *dérèglement*, but because it is a dysfunction in the domain of social life governed by moral rules, it is closely related to moral anomie. Outside this classification scheme lies a host of further possible social pathologies, many mentioned above,[16] that are not best construed as instances of *dérèglement* or anomie. Finally, soon after the publication of the first edition of the *Division of Labor*, Durkheim focused almost exclusively on moral forms of *dérèglement*, that is, on moral anomie and related pathologies, such as hypernomie.

Among the specifically moral forms of anomie treated by Durkheim it is possible to distinguish three subclasses, according to the *social domain* within which the relevant rules are absent or inefficacious. Unsurprisingly, this threefold division tracks the three institutions Hegel took to be distinctive of modern society – the family, civil society, and the state – and so moral anomie for Durkheim comes in three principal guises: domestic, economic, and political. Of these, the economic varieties, at issue throughout the *Division of Labor* and *Suicide*, are the most prominent, reflecting his view that the "juridical and moral anomie in which economic life currently finds itself" (DLS: xxxi–xxxii/ii) is the most urgent social pathology of his day. Domestic forms of moral anomie are discussed in *Suicide*, primarily in its treatment of the relation between divorce and suicide;[17] they are clearly important to Durkheim, if also less so than economic anomie. Finally, although Durkheim has resources for conceiving of anomie within political life, only hints of this variety of the pathology are found in his major texts. This may reflect Durkheim's own personal interests, but it may also be a reaction to Comte's overemphasis of the state's role as the conscious, overarching agent of social

[16] Some of these – hyperactivity, hypertrophy, cancerous growth – are possibly construable as forms of *dérèglement* and hence as anomie, but others – maladaptation, ossification, reproductive dysfunctions – are not.

[17] See also Durkheim's brief description of the family at DLS: xliv–xlv/xviii–xix.

cohesion in his vision of the cure required for the social pathologies of nineteenth-century Europe.

Nonmoral Anomie

Although moral anomie is the most fruitful and innovative part of Durkheim's account of social pathology, it is worth first taking a look at his discussion of nonmoral anomie. As with moral anomie, the most prominent nonmoral forms of anomie come from economic life. The first that Durkheim points to are "industrial or commercial crises" in which production and consumption fail to adjust to one another, resulting in widespread bankruptcies and recurring cycles of boom and bust. Clearly, these phenomena "attest to the fact that at certain points in the organism certain social functions are not adjusted to one another" (DLS: 292/344). The problem here, according to Durkheim, is not some defect in the consciousness of society's members but in the spontaneous consensus of its economic activities (DLS: 297/351). As noted above, by this term Durkheim understands coordination established not by a central conscious agency but "organically," or bottom-up, through processes of mutual self-adjustment among society's quasi-independently operating "organs." Adam Smith's invisible hand is an obvious model of such coordination, but for Durkheim spontaneously achieved consensus is a more general phenomenon (of both social and organic life) and not restricted to productive activities mediated by the free market. For him the main condition making spontaneous consensus possible in social life is not some agency similar to a central nervous system but social density, close and constant contact among society's "organs" and subsystems (DLS: xxxiv/v). This view is fundamental to Durkheim's sociology, marking an important divergence from Comte, who, in Durkheim's view, exaggerates the need for a central conscious agency, in the guise of the state, to establish and regulate cooperation of all kinds. Durkheim, in contrast, attributes far less power to a universal consciousness in establishing a healthy balance among society's parts. For the most part, the conscious unity that is possible in society depends on there already being a substantial degree of spontaneously (unconsciously) achieved equilibrium among its parts:

> Underneath this general and superficial life [of conscious coordination] there is an intestinal life, a world of organs that, without being completely independent of the former, nevertheless functions without the former's intervention and, at least in the normal condition, even without consciousness of it .... What constitutes the unity of organized societies, as of all organisms, is the spontaneous *consensus* of its parts. This internal solidarity

> is not only just as indispensable as the regulative activity of the superior central [organs of consciousness]; rather, it is itself the condition of that activity. ... The brain does not create the organism's unity but expresses and crowns it. (DLS: 297/351)

In other words, no amount of conscious direction from above can establish a healthy coordination of society's parts if it is not already largely present as a result of spontaneous self-adjustment from below. Even in Durkheim's treatment of moral forms of anomie, the remedy for which depends on some type of conscious regulation, the consciousness at issue is rarely universal – requiring an overview of society's processes as a whole – as it is for Comte. (At the same time, the differences between the two thinkers should not be overstated. As the brief discussion of Durkheim's socialism below indicates, the state has a large role to play in creating the legal and economic conditions required if anomie is to be eliminated or reduced. Moreover, as the passage above shows, he does not deny the need for some measure of general oversight and guidance of social life on the part of the state, but, first, this is much less extensive than the central planning envisioned by Comte, and, second, its efficacy depends on an already existing spontaneous consensus in society's "inferior organs."[18])

Durkheim's treatment of economic crises is mostly unremarkable – it is far less sophisticated than Marx's – and he has little to say about what a remedy for these pathological coordination failures would look like beyond the general thought that it requires society's economic "organs to be in sufficient, and sufficiently prolonged, contact" with one another (DLS: 304/360).[19] This prescription is but a specific instance of Durkheim's general view that substantial spontaneous consensus among a society's parts can come about merely through contact and local communication among them without passing through a regulative agency of the whole (DLS: 304–5/360–1).

What is most interesting about Durkheim's treatment of economic crises is its brevity. After making the points mentioned above in a single paragraph, he moves immediately to a different topic that, perhaps because it is more closely related to moral anomie,[20] is clearly of greater interest

[18] For Durkheim on the state, see Karsenti 2006b: 25–9.

[19] It is unclear how this proposed solution is compatible with increasingly international economic markets, of which Durkheim was aware (DLS: 305/361–2).

[20] In the section of the *Division of Labor* devoted to nonmoral anomie, Durkheim cannot avoid referring to the moral elements of class conflict even though, as he admits, those considerations belong in the following section of the text (DLS: 293/346).

to him, namely, class conflict, or "the antagonism between labor and capital" (DLS: 292/345). Because he is still discussing nonmoral anomie, Durkheim restricts himself here to treating class conflict as the result of increased specialization that places the activities of workers and capitalists at such a great physical and social distance that the ongoing contact and communication needed for spontaneous consensus cannot occur (DLS: 304–6/360–2). Strangely, even though class conflict is the social pathology at issue, economic class in Marx's sense does not figure fundamentally in Durkheim's explanation of it. Insofar as class conflict is regarded as a nonmoral form of anomie, its ultimate source is taken to be insufficient social density, not an objective clash among economic interests built into the class structure of capitalism. Viewed from this angle, the problem that gives rise to conflicts between workers and capitalists is no different in nature from the factors that might cause disorders of production among members of the same class (laborers) who are not working in sufficient proximity to their immediate cooperators. As if to confirm this point, Durkheim moves immediately, and without noting the change in topic, from class conflict, allegedly caused by laborers not working in sufficient proximity to capitalists (however this odd idea is to be understood), to a problem that arises because in contemporary industrial production the distance among laborers themselves is too great.

To his credit, this is not Durkheim's final word on class conflict – the topic recurs in his more illuminating discussion of the (moral) pathology of hypernomie – but neither here nor there does he adequately appreciate the intrinsically opposed nature of class interests in capitalism articulated so compellingly by Marx. This deficiency in Durkheim's account of the consciousness-independent sources of class conflict is related to his ambiguous stance on whether – despite his avowal of socialism – his envisaged solutions to anomie are compatible with a capitalist order. His criticism of Comte's vision of the state as the principal coordinating agency of society implies a rejection of centralized economic planning, but he is equally suspicious of an unregulated market and the crises that accompany it (DLS: 303/358). Moreover, Durkheim fails to take a clear stance on whether private ownership of the means of production, especially when most individuals own none, is compatible with social health. His view appears to be that wage labor should be more thoroughly regulated by law but not eliminated (DLS: 303/359).

Durkheim is unable to complete his discussion of nonmoral anomie without, at the end, introducing an element of consciousness into his picture of how even nonmoral anomie is to be overcome. The consciousness at issue here does not involve following moral rules – which is why these forms of anomie are nonmoral – but a more general type of awareness that nevertheless possesses a certain moral flavor. Durkheim makes this point in addressing the by then familiar criticism that the highly specialized division of labor of modernity turns the laborer into a mere cog in a machine who, because "he does not know where the operations demanded of him are leading and does not attach a goal to them," is able "neither to take interest in [his labor] nor to understand it" (DLS: 306/363). Although Durkheim does not use the term, he is clearly addressing what Marx diagnoses as subjective alienation (although Marx locates its ultimate source in capitalist relations of production). Durkheim implicitly recognizes that socially generated subjective alienation constitutes a social pathology, and he, too, argues that it is not a necessary consequence of a highly specialized division of labor. The more interesting question is what leads him to regard such a condition as pathological, especially since, if function is understood narrowly, production that is alienating for laborers might be considered, from the perspective of the social "organism," not dysfunctional but efficient. Durkheim's answer to this question explicitly invokes a moral standard, namely, the "debasement of human nature" (DLS: 307/363) and the "disastrous effect [it has] on human consciousness" (DLS: 307/364): "If the end of morality is the full development (*perfectionnement*) of individuals, it cannot permit such a ruining of the individual, and if its end is society, it cannot allow the very source of social life to run dry; for this ill threatens not only economic but all social functions, however elevated" (DLS: 307/363).

Three points follow from this simple statement of the moral impermissibility of a system of production that generates widespread subjective alienation. The first is an expansion of Durkheim's conception of the moral so as to include not merely norms of freedom and justice, as indicated above, but also a perfectionist ideal that champions the full development of human talents and capacities. Second, morality is said to aim not only at collective goods but also at individual development, which in this case translates into a version of the liberal moral claim that collective ends may not be achieved at the expense of the basic interests of the individuals whose activities achieve those ends. Third, and most important, healthy functioning in human societies is to be understood not as mere efficiency in the economic sense but broadly enough to include the achievement of more elevated, moral ends. It is important, however, that Durkheim's claims are not *deontologically* moral. It is not as if there were for him an independently available conception of the moral that placed external permissibility constraints on social arrangements that might appear desirable from a narrowly functional perspective. His view, rather, is that in the case of human life – which is, of course, primarily social life – the (humanly) functional and the moral are so intertwined, even conceptually, that what is immoral "does not work" (it saps human energy) and what is functional includes what allows humans to realize their higher potential. This is a Hegelian, not a Kantian, view of morality that, as we will see in Chapter 12, tracks the core of Hegel's concept of spirit: in the case of human social existence there is no end of biological life that is not at the same time an ethical end and no ethical end that is wholly separate from the ends of living creatures.

Finally, it is worth noting Durkheim's proposed remedy for subjectively alienated labor for what it says about the relation between consciousness and even nonmoral forms of anomie. To his credit, Durkheim rejects the overly sunny view of some liberal philosophers that the solution for subjective alienation lies in educating workers so that, like their idle employers, they, too, can enjoy the higher pleasures of literature and art, which presumably would compensate them for long days spent carrying out routine, meaningless tasks. For Durkheim this solution remains too external to the activity that constitutes the problem. Picking up on the thought that the laborer is unable to understand or take interest in his labor because "he does not know where the operations demanded of him are leading and does not attach a goal to them," Durkheim concludes that the cure for this species of anomie requires that laborers acquire a certain *understanding* of the point of what they are doing, which alone can dispel the sense of meaninglessness their specialized labor takes on for them. Importantly, Durkheim recognizes that the requisite understanding must be more than a local awareness of how one's own labor technically complements the labor of one's immediate neighbors in the production process. (This much awareness seems to be required for the mutual adjustment among specialized productive activities that Durkheim calls spontaneous consensus. A kind of consciousness is involved when workers spontaneously adjust their labor to that of their co-workers, but it is not consciousness of a rule, nor does it have moral significance.) Durkheim characterizes the consciousness necessary for eliminating this species of anomie in terms of the workers sensing "more or less distinctly" that "they are collaborating on a single project" that is in "the service of some end" or "goal outside themselves" (DLS: 307–8/364–5). Although he has little more to say about his proposed remedy, two of its requirements are clear: if specialized labor is to be experienced as meaningful, those who do it must see it as an integral part of a larger project they carry out with others, and they must see the point of that project and affirm it as good. Unlike Marx, Durkheim does not further investigate the objective conditions to which the labor process must conform if those who carry it out are to see it as a collective project in service of the good. One could fashion a partial answer to this question by appealing to his later discussion of the conditions, including equality of various kinds, under which contracts must take place if they are to carry moral authority, but it seems doubtful that these formal conditions of justice, applied to wage contracts, would suffice to guarantee the affirmability of labor or its products.

Moral Anomie and Hypernomie

The ethical character of Durkheim's remedies for social pathology becomes even clearer when one turns to moral anomie and its close relative, hypernomie. Durkheim's principal example of the latter, the "forced division of labor," is a class or caste system in which individuals' locations are determined by custom or law rather than matched, perhaps through their own choice, to their particular talents and dispositions. The most interesting aspect of Durkheim's discussion of hypernomie is its account of why a forced division of labor is pathological, the basic ideas of which apply to his vision of social pathologies more generally. The most important reason is evident in his statement that the pathology represented by the forced division of labor has no analogue in biological organisms. In other words, it is a social pathology whose pathological character depends on the fact that the "organism" at issue is composed of human individuals. In contrast to the latter, "a cell or organ never seeks to take on a role other than the one allotted to it" (DLS: 310/367). Again, the pathological character of a division of labor in which individuals are assigned to places on the basis of heredity or their class at birth does not consist in the inefficient coordination of society's parts. It resides rather in the discontent and lack of freedom experienced by society's members. The division of labor that determines their lives is constrained or coercive – brought about by "force alone, more or less violent and more or less direct" (DLS: 311/369) – and is therefore not an arrangement those subject to it can affirm: "If the division of labor is to produce solidarity, it is not sufficient that each has his task; it is necessary, in addition, that this task be fitting for each (*lui convienne*)" (DLS: 311/368). The forced division of labor is pathological only because human individuals differ fundamentally from the cells of organisms in possessing consciousness and will, and hence capacities for freedom and contentment. In organized societies these capacities translate into moral demands on the part of individuals – claims to have a right – to be free (uncoerced) and subjectively satisfied in their social life.

That there is an ethical component to pathology in the social realm becomes even clearer when Durkheim discusses the conditions to be met if social members are to be content with their social positions: they must perceive the conditions under which they have ended up in those positions as just (DLS: lv, 314/xxxiv, 372), where this includes "absolute equality in the external conditions of [economic] competition" (DLS: 313/371) – or what Rawls calls "fair equality of opportunity" (Rawls 1999: 63).[21] This in turn requires abolishing the inheritance of wealth (DLS: 314/371) and establishing conditions under which the contracts regulating coordination are not merely formally free but substantively just, that is, where "the services exchanged have an equivalent social value" (DLS: 317/376) rather than involving exploitation (DLS: 320/379) in precisely the Marxian sense. Another condition of an unconstrained division of labor, perhaps also appropriated from Marx, is that the latter allow for "the free deployment of the social force that each carries within himself" (DLS: 313/370). These conditions are not unrelated to functional considerations in the narrow sense – a constrained division of labor tends to engender class conflicts that threaten stability and efficiency – but Durkheim's interest in these ethical goods is not reducible to the value of stability or efficiency. In organized societies, stability requires freedom and justice; organic solidarity can develop only under ethical conditions that Durkheim later calls socialism,[22] and the fact that freedom and justice produce stability and cohesion does not detract from – indeed, it is partly constitutive of – their ethical value. Health in human societies is always "moral health" (DLS: xxvii/xxxix).

Another way of putting this point is to say that the fact that social "organisms" are made up of thinking and acting individuals implies that a healthy

[21] Durkheim believes that these terms must also in fact be just if workers' beliefs that they are so are to be nonpathological, as well as that current class inequalities, including the propertied class's unlimited right to pass its wealth on to succeeding generations, must be significantly reduced if this condition is to be met. Moreover, the conditions of just economic cooperation must be not only based in shared moral principles but also backed up by law.

[22] Where socialism is understood as "a better, more intelligent organization of collective life whose aim and result would be to integrate individuals within social frameworks or communities invested with moral authority and capable of performing an educational function" (Aron 1970: 65).

organized society must also be good – that it must be just and substantively equal, but also able to accommodate the freedom and happiness of each of its members. In the case of a human society, the well-functioning of the whole requires that its individual members be able to find their own satisfaction and freedom by participating in social life. In fact, the connection for Durkheim between social pathology and the well-being of social members is tighter than even this: there is no pathology of social life – nothing that qualifies as a social dysfunction – that does not in some way impinge on the well-being of at least some of its members. This means that the well-being of individuals is at least partially constitutive of what is good or healthy for the social "organism" (DLS: xliii/xvii).

The main object of Durkheim's diagnoses in the *Division of Labor* is moral anomie in the economic sphere. The first variety mentioned is class conflict (DLS: xxxii, 292/ii, 345). As we have seen, Durkheim takes class conflict also to have sources other than the absence or inefficacy of moral rules, above all a failure of classes to adjust their respective activities spontaneously, due mostly to the largeness of modern industrial firms. This means that Durkheim regards class conflict as a complex pathology, having both moral and nonmoral causes that call, correspondingly, for moral and nonmoral remedies. With regard to the former, Durkheim has little to say in the first edition of the *Division of Labor*, beyond referring to the remedies he proposes later in the case of hypernomie: "We shall see that this tension in social relations is due in part to the fact that the working classes do not really want the condition they are given but are simply constrained and forced to accept it, not having the means to achieve another" (DLS: 293/346). It may seem strange that Durkheim conceives of this problem as a deficiency in the *rules* governing economic life, but this confusion dissipates when one understands that his proposed moral solution to class conflict consists in establishing laws that create the conditions of justice, material and legal equality, and fair equality of opportunity required if workers are to accept freely – regard as the just outcome of fair competition – the economically inferior positions they occupy.

Only in the Second Preface to the *Division of Labor* does it become clear what species of moral anomie Durkheim is most interested in: the absence of an articulated and legally established professional ethos (*morale professionnelle*), specific to each profession or branch of industry, that regulates "relations between office worker and employer, factory worker and boss, and among competing industries" (DLS: xxxii/ii). Among the issues such *morales professionnelles* would regulate are "the loyalty and commitment that employees of all sorts owe to their employers; the moderation with which the latter should exercise their superior economic power; a condemnation of competition that is too openly unfair; the overly glaring exploitation of consumers" (DLS: xxxii/ii) – in effect, all issues that give rise to the "constantly renewing conflicts and disorders," a veritable (and "morbid") anarchy, that characterizes the economic world of early twentieth-century industrial Europe (DLS: xxxii/iii). As noted above, Durkheim thinks of the rules that would restore order to economic life as laws and moral principles that limit the egoism of economic agents and hold in check the unconstrained war of all against all that he thinks would otherwise result from it.

It is an important part of Durkheim's remedy for moral anomie in the economic domain that even the professional mores that prevent conflicts and create cohesion – as opposed to the rules codified in law – be given an institutional form, which, echoing Hegel, he conceives of as a modern version of the legally recognized but state-independent corporations of ancient and medieval times. These organizations would enable economic agents not only to formulate and defend their shared interests but also to enforce – formally, if not necessarily legally – the *morales professionnelles* that contribute to social order and cohesion. By the time he prepares the second edition of the *Division of Labor*, Durkheim realizes that the corporations representing workers or factory owners in a particular branch of industry, *les syndicats*, can also play a significant role in mediating class conflict precisely because such bodies facilitate regular contact and communication between workers and employers.

Because of the emphasis Durkheim places on rules, constraint, the curbing of egoism, and social sanctions when discussing morality, it is important to remember that self-restraining discipline is for him only one side of moral life; as he repeatedly reminds us, social life has its attractions at the same time that it is constraining (DLS: xliii/xvii). With respect to these attractions, recall Durkheim's claim that morality is "a source of life … from which emanates a warmth that inflames and re-animates our hearts, opening them to a sympathy [for others] that melts our egoism" (DLS: lii/xxx). Moralized (or socialized) individuals (DLS: xliii/xvii) – those who have internalized the moral rules that bind them in their activity to others – typically have a subjective attachment to their fellow social members that enables them to submit willingly to the rules that demand the curbing of their egoism. This means that Durkheim regards the corporations in which he sees a remedy for anomie not as sites of austere moral discipline but as subjectivity-forming institutions within which, by engaging in a common project with others, members come to care about their corporations' collective interests and to take these into account in their own actions. Whether consciously or not, Durkheim avails himself of terminology from Hegel and Rousseau when characterizing this aspect of the dispositions of socialized humans: "This attachment to something beyond the individual, this subordination of particular interests to the general interest, is the very source of all moral activity" (DLS: xliii/xvii). This process of moralization not only produces individuals who are reliable followers of the rules required for coordinated activity and social cohesion but also "sustains in the hearts of [corporation members] a livelier sentiment of their common solidarity" (DLS: xxxix/xii), one that animates social life and "melts" rather than suppresses egoistic love of self. Consensual and nonrepressive social life is possible for humans only because "the individual [is also capable of] finding in it a source of joys" (DLS: xliii/xvii).

Holistic and Nonholistic Aspects of Social Pathology

I return now to the more general question of what makes the conditions that Durkheim classifies as anomie pathological. We know that anomie involves a disturbance in a society's vital functions, a failure to satisfy vital needs of social life, and that Durkheim regards the society that exhibits such a dysfunction as ill; at issue in such a dysfunction is "the general health of the social body" (DLS: lv/xxxiv). We also know that he considers illness to be both a normative and a prescriptive category, such that to call a society ill is to imply both that something within it is not as it should be and that something ought to be done to relieve the social body of its illness.[23] Why, though, is the health of the social body something we, its human members, ought to care about? Why should the diagnosis of a social illness move us to remedy it? To these questions Durkheim responds: "If anomie is an ill, it is above all because *society suffers* from it, not being able to do without *cohesion* and *regularity* in order to *live*. Thus, moral or juridical regulation expresses essentially *social needs* that only a society can know" (DLS: xxxv/vi; emphases added). Given what we have seen of Durkheim's position, this answer is unsurprising. Still, it is worth pausing to formulate it more precisely: anomie is a social illness that we are enjoined to remedy because cohesion, and therefore the rules that produce it, are essential to social life. To say that cohesion is essential to social life is to say that society

[23] "For societies as well as for individuals, health is good and desirable; illness, on the contrary is something bad that ought to be avoided" (RSM: 86/49); see also DLS: xxvii/xxxviii–xxxix.

suffers when it is lacking, where "suffer" must be assumed in the dual sense it has in both French and English: one suffers (consciously) when one is in pain, but it is also possible to suffer from (to be afflicted with) an illness without knowing that anything is wrong. To say that a society suffers from anomie is to say that it is in some way harmed by that condition, which may or may not be present to consciousness.

It is important not to be misled by Durkheim's apparently hypostasizing talk of the social body as something that has its own health, suffering, needs, and even consciousness. It is important to him to claim that the social body is susceptible to its own *sui generis* illnesses, but this must be understood in a way compatible with his ontological position that a social body is made up of no material substance or locus of consciousness beyond the material substance or consciousness of its human members (and the material substance of the physical objects integrated into social life). This is the view Durkheim means to embrace when insisting that socially *sui generis* phenomena, including social consciousness, imply "nothing substantial" and that his theory is not guilty of "ontologizing" society (e.g., RSM: 34/xi). Rather, the *sui generis* phenomena that are sociology's subject matter reside in the structural features of an *ensemble*, a whole that incorporates and puts into relation human bodies, the consciousness of human individuals, and ordinary physical objects.

There are two reasons it is coherent, and necessary, for the social theorist to treat the phenomena of social life as *sui generis* in Durkheim's sense. The first is that, as an *ensemble*, a society possesses properties, including capacities, that cannot be instantiated severally in its parts (precisely because those properties consist in relations among its parts). For the same reason that a four-legged animal can run without any of its parts severally being capable of running, so a society can be cohesive in a sense that cannot hold of any of its individual parts. Social cohesion is a *sui generis* property of societies precisely because, and in the sense that, it resides in a certain structure among their parts. Like biological organisms, societies can possess properties and capacities that none of their parts as such possess without any new material substance or locus of consciousness entering the world. By the same token, a society can be ill without any of its members being afflicted with that illness, and for the same reason: a social pathology is a structural property of an *ensemble*, consisting in the social body's activities being poorly coordinated or its members being insufficiently integrated.

The second, more important reason for treating the phenomena of social life as *sui generis* concerns not the ascription of properties to societies but the explanation of social phenomena (or social facts). As we have seen, Durkheim's sociology is motivated at its core by the conviction that the principal phenomena of social life can be explained only by invoking some notion of "collective forces" (RSM: 46/xxiii) that affect the behaviors and mental states of individuals and cannot be understood atomistically, as mere aggregates of the behavior or states of independently constituted individuals. We encountered one version of this view in Durkheim's mysterious doctrine of causally efficacious "social currents," but a similar explanatory holism is at work in his entire sociology, as can be seen in his most general definition of a social fact as "a way of acting ... capable of exerting an external constraint on the individual" (RSM: 59/14). The thought that social phenomena can be understood only by referring to ways in which the social body *constrains* individuals – where the whole is somehow explanatorily prior to its parts – is central to Durkheim's sociology. As we saw in his treatment of the moral facts of social life, however, this constraint of the whole on its members can be construed not causally but normatively, that is, as producing effects by means of rules that individuals take to impose obligations on them. Although Durkheim never worked out the precise details of this claim, he is committed to the view that the rule-governed activity in which moral facts consist cannot be explained atomistically. I return to this issue in Chapter 11 in discussing Hegel's doctrine of objective spirit, but the main idea can be stated in a preliminary way here: the social whole is composed exclusively of human individuals related to one another through commonly recognized rules that define their social institutions, govern their interactions, and provide the context within which their social activity first becomes possible. When individuals are joined together in normatively governed structures that constitute the contexts of those individuals' actions, no new substance or consciousness appears in the world, but such structures, because they make ways of acting in the world possible that otherwise would not be, count as a reality that possesses its own *sui generis* existence, namely, as *social* facts (or, in John Searle's terms, "social reality").

The first of these reasons for treating social phenomena as *sui generis* – because properties can be ascribed to a whole that do not pertain to its parts severally – implies that a society can be ill without any of its individual members being so. Indeed, strictly speaking, anomie is a condition that only societies, but not individuals, can be in. Individual social members can suffer the effects of anomie – anomic suicide is an example of this – but the bearer of the social pathology is society, not the affected individuals as such. This point is important, among other reasons, if we want to avoid pathologizing the affected members of sick societies; individuals who commit anomic suicide are not themselves ill. Nor are they morally blameworthy, even if the social rules that have failed to function properly are moral in character. It is more accurate for the social theorist to regard such individuals as the unfortunate casualties of social disorder.

This picture is complicated by the fact that – although Durkheim does not say so explicitly – his modest *ontological holism* is joined with an equally sober version of *normative individualism*, according to which society has no interests or good not bound up with the interests or good of its members (though these include interests and goods that members can have and achieve only in society). Because of this there can be, for conceptual reasons, no ill suffered by a social body without some of its members being harmed in some way as well. In other words, social dysfunction is bad – it counts as an illness that calls on us to remedy it – only because it is bad for some of the individuals who make up the ill society. When a society is ill, some of its members necessarily suffer harm, even though this harm need not itself be an illness. Durkheim is at pains to point out that whenever a society suffers a pathological deficiency of cohesion, its members are vulnerable to a variety of ills: anarchy is "painful" for the individuals who experience it (DLS: xliii/xvii), and the same is true of the state of war that moral rules ought to prevent. Anomie detracts from individuals' ability to enjoy the inherent goods of community, and abnormal rates of suicide or crime – economic crises, too – count as social pathologies only because they affect the well-being of social members; the latter is a necessary condition of their society's being in a pathological state.

One aspect of the effects moral anomie can have on individuals deserves special attention, in part because it is central to the close relation between social health and morality in Durkheim's thought and in part because it brings out an important similarity between his conception of social pathology and more familiar versions of the idea articulated by Plato, Rousseau, and Marx. The view in question is expressed in Durkheim's claim that one effect an anomic society has on its members is that "their [own] activity is *déréglée*" – dysfunctional because of the absence of rules – "and they suffer from it" (S: 219/288). One feature of anomie, then, is that the moral rules that normally function to curb egoism and promote action oriented to a collective good lose their grip on individuals. The result is not only that discipline falters and collective projects become more difficult but also that the individuals whose activity is *déréglée* suffer from this condition, and suffer subjectively since their own well-being is affected by it. The phenomenon Durkheim has in mind has its source in the world of commerce and industry, especially in the exceedingly rapid transformations that world has recently undergone (S: 215/283). (It is typical of Durkheim that he fails to understand these rapid transformations and the *dérèglement* accompanying them as consequences of specifically capitalist commerce and industry.[24]) The rapid pace of change in this domain – the liberation of European economies from the dictates of Church and state – has severely weakened the moral and political forces that in earlier eras held economic activity in check: "Industry, instead of continuing to be regarded as a means in relation to an end beyond it, has become the highest end of individuals and societies" (S: 216/284).

The crucial point is that these developments have unfettered, or inflamed, not only economic growth but also the desires of those who participate in economic life. When freed from the constraint of moral ideals, desires for goods – and ultimately for money – tend to become endless and for that reason unsatisfiable. (As we saw in Chapter 3, Marx offers a more sophisticated account of this phenomenon in analyzing the nature of capital.) Following Plato and Rousseau, Durkheim signals the pathological character of a society that generates such desires by characterizing them as feverish (S: 217/285), and the passages in which he describes their deleterious effects on humans are nearly indistinguishable from Rousseau's descriptions of the same phenomenon:

> This is the source of the turmoil that reigns in [commerce and industry] and that has spread from there to other parts of society. A state of crisis and anomie is constant there and, so to speak, normal. From the top to the bottom of the social ladder desires have been stirred up that cannot be definitively attached to specific ends. Nothing is able to calm them because the goal they seek is infinitely beyond anything they can achieve .... We thirst after new things, unknown pleasures, unnamed sensations, which, however, lose their flavor as soon as they are known .... When this fever recedes, we perceive how sterile this tumult was and how all these new sensations, accumulated without end, were not able to establish a solid stock of happiness .... Fatigue by itself suffices to produce disenchantment, for it is difficult not to perceive finally the uselessness of a pursuit that has no end. (S: 216–17/284–5)

In addition, "this evil of the infinite, which anomie brings with it everywhere" is capable of infecting the domestic realm as well (S: 234/304), producing unlimited and unsatisfiable sexual desires that the rules of marriage, when functioning well, allow human beings – or men, as Durkheim

[24] However, he recognizes the role the market's geographical expansion has played in making desires infinite (S: 216/284).

sees it – to avoid. The importance of this aspect of Durkheim's account of the harmful effects of anomie – his conception of "healthy discipline" (S: 217/285) – is that it reveals how the moral rules central to his conception of a well-functioning society perform not only social functions but also serve the interests of individual social members, in this case by making an achievable happiness compatible with the social membership they need in order to satisfy their biological (and spiritual) needs.

Before proceeding to Durkheim's understanding of his project as a science of morality, I summarize several of his claims that deserve to be retained in a contemporary account of what human societies are and how they are vulnerable to falling ill:

- Societies differ from biological organisms in that the individual members of the former are endowed with consciousness. This enables some of society's coordination of specialized functions to be achieved consciously, and it underlies the normatively relevant aspiration of individuals to grasp and affirm their contributions to the tasks of the whole or, more typically, to its subsystems.
- The essential functions of human societies include not only material reproduction but also achieving ethical ends, including (in the modern West) justice and freedom (in several senses). Social health is moral health.
- Integration and solidarity, achieved in part by social members' acknowledgment of the authority of moral rules, is a vital need of human society, although in organized societies solidarity depends more on individuals' integration into society's subsystems than it does in segmental societies, where individuals are attached directly to the whole via a strong, normative collective consciousness.
- Solidarity in organized societies relies on coordination in which parts or subgroups adjust themselves "spontaneously" to the activities of others without that coordination being governed by moral (or other) rules.

CHAPTER 10

Durkheim: A Science of Morality

One of Durkheim's most intriguing claims is that social health and moral health are the same condition and that sociology is therefore a moral science. Sociology is a moral science not in the loose sense in which philosophy and history are sometimes said to be moral disciplines; rather, sociology is a science *of* morality (DLS: xxv/xxxvii), where morality (or ethics) has to do with right and wrong (and, ultimately, the good) in human conduct and social organization.[1] Sociology is a moral science because morality is central to social reality, and not only in the sense that moral rules are the necessary glue that holds societies together and allows them to function but also in the more robust sense that the characteristics of a well-functioning society are the same as those of one that is just and morally good. Durkheim puts the latter point even more strongly: "There exists a state of moral health that [social] science alone is competent to determine" such that knowing how "we ought to orient our conduct," far from being the province of a priori moral philosophy, requires an empirically grounded science of society (DLS: xxvii/xxxix). In this chapter I reconstruct this aspect of Durkheim's position.

Durkheim's science of morality has three related aspirations: first, to orient our conduct, or to tell us what we ought to do (via an account of a morally healthy state of society); second, to illuminate the connection between social functioning and morality (to explain why moral and social health are the same condition); and, third, to "seek out the laws that explain [moral facts]" (DLS: xxv/xxxvii) – or to explain why the specific moral ideals that animate a given society do so and how they vary with changing social conditions. The second and third of these aspirations make

[1] When Durkheim discusses the relation between morality and social functioning, he often associates the former with justice rather than the good. Careful reading reveals, however, that his conception of morality involves elements of the good as well, such as self-transparency and the absence of alienation. For more on Durkheim's science of morality, see Karsenti 2006a: 69–91.

Durkheim's moral science similar to Marx's historical materialism, insofar as it, too, draws a tight connection between morality and social functioning (such functioning need not, however, be healthy) and claims to explain why, for example, "bourgeois" morality holds sway in capitalism. The most important respect in which Durkheim's project differs from Marx's is that Durkheim does not regard his explanation of why different moral rules hold in different kinds of societies as undermining the legitimacy of such rules or as demonstrating their ideological character (in the negative sense of that term). His science of morality leaves the moral authority of the rules it explains largely intact. While Durkheim's accounts of specific social pathologies imply a critique of certain social rules, they do not discredit the fundamental ethical norms at work in the societies he studies.

After reconstructing Durkheim's science of morality, I argue that despite its virtues it falls short of fully elucidating the connection between social functioning and morality and that, relatedly, it also fails to provide an adequate moral grounding of the conception of social health it employs. At the end of the chapter I suggest that some of the deficiencies of Durkheim's science of morality can be addressed by adopting aspects of Hegel's account of that same connection implicit in his conception of rational (or healthy) social institutions as guided by and systematically productive of the moral good. I argue that Hegel's conception of the good that governs human social life – a "spiritual" unity of ideals of freedom and the goods of life – offers more resources than Durkheim has for grounding the moral aspects of an account of social health. This difference between the two positions can be traced back to their respective understandings of social transformation, in which context Hegel, in contrast to Durkheim, conceives of the moral ideals of (later) societies as rational responses to problems or crises, both functional and moral, encountered by earlier societies, where the idea of a rational response plays a key role in justifying the (new) moral ideals under consideration. This complex idea, however, can be articulated only after examining the details of Durkheim's moral science.

Morality and Healthy Social Functioning

I begin with Durkheim's central claim that social health is also moral health. As we have seen, there are many respects in which for Durkheim the moral is bound up with healthy social functioning. This can be seen in the variety of ways in which the hallmark of social health, *solidarity*, converges in an organized society with moral ends:

- It reduces social conflict;
- it consists in cooperation coordinated by normative rules taken to have moral authority by social members themselves, thereby realizing a kind of freedom in which individuals determine their activity in accordance with rules they understand and affirm;
- it persists only if individuals recognize their social activity as satisfying their own interests, including a felicitous development of their capacities;
- it requires that social members determine the basic contours of their lives (at least the economic roles they play), the absence of which is hypernomie, a pathological condition in which social rules are overly constraining and the individuals they govern unfree; and
- it depends on a just arrangement of institutions, including enough economic equality to allow for equality of opportunity in competition for society's advantaged positions.

It is not difficult to see that moral considerations figure centrally in Durkheim's conception of social health. It is clearly his intention to develop a science of society that shows that the functional and the moral are not only compatible but also mutually dependent – where there is no well-functioning of society without morality, and no realization of morality in the absence of social institutions that reproduce society and make moral ends the standard outcome of social functioning. In the domain of social life the ethically repugnant is existentially incompatible with the humanly functional. (Recall Durkheim's claim that "morality is … the daily bread without which societies cannot live" [DLS: 13/14].) How, though, is the connection between the socially functional and the moral to be understood?

It must be emphasized that morality for Durkheim comprises something more than whatever rules a society needs to ensure its reproduction, a view that would *reduce* morality to (mere) functionality rather than unite the two.[2] More precisely, morality's content cannot be derived from some quasi-biological or narrowly economic conception of what is required for the social "organism" to reproduce itself efficiently and smoothly. Contrary to the sociological functionalists referred to in Chapter 1, morality cannot be reduced to whatever best contributes to social stability, homeostasis, survival, and adaptation to the environment, defined in nonmoral terms.

[2] "We do not … fall back upon the formula that defines morality as a function of society's interest" (DLS: 332/394).

This is Durkheim's point when distinguishing his science of morality from the views of other self-described scientists of morality – he is thinking especially of Spencer – "who deduce their doctrines … from propositions borrowed from one or more positive sciences such as biology, psychology, or sociology. … We do not want to extract morality from science but to create a science of morality, which is something quite different" (DLS: xxv/xxxvii).[3] Durkheim's view, then, does not begin with a nonmoral conception of social functioning from which it then derives morality's content, nor does it regard morality as of merely instrumental value in relation to social functioning. Instead, his science of morality – a version of what might be called "moral functionalism" – aims to *integrate* the moral and the functional without reducing either to the other. Explaining what this means is the main task of this chapter.

If Durkheim does not want to reduce the moral to the socially functional, he also does not seek a reduction in the opposite direction. That is, he does not begin with a settled conception of justice or the moral good and then count as socially functional whatever social relations best promote those ideals. Rousseau might be taken to adopt such a position, if one sees as his conception of healthy social life as determined by an a priori principle of justice applied to the circumstances of human nature. If Durkheim aspires to follow neither Spencer nor Rousseau, what path does his science of morality take? I suggest below that the *Wealth of Nations* is a closer relative to Durkheim's science of morality than the positions of Spencer or Rousseau, with the important difference that Durkheim emphasizes more than Smith that the ethical standards ultimately informing his conception of healthy social life are valid not eternally but only relative to basic facts about the kind of society under consideration. In referring to the ethical standards that "ultimately" inform Durkheim's conception of social health, I mean to note that these standards come fully into view only at the end of (empirical) inquiry rather than being articulable from the start. For this reason Durkheim's science of morality could be considered a form of non-a priori *constructivism*, and in this respect – strange as this may sound at first – it has affinities with both the *Wealth of Nations* and Hegel's account of rational social life. For Hegel's moral "science," too, is a kind of constructivist account of the conception of the moral good appropriate to Western modernity that, contrary to what

[3] "If the laws of life are also found in society, they possess new forms and specific characteristics that cannot be surmised from the analogy [between social and biological life]" (SP: 37/3–4).

many believe, depends on non-a priori accounts of its basic social institutions and how they actually function.

An implication of Durkheim's intention to avoid reducing the moral to the functional, and vice versa, is that there must be some *conceptual* difference between the two bare ideas such that their identity rests on more than a logical truth: social health and moral health are the same condition but not because the two concepts are synonymous. If Durkheim wants his claim to be more than tautological, it must be possible to give an abstract, preliminary characterization of the idea of the socially functional that makes no reference to moral goodness. If, in other words, his moral functionalism is to claim something contentful about the functional character of the moral, then we must have some initial, independent grasp of what the functional adds to the idea of the moral, and vice versa. In order to do so, we can posit *efficient and stable reproduction* as essential to the idea of the socially functional and, for that reason, as a necessary component of any conception of healthy social functioning. On this basis one might even compile a rough list of that concept's essential marks, including (in some form) stability, cohesion, flexibility, vitality, well-coordinated functions, and an ability to meet the material and psychological requirements of social reproduction. Using Rawls's terminology, we could say that the (bare) *concept* of the socially functional is definable independently of the (bare) concept of the moral, but that an appropriate (determinate) *conception* of the socially functional cannot be had without in some manner bringing it into relation with the concept of the moral (and that the converse holds as well).[4]

Morality, in contrast, concerns at its core "everything that obligates man to take others into account, to regulate his movements on the basis of something other than the promptings of his egoism" (DLS: 331/394). On this view, the domain of the moral comprises rules and ideals that regulate social cooperation via notions of obligation that command individuals "to take others into account" and to act in ways different from what narrow self-interest alone would dictate. In *organized societies* morality consists primarily in a form of justice – although considerations of the good enter the picture, too – that regulates human relations in accordance with ideals of freedom and equality. That the socially functional and the moral are not conceptually identical can be seen in the fact that it is possible to *conceive* of forms of stability or coordination that lack freedom or equality, as well

[4] For the distinction between a (bare) concept and a (more determinate) conception, see Rawls 1999: 8–9.

as versions of justice that impede social stability, vitality, or reproducibility. Conceptions of justice of the latter type might be characterized by the slogan *fiat justitia et pereat mundus*, whereas Durkheim holds that a conception of justice compatible with the world's perishing is a "false" moral ideal – and, of course, that social reproduction achieved without justice (if even possible) would not be healthy social functioning. (The fact that we can conceive of unjust, stable societies does not imply that the conditions of the world are such that there could actually be one. This is a trickier question than it first appears,[5] but the view I am both defending and attributing to Durkheim asserts that, for any human society, justice contributes to stability, cohesion, coordination, and so on, and that injustice detracts from those same conditions. Since all these properties – both moral and narrowly functional – are realized in actual societies to varying degrees, and since stability makes up only part of the socially functional,[6] one cannot rule out the possibility of a relatively stable society – which might, however, be lacking in flexibility, vitality, and so on – that is also in significant respects unjust. The real world, always more complex than theory, seems rife with such examples.)

The claim that the moral and the socially functional are distinct concepts suggests one way of understanding Durkheim's project of integrating the two that nevertheless leads in the wrong direction. One might think that his science of morality integrates the two concepts by taking each to place external constraints on the acceptable content of the other in the following sense: What in the end will count for sociology as genuine well-functioning – the specific versions of stability, cohesion, and so on, that qualify as healthy – must also satisfy standards of justice and the good; and the versions of freedom and equality that count as morally salutary or authoritative must be compatible with social stability, cohesion, and so on. On the most complex version of this view, "constructing" the conceptions of moral goodness and functionality that social theory is to use would be a matter of making the respective contents of two independently definable concepts not merely logically compatible but also compossible, or

[5] One source of complexity is that the content of justice varies in different types of societies; in segmental societies the absence of modern justice is not dysfunctional.

[6] The thesis that morality and social functionality are realized only jointly may appear implausible if we think only of one dimension of social well-functioning: it seems unlikely that a moderate degree of stability, for example, requires good or just social arrangements. The thesis looks more plausible, however, if we take it to claim that realizing the various elements of social functionality *jointly* – stability, cohesion, flexibility, vitality, coordinated functions, and social reproduction – depends on society's being just or good.

jointly realizable.[7] Although it seems likely that Durkheim would accept that an adequate conception of justice must be logically and existentially compatible with both the conceptual and real conditions of efficient social reproduction – and, conversely, that healthy social functioning must be logically and existentially compatible with justice and moral goodness – this way of construing the method of his science of morality does not adequately capture his position.

One reason for this is that, as indicated above, Durkheim's view that social and moral health are the same condition is more robust than the claim that acceptable conceptions of each must be mutually compatible. Saying that social and moral health are the same condition implies not only that what counts as "true" (genuinely authoritative) morality must be compatible with the core elements of social functioning, and vice versa, but also, and more robustly, that each causally depends on the other: "true" morality contributes to social stability, cohesion, flexibility, vitality, coordination, and reproducibility – which is why there are aspects of "moral health that [social] science alone is competent to determine" (DLS: xxvii/xxxix) – and, conversely, healthy social institutions are constituted such that justice and the moral good are the standard outcomes of their normal functioning. In other words, the functional and the moral are mutually dependent not only because they must be conceptually and existentially compatible but also because, more substantially, what is moral promotes functionality and the functional produces just social conditions; their relation is one of real *causal interdependence*.

These considerations make clear that the identity Durkheim posits between morality and social functioning is a substantive, partially empirical assertion about the world rather than a conceptual claim or an a priori normative demand his social philosophy imposes on the world as a condition of its being found rational or good. At the same time, his position is a form of (empirical) moral constructivism that aims ultimately to "orient our conduct" by determining the specific content of the conceptions of justice and the moral good appropriate for modern societies. His science of morality, then, is an empirical undertaking that intends to teach us

[7] An example of logical incompatibility might be that slave-based production is ruled out by the very concept of (modern) justice. Real, or existential, incompatibilities come into view only when basic empirical conditions of the world and human nature are brought into the picture: Strict economic equality might be ruled out by the real conditions of satisfying the social need for vitality and a substantial level of social wealth; and moral rules requiring a degree of altruism such that most humans are able to follow them only when subject to a costly, energy-draining penal apparatus might be ruled out by the social need for a sufficient degree of stability and efficiency.

something about both how existing social institutions function and what conceptions of justice and the moral good are appropriate for our world and worthy of our endorsement, where each task can be carried out only in conjunction with the other. How, though, is a project of this kind to be understood?

It may help to recall Hegel's remark in the *Philosophy of Right* that it is not the business of social philosophy to *prescribe* moral ideals to reality (PhR, Preface) but rather first to find them at work in the existing social world and then rationally to reconstruct them, in part in light of what we find out about how institutions governed by them actually function, where such functioning includes promoting stability, social reproduction, and so on. Something similar holds for Durkheim's science of morality, and for this reason it is more accurate to describe its method not as constructivism but as (empirical) *re*constructivism. Durkheim's approach, like Hegel's, is to begin by looking at the social world he finds before him, including its dominant moral ideals, and to investigate empirically both how those ideals, actually embraced and already at work in social life, jointly can or do promote social functioning and, conversely, how the normal functioning of existing institutions realizes or gives expression to the same ideals.

Since the method of the *Philosophy of Right* is famously complex and obscure, it hardly makes sense to attempt to illuminate Durkheim's method by reference to Hegel's. More promising is a comparison to Adam Smith, whose method, as I sketch it briefly here, bears similarities to Hegel's approach while avoiding its deepest metaphysical pretensions, including its picture of human history as teleologically organized. With respect to method, Durkheim's science of morality might be understood as expanding on the general approach at work in Smith's account of the functional role played in commercial society by a historically specific conception of freedom, the legally codified and highly prized right to dispose of one's labor power as one sees fit. To simplify considerably, Smith legitimizes this conception of freedom, in part, by revealing its functional necessity for the workings of commercial society, insofar as it makes possible the principal social relation on which production in such a society depends: the "cooperation" between wage laborer and capitalist. In taking this functional claim to contribute to a moral justification of this feature of commercial society, Smith relies on our ability to recognize it as a form of *freedom* – as a form of not being subject to another's will – and hence as *prima facie* morally desirable. (Up to this point Marx would agree since he, too, recognizes the wage contract as an advance in freedom over feudal serfdom, even if – still following in essence the *method* of Smith described below – he also points out the one-sided nature of this fact in light of other morally objectionable consequences of the mode of production based on such freedom.)

Of course, other conceptions of freedom are imaginable, too, and considered independently of how they function in a real social system each might appear *prima facie* as of equal moral value to the one highlighted by Smith. What marks out the right to dispose of one's labor power as especially valuable for him is the central role it plays in the functioning of an entire social system that, once his argument is complete, reveals itself to be efficient, just, and productive of the good. In order to complete this larger argument, two kinds of claims must be made beyond merely pointing out that this right is a form of freedom and for that reason of moral value. First, Smith supplements his argument for the importance of this type of freedom by showing how the economic system based on it systematically realizes other moral ends as well, such as personal independence and a morally acceptable distribution of the social product. (The latter claim constitutes the moral significance of Chapter 8 of the *Wealth of Nations* – an application *avant la lettre* of Rawls's difference principle – which argues that increasing social wealth produces an upward pressure on wages, guaranteeing improvements in the conditions of the worst off.) Second, Smith argues from the opposite direction as well, showing how the economic system that realizes such moral ideals is also socially functional in the narrow sense of promoting stability, social reproduction, and general improvements in the conditions of life.

Smith's importance for understanding the method of Durkheim's science of morality lies in his sense of how the functional and the moral are intertwined in social life; normatively speaking, he refuses to reduce either *desideratum* to the other. Smith is no more explicit about this than Durkheim, but it is clear that his finding the freedom to sell one's labor power a significant moral advantage of commercial society is not reducible to the fact that the functioning and reproduction of that form of society require it. Beyond this, what makes Smith's form of moral justification similar to those of Durkheim and Hegel is its holistic character, where arguing for the importance of a specific sort of freedom involves showing both that it is systematically bound up with the realization of other moral values and that, integrated into commercial society as a central element of that system, it simultaneously contributes to realizing the more mundane ends of social life. In other words, part of what makes this particular version of freedom so valuable to Smith is its elective affinity with an economic system that functions impressively with respect to not only justice and the good but also stability, cohesion, flexibility, vitality, coordination, social self-maintenance, and the production of wealth. Although it is far from obvious, Durkheim's defense of the moral ideal characteristic of modern organized societies – its valorization of individualism, discussed below in this chapter – rests on an argument of this same type, and his general claim that social and moral health are the same condition is based on the conviction, backed up by various kinds of evidence, that similar arguments can be made for human societies more broadly.[8] Smith's analysis of the role of freedom in commercial society is a fitting example of Durkheim's concept of "moral reality" and his claim that "morality is … a system of realized facts bound up with the total system of the world" (DLS: xxviii/xli).

One cannot overemphasize that Durkheim's approach to specifying what morality and social well-functioning consist in is not a form of a priori constructivism in which philosophical reflection on the two concepts shows how they can be unified in some conception of moral health. It relies instead on empirical knowledge specific to time and place in order to see how specific forms of freedom and equality actually do or could promote social stability, cohesion, and so on, in a given society (and vice versa). What makes this approach similar to constructivism in moral theory is that it presupposes only very abstract concepts of justice and the good and proceeds to give them determinate content by observing how specific versions of those ideals, already at work in the world, both are complemented by other realized ideals to form a coherent moral landscape and are bound up with processes satisfying the more mundane requirements of social life. (That there are purely material requirements of social life – food, shelter, clothing, clean air, and so on – does not imply that they are met through purely material social activities lacking in moral significance. For humans, eating is always more spiritually significant than merely ingesting nutrients. This does not mean, however, that eating is spiritual "all the way down": however much it may injure our pride to admit it, biology imposes real, permanent constraints on the forms of life available to us.) Durkheim should be understood not as simply presupposing that social functioning and morality are intertwined in this way but as looking at the world with an eye to *finding* that relation in the empirical reality before him.

The supposed identity of the moral and the socially functional, then, guides Durkheim's empirical inquiry as a regulative ideal of the same sort

[8] See his claims about Rome in the Second Preface and Book I of *DLS*, as well as his statement that certain "ideas were incompatible with the Roman city" (DLS: xxvi/xxxviii).

operative in his basic assumption that societies are functionally organized and that it is therefore reasonable to ask what functions their enduring features serve. If there is an element of apriorism in his approach, it can be traced back to a basic (and plausible) anthropological assumption: that human nature and the fundamental conditions of the world are such that every society can be expected to face – and, if viable, to solve at least moderately well – two independently definable tasks (that are, however, always intertwined in reality): efficient, stable material reproduction and normative regulation of the cooperative relations under which such reproduction is accomplished. Further details regarding how either task is accomplished by a given society, or precisely how the two are interfused, must be ascertained empirically.

It is not surprising, then – as befits an empirically oriented science of morality and society – that there is no single fact or principle that explains for Durkheim why in general morality contributes to social well-functioning. One way of partially explaining the connection is to invoke a broad claim about human nature, mentioned in Chapter 9, according to which humans bring vitality and energy (DLS: 63/76) to their social activity only when they take their society to be fundamentally just or in some other way morally affirmable. Durkheim clearly endorses this claim, but his position is even stronger, insofar as he regards the actual justice or goodness of society, and not merely perceptions of its justice or goodness, as essential to social well-functioning; his claim is that morality, not the mere appearance of morality, is society's daily bread. When Durkheim endorses inheritance laws, greater economic equality, fair equality of opportunity, and even socialism, there is no reason to think he is interested only in society's appearing to be just; he is interested, more fundamentally, in its *being* just, and his claim is that the latter is indispensable to social well-functioning.[9]

An example of how Durkheim takes justice, and not merely its appearance, to contribute to social well-functioning can be seen in the measures that promote fair equality of opportunity. Such measures, independently of how they are perceived by social members, are almost certain to unleash talents and energies of individuals from disadvantaged social positions that society would otherwise be unable to employ for the benefit of all. One might also appeal to general considerations in support of the claim that justice promotes social well-functioning. There is a direct and plausible argument for this claim (with respect to the modern world), even though

[9] "The mere existence of [coordinating] rules is not sufficient; they must also be just" (DLS: 338/403).

Durkheim never articulates it as such.[10] The basic thought is that standards of justice track the fundamental interests of individual social members – in freedom, bodily and psychological well-being, and so on – and that a necessary condition of social cooperation's going well is that individuals perceive it as promoting their fundamental interests. (Following Rousseau and Hegel, Durkheim thinks of these generally defined interests as highly malleable; what *specifically* freedom and well-being – or "humanity" itself [DLS: 330/392] – are taken to consist in varies widely with historical and social circumstances.) To this thought must be added the liberal presupposition that modern individuals generally recognize these interests as such, regard them as fundamental to their well-being, and, despite the possibility of ideological distortion, for the most part know when those interests are being violated. On this view, society's being just is functional because grievous forms of injustice tend to be perceived by those subject to them as contrary to their fundamental interests, with the result that they participate only lethargically or grudgingly in social life or demonstrate outright hostility to its demands. This argument differs from the one sketched immediately above in that it requires both that society be just (promote the fundamental interests of all) and that it be perceived as such by its members (where this perception follows more or less directly from society's being just).

It is reasonable to think that something like this argument underlies Durkheim's thought that morality is the daily bread without which societies cannot live. Yet there are two important reasons this cannot be exactly his view (nor, for similar reasons, Hegel's). In the first place, Durkheim does not suppose that there are ahistorically valid moral concepts that, for example, define justice in terms of the fundamental interests of social members, and yet he wants to construe the connection between morality and social well-functioning as historically invariant. Certainly, the religious origins of morality – where what commands reverence is something greater than individuals – speak for him against the modern prejudice that justice is always to be defined in relation to individual interests, but so, too, does his view that moral notions vary (appropriately) with different forms of social organization. It is plausible that Durkheim regards the connection between morality and the fundamental interests of individuals as typical of organized societies, but he cannot believe it to hold in earlier, segmental societies, where processes of individualization have not taken

[10] This paragraph is inspired by Joshua Cohen 1997: 91–134.

place. And yet there is no sense in Durkheim's texts that the moral practices of the former societies are somehow less justified than our own, given their circumstances, or that they should be criticized for not measuring up to modern moral standards. This means that when Durkheim says, of all societies, that morality, and not merely its appearance, is society's daily bread, he is operating with a conception of morality more robust than merely whatever a group of humans might take morality to be and that at the same time, because it allows for substantial historical variation, does not claim eternal validity. The moral standards of a given society are at once something more than an arbitrary construct and something less than what, independent of time and place, morality "truly" is.

The second reason the connection sketched above between morality and social well-functioning – that allegiance to social norms depends on individuals seeing how their fundamental interests are promoted in a just society – is subtler than the fact that the tight connection between justice and individual interests holds only for modern societies. It derives from Durkheim's view that morality has its origins in, and always retains vestiges of, religious practices and the form of collective consciousness intrinsic to them, the characteristic feature of which is shared reverence for the sacred, defined in opposition to the profane world of everyday life (Durkheim 1995 [1912]: 34–8; Karsenti 2006a: 78–82). For Durkheim this implies that our allegiance to moral rules includes an element of reverence for the sacred that goes beyond what is merely useful or what we take to serve our own interests, however fundamental the latter might be. If a society is to be viable, its members must have some noninstrumental allegiance to the rules governing their cooperation – or, more precisely, some shared sense of the sacred that secures the moral authority of those rules. This view is connected to Durkheim's claim that "there are no genuinely moral ends except collective ones and no truly moral motivation other than attachment to a group" (ME: 105/92).[11] Morality, in other words, extends beyond what is good for me in particular, and allegiance to its rules requires reverence for something I take to be of higher value than myself. This view is compatible with the one described above, according to which in organized societies individuals bring more energy to their social activity when they see social rules converging with their own interests,

[11] With regard to Durkheim's holism, it is worth noting that he immediately goes on to say that "when one is attached to society …, it is psychologically impossible not to be attached, as a consequence, to the individuals who compose it and in which it is realized" (ME: 105/92).

even though individuals whose allegiance to those rules were based *only* on such a convergence would not be acting morally.

Durkheim's position is interestingly complicated by his theses that in organized societies collective consciousness is significantly less forceful than in their predecessors and that in the collective consciousness of such societies some ideal – indeed, a "cult" (DLS: 122, 333/147, 396) – of individuality tends to replace, though never completely, divinities or collectivities as the bearers of the sacred. What makes this position complex, even paradoxical, is that in modern societies, collective consciousness – precisely what is supposed to produce social cohesion – deifies the individual, which would seem to promote the opposite of cohesion, namely, egoism and a longing to be free from social constraints. Although the modern idealization of individuality explains in part why anomie, involving insufficient attachment to collective ends, is a persistent danger in organized societies, it is also Durkheim's view that in organized societies a shared commitment to the "dignity of the person" is the aspect of collective consciousness most responsible for producing social cohesion (II: 23–4/268).[12] This view is plausible if social members think of securing the conditions of individual dignity as exactly what it is: something that can be achieved only cooperatively, as a collective project in which all members participate and in which the dignity of all is secured. In this context it is important to bear in mind that the individualization characteristic of organized societies is from the very beginning a social achievement that emerges only with an increasing division of labor, itself a response to changing social conditions. Since increased specialization means increased dependence, the forging of social bonds in organized societies and the creation of modern individuals are merely two aspects of the same process; specialized individuals and social solidarity are equally primordial,[13] and they are no less a collective project than creating the social conditions that realize the ideal of modern individualism.

It is not obvious what Durkheim takes the modern ideal of individuality to consist in beyond an often vaguely defined "dignity of the person." If we attend to his description of that ideal more carefully, however – especially in his lesser-known works – we find another reason the ideal of individuality is more compatible with social cohesion than it might initially seem to be. Although his treatment of hypernomie suggests that freedom of choice

[12] On these points I am indebted to Lea-Riccarda Prix.

[13] I am indebted to Rahel Jaeggi for directing my attention to this point.

plays some role in this ideal – as do individual rights, sympathy for the suffering of all persons, and the celebration of individual difference (II: 24, 26/268, 271) – his explicit discussions of individuality emphasize *autonomy* (ME: 130/112),[14] characterized (allegedly in line with Kant and Rousseau [II: 20–1/263–4] but actually in closer proximity to Hegel) as "willing what is rational" (ME: 138/118), or "freely accepting the rules that prescribe one's acts," where "free acceptance is nothing other than acceptance based on enlightenment" (ME: 143/122). In order for compliance with such rules to be truly autonomous, however, more is required than rational insight into their validity. If individuals are to see their social activity as coming from themselves, they must also experience it as expressive of norms that reside at the very core of their being; they must experience themselves as "penetrated [by the social world and its laws] from all directions" (ME: 142/121). It is this conception of freedom as self-determination or autonomy, more than liberal conceptions emphasizing rights-based forms of negative freedom, that Durkheim has in mind when articulating the modern ideal of individuality.

Why Moral Ideals Vary among Social Types

One can gain a further perspective on Durkheim's reasons for subscribing to the real mutual dependence of morality and social well-functioning by looking at how he proposes to fulfill the third aspiration of his science of morality: showing how particular moral ideals of societies respond to, and can be explained by, changing social conditions. The fundamental principle here is expressed in Durkheim's statement that "[morality] develops within history and subject to the reign of historical causes; it has a function in our temporal life" (DLS: xxvi/xxxviii). As this suggests, Durkheim's explanation of why societies have the moral ideals they do is functional: moral transformations occur when changing social conditions make existing ways of generating solidarity obsolete – "changes … in the structure of societies [make changes] in *mores* necessary" (DLS: xxvi/xxxviii)[15] – and the moral ideals that replace old ones do so because they respond more effectively to the new social conditions under which solidarity must now be produced. An explanation of this sort is functional – it is an existential functional explanation of a thing's persistence but not its origin – because

[14] "This cult of man has for its first dogma the autonomy of reason" (II: 24/268).

[15] Similar remarks are found throughout the First Preface.

it appeals to a putative function of the *explanandum* to account for, not how moral ideals first come into the world (in the minds of a Confucius or Socrates, for example), but why, once they do, they take root in and rule the society in question.

Reconstructing Durkheim's explanation of why moral individualism became the dominant "religion" of modern societies requires referring back to his causal account, discussed in Chapter 8, of the origin of the specialized division of labor characteristic of European modernity. Durkheim's argument appeals most fundamentally to social changes – population growth and increasing social density, both moral and material[16] – that effaced the boundaries of existing segmental societies, filled up the spaces between them, and merged existing segmental societies into a much larger but still homogenous unit. Increased moral and material density allowed for more frequent interactions among social members, but it is primarily the increased difficulties of survival – more living individuals with roughly equal resources – to which he appeals in explaining the ensuing development of the division of labor: the need to survive forced individuals to differentiate themselves because developing different needs and means to satisfy them put those individuals in less direct material competition. Parallel to increasing differentiation occurred an independent process, the weakening of collective consciousness due to society's increased size and the consequent decrease in individuals' ability to identify with the larger unarticulated social whole of which they now formed a less significant part. This, too, led to increased differentiation among individuals by giving them more normative distance from the social norms that had previously governed every aspect of their lives with great force. This increased differentiation, together with the need to survive under more competitive conditions, resulted, according to Durkheim, in a substantially extended division of labor in which differences among social members' characters, capacities, and ways of life became relatively fixed, making them specialized individuals who could no longer move easily from one functional role to another. This development represented both an increase in freedom – from the repressive authority of premodern collective consciousness – and an increase in material dependence (since specialized individuals are less self-sufficient than those who can assume multiple roles in social reproduction).

[16] Moral density refers to the number of possibilities for interaction and influence among social members, material density to the spatial distance between them.

The key point in Durkheim's explanation of why a valorization of the individual became the dominant element in the morality of organized societies – of why a change in social structure made a change in morality necessary – is that the erosion of collective consciousness produced a functional problem for society. Since previously it was the collective consciousness that guaranteed social cohesion and solidarity, its weakening left a gap that needed to be filled by a new form of moral consciousness. Here we encounter an easily missed complexity in Durkheim's position. On the one hand, he holds that the new division of labor – with increased specialization and dependence – was itself, independently of any form of consciousness, a new source of solidarity; specialized, non-self-sufficient elements *must* find a way of cooperating if they and their society are to survive. On the other hand, he also holds – this is the point it is easy to miss – that solidarity can never be achieved without corresponding forms of consciousness, including some degree of collective consciousness that binds social members through a common allegiance to what is taken to be sacred or of unconditional value. Despite the danger of egoism it brings with it, the "cult" of individualism appears well suited to filling the functional gap left by the erosion of collective consciousness: it unites society in an allegiance to something thought to have unconditional value, and its content fits perfectly with the requirements of the new conditions of social existence in which individual differentiation is necessary. The idea, then, is that although every society must be bound together by some conception of the sacred, many such conceptions are possible, and some are more suitable to given social conditions than others. A morality of individualism corresponds to the new social conditions – it serves a "vital function" of society – because it promotes, and gives moral meaning to, the very phenomenon that organized societies rely on to maintain themselves. This explanation exemplifies Durkheim's claim that, like a plant, a moral ideal can live only "in a soil capable of nourishing it" (DLS: xxvi/xxxviii).

I say that a morality of individualism "corresponds" to the new social conditions because this is the term Marx uses when describing the relation between social consciousness and the relations of production (MER: 4/MEGA: XIII.8–9). The allusion to Marx's historical materialism is apt because for both thinkers the relation between moral consciousness and social conditions is functional, and in a similar sense: a form of consciousness corresponds to social conditions when it promotes phenomena required for the continued existence and functioning of those conditions.

Another similarity between the two theories is that in both cases readers have been confused about how strong the theories' explanations of new forms of moral consciousness claim to be, that is, whether they argue that social conditions determine, or merely condition, new forms of consciousness. Although Durkheim frequently encourages the stronger reading,[17] neither he nor Marx is best understood as taking that view. It is surely enough for a science of morality of Durkheim's type to show that a given moral system is well suited to play the functional role social conditions require of it without making the stronger claim that it is uniquely suited to do so. Here, again, demonstrating an elective affinity is a sufficiently ambitious aim.

Most important, however, is a further (fateful) similarity between Marx's and Durkheim's functional accounts of the correspondence between morality and social conditions. Both thinkers articulate their versions of this claim in the form of a historical narrative in which a change in social or material circumstances poses a (narrowly) functional problem, the solving of which requires a transformation of (among other things) dominant moral ideals. In Marx's case,[18] the underlying functional problem is always a clash between newly developing productive forces and existing relations of production (society's economic structure), where the latter have ceased to function in the sense that they no longer promote the growth of productive forces. New relations of production are then required, which themselves demand new forms of moral consciousness for their legitimation, but the underlying crisis that calls for these changes is functional in a narrow sense that can be captured without moral concepts. The morality of society must undergo change in order for the crisis to be resolved, but the crisis is not itself of a moral character, which is one reason interpreters generally agree that morality for Marx is epiphenomenal.

The fatal similarity between Marx's and Durkheim's positions consists in the fact that a similar "vulgar functionalism" is at work in the latter's account of the epochal transition from segmental to organized societies. While the transition ultimately requires a new form of moral consciousness – the "cult" of individuality – corresponding to new social conditions, the crisis that occasions the transition is narrowly functional: the relevant dysfunction results from increasing growth and social density that makes

[17] "If [morality] is as it is at any given moment, it is because the conditions in which men are living then do not permit it to be otherwise" (DLS: xxvi/xxxviii).

[18] This is the orthodox reading of his position (MER: 4/MEGA: XIII.8–9; G. A. Cohen 1978). But see the following note.

the homogeneity of segmental societies no longer viable, but there is no suggestion that the crisis in question is at the same time moral in character.[19] I return to this aspect of Durkheim's position below, when considering the resources available to him for providing a rational grounding of a society's moral ideals that goes beyond pointing out their narrowly functional character.

Sociology as Normative Moral Theory

This comparison with Marx raises two questions that return us to Durkheim's view that a science of morality not only explains moral reality, as does Marx's, but also tells us how "we ought to orient our conduct" (DLS: xxvii/xxxix): Why does Durkheim's science of morality, in contrast to Marx's, not undermine but instead reinforce the authority of the moral beliefs it explains, and how, more generally, can an *explanation* of moral ideals contribute to *justifying* them? The easy (and partially correct) response to the first question is that for Marx the social system sustained by modern morality is plagued by ineliminable dysfunctions that make the system's abolition, rather than its cure, the appropriate practical response, whereas the same is not true of contemporary society for Durkheim. As we know, Durkheim takes the form of society he investigates to reveal itself to be something we ought to maintain (and remedy where necessary), and this judgment is grounded in something more than a claim that that society is, or can be reformed so as to be, functional in a narrow sense. Rather, contemporary society achieves for him forms of stability, cohesion, and so on, and its way of doing so realizes a modern version of justice (and goodness) that accommodates the autonomy, rights, and other interests of its members. As noted above, this ethical justification of organized society – a form of reflective affirmation – differs little in form from Smith's ethical-functional defense of commercial society.

For Durkheim, then (as for Hegel), the moral notions that prevail in societies are both conditioned by social reality and genuinely authoritative; the latter, however, does not imply eternal validity in a sense that Plato,

[19] I do not mean to rule out the possibility of there being resources in Marx's and Durkheim's theories for ascribing to them more sophisticated accounts of social change. Indeed, the most plausible version of Marx's account of the transition from capitalism to communism conceives of the crisis of capitalism in moral-functional terms – insofar as subjective alienation and the perception of exploitation play a role in it – but it is precisely because this seems to be in tension with his "official" formulation of historical materialism that so much ink has been spilled over the extent to which communist revolution depends on its participants' awareness of capitalism's ethical failings.

Rousseau, or Kant would demand. One might say that for Durkheim there is a kind of truth content, or objective validity, to social morality that is bound up with, but not reducible to, the requirements of social reproduction. How, though, is such a position philosophically defensible? What, if the conditions of social reproducibility narrowly understood cannot, lends *moral* legitimacy to modern justice and makes a modernly just society morally superior to other possible ways of achieving social stability, cohesion, and so on? Although Durkheim does not provide a fully satisfying answer to this question, before concluding that no such answer is available, let us see whether more can be teased out of his position that brings us closer to that goal.

Whatever meta-ethical position it makes sense to ascribe to Durkheim, it will not be foundationalist. His dismissals of a priori moral philosophy that locates the ends of morality in a transcendent realm, together with the historically relative and empirically contingent nature of his science of morality, make that indisputable. This might lead one to think that he adopts a Hegelian perspective instead, but he also rejects Hegel's historically sensitive but still rationally demanding strategy for answering this question, which, in its most robust version, involves showing the moralities of historically diverse societies to be mutually dependent parts of a single humanity-wide project aimed at unifying the demands of life with standards of justice and the good, in the context of which later societies, building on the achievements and failures of earlier ones, devise ever more adequate responses to their species-unifying task of spiritualizing the ends of life and vitalizing those of spirit. Because unifying life processes with justice and the good counts for Hegel as a kind of emancipation, his picture of ethical progress is also one in which all human societies share the same generally conceived *ethical* end – defining and realizing freedom – and in which an ethically more advanced society is one that develops and institutionally realizes more adequate and comprehensive conceptions of freedom than less advanced societies. Moreover, the superiority of a more comprehensive conception is in principle rationally accessible to the inhabitants of less advanced perspectives precisely because it can be shown to resolve "contradictions" internal to those perspectives. On this model, then, rationally justifying a social formation, including its moral conceptions, consists in showing it either to offer a *more successful* solution to the species-specific task just described – one less susceptible to self-generated contradictions – than its predecessors' or, in a stronger version of the strategy, to offer a *finally and decisively successful* solution to the same task that generates no further contradictions that push historical development to a new level.

Durkheim's intention to reject Hegel's meta-ethical position is clear, but it is worth noting why he does so: "It is no longer possible today to believe that moral evolution consists in the development of a single idea that, confused and indeterminate for primitive man, gradually clarifies and determines itself more precisely through a spontaneous progress in enlightenment" (DLS: xxvi/xxxviii). Apparently, what leads Durkheim to dismiss Hegel's philosophical strategy is that he takes it to ascribe a kind and degree of rational coherence to human history that by the end of the nineteenth century no longer appears tenable. In effect, Durkheim allows for more radical forms of moral diversity than Hegel since he does not posit a continuity in content – resulting from a process of spiritual supersession that always preserves a part of the old – among historically different moral systems. A major aspect of the rational coherence of history rejected by Durkheim is Hegel's belief that social transformations result from the emergence of quasi-conceptual[20] contradictions that are internal to a society's moral conceptions and therefore in some sense present within them from the start; social change for Hegel is in reality social *development* in accordance with a logic intrinsic to what from the very beginning human society, including its ideals, implicitly is. This means that, unlike Hegel, Durkheim does not conceive of history as a continuous, rationally reconstructible process in which major innovations are shown to be rationally necessary (or appropriate) responses to the failings of earlier developmental stages. This can be seen in Durkheim's explanation of why moral individualism comes to prevail in organized societies: the events that trigger the described change – population growth and increasing social density – are contingent and morally meaningless, and the moral transformation they make necessary is a solution to a new functional problem the moral content of which need not be continuous with that of previous moral systems. This means that modern moral individualism is a fitting response to the functional needs created by the decline of segmental societies, but, strictly speaking, it cannot be regarded as a moral *advance* (DLS: 330/393)[21] over earlier moral systems. Each form of morality is a fitting response to a

[20] These are not logical contradictions but contradictions between a specific conception of freedom and the requirements of its institutional realization: a society of lords and bondsmen purports to realize freedom, but the impoverished conception of freedom on which it is grounded, when realized in the world, produces alienation and domination instead. Such contradictions are quasi-conceptual because they represent non-contingent defects of the relevant conceptions of freedom that become manifest only when realized in social practices.

[21] Alternative, pragmatist accounts of moral progress that might help Durkheim out of his problems can be found at: Kitcher 2014: ch. 6; Kitcher 2021: 73–82; and Jaeggi 2019: 221–36, 290–313.

specific social need, but it is hard to see what could make one such system morally superior to another.

None of this prevents Durkheim from referring to organized societies as "more advanced" than those of "lower peoples" (DLS: 329/391), and, as we will see below, "more advanced" includes moral advancement. If we take the points of the previous paragraph seriously, this means that there is an unresolved tension in Durkheim's position: his philosophical resources are insufficient to explain the moral advances – and hence the (conditioned) truth content of social morality – that he nevertheless wants to posit. One sometimes gets the impression that he means to borrow the notion "advanced" from biologists' talk of the higher animals, where "higher" denotes more complex forms of organization that provide such animals with expanded capacities in comparison with their "lower" counterparts. Organized societies could be said to be higher than segmental societies in the same sense, but the quasi-biological criterion appealed to here does not help with the question of how ethical ideals for Durkheim could be anything more than reductively functionalist properties of societies or how one could distinguish morally better from morally worse forms of equally complex functioning.

That Durkheim wants the form of justice appropriate to organized societies to be "truer," or more morally authoritative, than a reductively functionalist conception of it is clear in one of the concluding passages of the *Division of Labor*:

> What characterizes the morality of organized societies compared to that of segmental societies is that the former has something more human, hence more rational, about it. It does not make our activity depend on ends that do not touch us directly; it does not make us servants of ideal powers that have a nature completely different from ours and that follow their own paths without concerning themselves for the interests of men. … The rules that constitute this morality do not have a constraining force that stifles free inquiry, but, because they are in addition made for us, and in a certain sense, by us, we are freer with respect to them. (DLS: 33–40/403–4)

What Durkheim appeals to here to explain the ethical superiority of modern morality, and the reason the latter has the authority to "orient our conduct," is ultimately an anthropological thesis concerning the general characteristics of the mode of existence most appropriate to human beings, given the kind of being they are – or, expressed in Hegelian language, given the spiritual characteristics that give them the potential to lead lives that possess a dignity not capturable in biological terms. The "human" – and not merely modern – ideal expressed in this passage is a complex ensemble of rationality, self-transparency, absence of alienation, and, perhaps most salient, freedom (of a specific sort). There is an undeniable power to Durkheim's view, but it lacks the degree of articulation and justification that philosophers expect of a well-founded meta-ethical position. Perhaps more relevant, his appeal to "the human" has more than a hint of the foundationalism that sits uncomfortably with his empirical science of morality.

There is a further aspect of Durkheim's mostly implicit anthropology that, although it does not decisively solve the philosophical problems just described, adds an interesting element to his meta-ethical view that pushes his position in the direction of Hegel's. This aspect of Durkheim's view becomes more visible in later texts that expand on the nature of morality and, influenced by *The Elementary Forms of Religious Life* (1912), reflect further on the implications for morality of its origin in, and enduring inseparability from, the experience of the sacred (SP: 83/57). In these texts Durkheim draws our attention more forcefully than before to the *dual* nature of the sacred, namely, that it is experienced not only as issuing inviolable commands that constrain our egoism in the manner of Kant's categorical imperative but also as drawing our desire toward it as something we aspire to be, or to be in connection with. Durkheim calls this aspect of the sacred its "desirability," adding that "it is this *sui generis* desirable something that we commonly call *the good*" (SP: 71–2/42–3; emphasis added). Both the sacred and the moral, in other words, involve a synthesis of two moments, duty and the good.[22]

If we add to this that for Durkheim the sacred is inseparable from the social – our awe for the sacred is but a mystified form of the reverence we owe to society itself, on which nearly everything we are, including our very life, depends – what results is a view of human social life as incorporating two conceptually distinct but in reality inextricable projects. The first, and more obvious, is the task of social reproduction; the second derives from the fact that such reproduction, even in "less advanced" societies, depends on forms of cooperation that require for their success a shared conception of something taken to have a value greater than mere survival – Hegel's "absolute" – that imbues the rules governing cooperation with a moral, supra-utilitarian authority. The source of this component of social life is the sacred, including its image of the good, or of that which is objectively

[22] This view is already implicit in his earlier, already cited claim that, beyond constraining us, morality "emanates a warmth that inflames and re-animates our hearts" (DLS: lii/xxx).

more desirable than the ends of our (as Durkheim sees it) natural egoism. Most fundamentally, then, social life is rule-coordinated cooperation that at the same time expresses shared values not reducible to the mundane aspects of social reproduction. As suggested above, social participation on this view could be thought of as a form of Weberian "value-rational" action, which further suggests that the moral aspect of social cooperation is what is capable of imbuing our everyday lives with a meaning more substantial than pleasure or utility. It is not implausible, then, to think of there being for Durkheim two fundamental human "needs" to which social life must respond (even if the same activities that address the one typically address the other as well): not only the needs bound up with the maintenance of individuals and society but also a need to find what one does and who one is – together with the larger social life within which both are made possible – "spiritually" meaningful and affirmable as good.[23]

I have already noted that this addition to Durkheim's view does not decisively solve the philosophical problem of how a moral system can have a (historically specific) truth content, or objective validity, grounded in something more than the narrowly functional role it plays in social reproduction. What this addition suggests, however, is a possibility for enriching Durkheim's picture of social transformation that reconfigures the philosophical problem to be solved: if human social life is cooperation for the purpose of survival and reproduction that is at the same time meaning-endowing and expressive of the good, then Durkheim's account of the advent of modern moral notions appears grievously impoverished, even by his own lights. For it explains the succession of one moral system by another as a functional response to purely material, morally meaningless changes – population growth and increasing social density – that make the transformation from segmental to organized societies necessary. The ultimate motor of social, and therefore moral, change, on this view, is problems or "crises" (DLS: lv/xxxiv) located at the material level of social structure. If one takes seriously, however, the more complex view that the functional and the moral are inextricably linked in human social life, then one should expect that the problems that generate social transformation are neither purely functional nor purely moral but both at once. Think

[23] In discussing Durkheim, Habermas cites Peter L. Berger's *The Sacred Canopy*: "The anthropological presupposition … is a human craving for meaning that appears to have the force of instinct" (Habermas 1975: 118). Nietzsche is the follower of Hegel who most emphasized this "need for meaning," although he thinks of it not as an eternal need of humans but as one that arose in specific historical circumstances, even if human existence is now, for us, unthinkable without it (Forster 2015).

again of the crisis of the patriarchal nuclear family in the late twentieth-century West, where broken homes and gender injustice are impossible to hold apart – and where the moral crisis in question is occasioned in part by material conditions: new economic opportunities for women made necessary by World War II. And the same is true for Hegel's tale of bondsman and lord (examined in Chapter 13), who engage in material and moral cooperation plagued by instability and lack of cohesion, the source of which is partly moral: unsatisfying patterns of recognition that fail to grant equal dignity to both poles of the relation.

If these two essential pillars of social life are taken into account in understanding social transformation – if social crises are always also moral crises – then successful solutions to social crises must be both moral and functional. But, as Hegel sees more clearly than Durkheim, the form a moral solution to a social crisis is most likely to take is not a creation of new moral values *ex nihilo* but a revision of prevailing notions of justice or the good that, in contrast to Durkheim's example, involves some continuity in content among old and new moral conceptions; generally speaking, moral innovations do not arise out of thin air but involve transformations of earlier ideals, the results, to speak with Hegel, of spiritual processes that both negate and preserve the legacy of the past. This modification of Durkheim would reintroduce into his picture of social change an element of the moral continuity upheld by Hegel without an exaggerated view of the unity and teleological coherence of human history that the latter seems to espouse, including the idea that history is a rationally necessary series of developments, generated by contradictions internal to specific forms of social life, in which the human race achieves full clarity with respect to – as well as complete realization of – its governing ideal: freedom.

The picture of social and moral change that I am suggesting can be attributed to Durkheim without doing violence to the essential elements of his position does not provide a foundationalist justification of the content of modern morality, but it does offer a different, neo-Hegelian way of understanding justification in the moral domain, insofar as it makes it possible for moral innovations to possess a kind of rational superiority in relation to their immediate predecessors. If the problems to which social change must respond are inextricably functional and moral, then their solutions must address the old society's moral failings, too (Jaeggi 2019: 109–13). Both examples of social crises mentioned above – in the patriarchal family and in a society of bondsmen and lords – suggest that social crises are solved through not only changes in social structure but also, inseparable from these, reinterpretations of conceptions of justice and the good that present themselves as more adequate versions of the values animating earlier institutions, for example, by reinterpreting the values of family life to accompany the moral "discovery" that rigid gender roles are unjust because they are incompatible with all family members having equal possibilities for self-realization. The new conceptions of justice and the good acquire, then, a kind of legitimation that, though weaker than that aimed at by foundationalist moral theories, may be the best that can be achieved in the domain of social philosophy. In the following two chapters I examine how Hegel's conception of spirit articulates more consistently and precisely the moral-functional character of social life, as expressed in the claim that in human societies there is no end of biological life that is not at the same time an ethical end and no ethical end that is wholly separate from the ends of living creatures.

CHAPTER 11

Hegelian Social Ontology I: Objective Spirit

Unlike Durkheim, Hegel (almost) never speaks of social pathology or illness, and yet his understanding of human society remains the single most important resource for contemporary theorists of social pathology. One implication of this is that even though Durkheim discusses social pathology more explicitly and systematically than Hegel, his account of what human society is and of the problems it can run into – already Hegelian in many respects – can be enriched by incorporating even more of Hegel's conception of the rational (modern) society. One reason Hegel's position is crucial to a theory of social pathology lies in the pre-eminence it accords to *life* in understanding both nature and human existence, as manifested in the fact that life serves as the model for the central idea of his philosophy as a whole, that of *human* being as spirit, or *Geist*: if human modes of being are grounded in or structurally similar to those of living phenomena, it is reasonable to expect that, like living beings, human societies, too, are susceptible to something analogous to organic disease. Beyond its close connection to life, however, Hegel's idea of *Geist*, in which the ends of life are to be reconciled with those of freedom, provides us with an ethical conception of the good – as the supreme object of *Geist*'s striving – that integrates the functional and moral aspects of social life more satisfactorily than Durkheim does and therefore better explains how the dominant ethical ideals of societies can claim a moral authority beyond the functional value they have for social reproduction.

The next three chapters are organized as follows: Chapters 11 and 12 set out the principal elements of Hegel's social ontology by examining his characterization of social reality in terms of *objective spirit* and *the living good*, respectively. If these discussions are successful, they will illuminate complementary aspects of Hegel's social ontology.[1] Chapter 13 builds on

[1] Pinkard (2012) undertakes a similar project. His focus is broader than mine, but there is significant overlap between our interpretations.

these chapters to consider implications of Hegel's social ontology for conceiving of social pathology.

Numerous attempts to explain Hegel's concept of objective spirit can be found in the secondary literature, and many are worth consulting. However, because I am interested not only in what Hegel means by this concept but also in its relevance for contemporary social theory, I pursue here an indirect strategy for elucidating the idea of objective spirit. Rather than focusing on interpreting Hegel's texts, I turn first to the most compelling contemporary account of the ontology of social institutions in terms of the very concept introduced by Hegel, objective spirit. This is Vincent Descombes's project in his path-breaking work *The Institutions of Meaning*. Descombes's project is important for mine because his aim is not Hegel exegesis but a philosophical articulation and defense of the core insights Hegel (and others) had, or came close to having, in conceiving of social life as objective spirit. In the end, my appropriation of Descombes's work – and, to a lesser extent, of similar views articulated by Durkheim and John Searle – pushes his position closer to Hegel's than Descombes might have intended, but there is much to be gained from looking first at Hegel's position (or one similar to it) from afar, through the eyes of a philosopher interested more in finding the truth than in explicating Hegel's texts, and only then returning to those texts to see whether the view from afar has overlooked elements of Hegel's position we might want to retain.

The present chapter is divided into four sections. After some remarks in the first on the concept of objective spirit in general, I consider in the second how traces of the concept can be found in Durkheim, even though he does not use the term. In the third, most substantial section, I reconstruct Vincent Descombes's account of objective spirit. Finally, I consider aspects of Hegel's account of objective spirit missing from Descombes's that must be brought into the picture in order to undergird Hegel's ethically robust project in social philosophy and to furnish the resources needed for an appropriately rich understanding of social pathology.

Objective Spirit: Preliminaries

Hegel does not use the term "social ontology," but the question of how social reality differs from other kinds of reality – inorganic nature, living beings, subjective mind, culture – lies at the heart of his systematic philosophy. This is visible in the very structure of his *Encyclopedia of the Philosophical Sciences*, where each section is named for what Hegel takes to be a distinctive type of reality: nature, subjective spirit, objective spirit, and absolute spirit. The location of his social (and simultaneously moral and political) philosophy in the third of these sections points to the main ontological claim Hegel makes about human social life, namely, that it belongs to the domain of *objective spirit*. In the *Philosophy of Right*, social (or ethical) life (*Sittlichkeit*) is characterized as the highest form of objective spirit,[2] and in the *Phenomenology of Spirit* we are said to encounter spirit for the first time precisely when what we have before us are distinctively social phenomena: relations between recognition-seeking, living subjects that take the form of interactions between bondsman and lord. It is also at this point in the text where, further characterizing objective spirit, Hegel refers to it as "an *I* that is a *we* and a *we* that is an *I*" (PhG: ¶177; Hegel's emphasis). After discussing Durkheim's and Descombes's holistic conceptions of social reality, I will return to this phrase since only then will we be in a position to understand Hegel's thought that an *I* is possible only as a member of a *we*.

Hegel characterizes objective spirit in two ways: first, as "the concept of freedom that has become the world before us" (PhR: §142) and, second, as spirit "existing in the form of *reality* as a *world* it has produced" (PhM: §385; Hegel's emphasis).[3] Each characterization points to a distinct aspect of objective spirit.[4] The first is most naturally taken to mean simply that objective spirit is identical with rational social life, the rationality of which consists in the fact that the institutions composing it realize the freedom of social members. This is a claim about the goodness of the social order described in the *Philosophy of Right*, and it belongs squarely to the project of normative social philosophy. It can also be formulated in teleological language: the institutions of ethical life are rational because only within them are humans able to realize their telos (or essence) as self-conscious and self-determining beings. There is nothing wrong in construing Hegel's doctrine of objective spirit in this way, and elsewhere I have examined in detail how the claim that rational social life realizes freedom is to be understood.[5]

[2] I ignore world history, which is a more complete form of objective spirit than any single rational state (PhR: §§341–60).

[3] That objective spirit is produced by spirit means that the natural world must first be transformed by human agency – in the course of history – before it can accommodate the realization of freedom.

[4] Alznauer 2016b: 209–31.

[5] Neuhouser 2003: chs. 1, 3, 4, 5.

Yet, as the second quotation indicates, there is more to Hegel's characterization of objective spirit than this normative claim, for it is also said to be a way in which spirit (or mind[6]) acquires an objective existence in the real (spatial, temporal) world. To say that mind exists objectively is to say that, in a sense specified below, it exists external to the consciousness of individual social members. Durkheim, who is also committed to the reality of something akin to objective spirit, famously articulates this point by claiming that social phenomena exist, and are to be treated by sociology, as *things* (RSM: 35–7, 60, 69/xii–xiii, 15, 27–8). The claim at the heart of Hegel's second characterization of objective spirit is not only that mind exists externally to the consciousness of individuals but also, and more strongly, that it must do so if individuals are to be capable of belief and action. This claim is not normative but transcendental: minded individuals can exist only if they inhabit a social world that is also minded (in a sense to be explained); or, expressed in Hegelian terms, objective spirit is a necessary condition of subjective mind's capacity for belief and action being realized.[7] It is this aspect of Hegel's doctrine of objective spirit that Descombes attempts to render plausible rather than the normative claim that rational social institutions are the sites of freedom realized in the world.[8]

One familiar sense in which mind can exist outside consciousness is when subjective acts leave behind material traces, such as prehistoric paintings on a cave wall. This phenomenon – sometimes called *objectified mind* (IM: xxiv) – is not, however, what Hegel and Descombes have in view when speaking of objective spirit. The latter is not simply a material trace of a mind that used to exist but no longer does. Rather, objective spirit is (somehow) mind that is still at work but in a realm external to individual consciousnesses. If, as both Hegel and Descombes believe, there is no locus of consciousness above and beyond individual consciousnesses, then it is initially a mystery as to how there can be such a thing as "mindedness" (Pippin 2008: 17, 104) operating in a realm of things external to individuals. It is this ontological claim – that social reality is externally

[6] English speakers will find it odd to equate "spirit" and "mind," but in the languages of both Hegel and Durkheim the two meanings are joined in a single word: *Geist*, like *esprit*, commonly means both.

[7] As Descombes points out, a similar claim can be attributed to Wittgenstein, namely, that belief and action presuppose membership in a linguistic community, which requires a specific (social) form of life (IM: xvi–xvii).

[8] Descombes, too, hints at ethically normative implications of his account of objective spirit, as in his claim that objective mind "can take place only within an order of justice, which can be established only among autonomous people" (IM: 243). For further connections between institutions and their participants' freedom, see IM: xxvii, 9, 245, 306.

existing mind, or objective spirit – that Descombes explains and defends. As we will see, Durkheim, too, struggles to articulate a similar thought in addressing the nature of the specific realm of reality taken by sociology as its subject matter.

Before turning to Hegel's successors, I must introduce one more confusing feature of his characterization of objective spirit, namely, that the latter contains both subjective and objective aspects. The objective aspect of objective spirit is said to reside in "*laws and institutions that exist in and for themselves*" (substantially or absolutely) and that are also characterized as "*ethical powers* that rule individuals' lives" and find in those lives "their phenomenal form and actuality" (PhR: §§144–5; Hegel's emphasis). The subjective aspect of objective spirit, in contrast, consists in the conscious, or subjective, relation social members have to their laws and institutions, which itself has two aspects: individuals recognize laws and institutions as having normative (moral) authority over them, and they regard laws and institutions not as foreign powers but as authorities that secure the conditions under which they realize themselves as the beings they take themselves essentially to be (PhR: §§146–7): spiritual, living beings whose essential interests include freedom, recognition, and well-being.

Thus, Hegel's description of objective spirit contains various claims in need of explanation: i) There is a form of mindedness that is real, or that *exists objectively*, outside the consciousness of individuals; ii) this externally existing mind, embodied in social institutions, is in some sense *prior to the individuals* whose lives they rule (or, as Hegel says, those institutions constitute the "substance" of which individuals are the "accidents" [PhR: §§144–5]); iii) social reality is *normatively constituted* in that its existence depends on (to use Searle's language) the collective acceptance of institutions' constitutive rules or norms; and iv) the rules or norms that social members regard as authoritative not only constrain what they do but also *expand their practical possibilities* (and therefore promote a kind of freedom) by enabling them to act in ways that would be impossible outside social life and that enrich their lives. Durkheim and Descombes endorse versions of all four of these claims. Because the latter presents his doctrine of objective spirit as a reworking of the former's social ontology, I will briefly consider the Durkheimian sources of Descombes's view.

Objective Spirit in Durkheim

The key to understanding the concept of objective spirit is figuring out in what sense institutions qualify as *objective* mind. Unfortunately, "objective" has several meanings in this context that even in Descombes's scrupulous account are not always held apart. Surely this is in part because Descombes orients himself on this issue in relation to Durkheim's social ontology, which is centered around the similarly polysemous claim that "social facts" (social phenomena[9]) are to be regarded by sociology as things. Since much of what makes institutional reality objective for Descombes is related to what makes it thinglike for Durkheim, examining the latter claim will put us a position, in the following section, to untangle the former.

The central feature of things in the context of Durkheim's claim – think of the book or computer in front of you – is that they exist independently of consciousness. One meaning this has for Durkheim is that in the typical case, *institutions temporally precede and survive their individual members*, and in this sense they exist outside the consciousness of individuals (RSM: 44–5, 50–1/xxii, 3–4). The point is that, although there could be no social facts without conscious individuals who recognize the authority of institutional rules, such facts do not depend on the existence of any specific individual consciousness, and such facts can survive the death of all individuals who existed at a prior time, provided they are replaced by a later generation. A related point is that *institutions have an enduring existence independent of any particular instance of the practices they define*:[10] The institution of money continues to exist regardless of whether I use mine today to buy lunch or store it instead under my mattress. This is also part of what makes social reality objective for Descombes (IM: 93). At the same time, calling social facts thinglike is potentially misleading since the existence of nonsocial things does not depend on there being conscious subjects at all. In other words, social facts have a reality all their own, which in some respects is like that of physical things and in other respects not.

A further aspect of social facts' thinglike character for Durkheim is that, in contrast to mere ideas or mental states (RSM: 60–1/15–16), they are *real* – or, as he also says, they possess "a degree of reality at least equal to that which everyone recognizes [the realities of the external world] as having" (RSM: 35/xii). Here Durkheim is careful to say that social facts

[9] Social *facts* for Durkheim are not facts in the ordinary sense – e.g., in the sense in which Searle takes "Obama was the 44th President of the United States" to be a social fact, a "fact involving collective intentionality" (CSR: 26). Durkheim's *faits socials* are more like acts or realities than facts (although all three terms are possible translations of *faits*). Social facts are social *phenomena* (RSM: 31/vii) and include institutions of kinship and marriage, money, language, political organizations, etc. (RSM: 50–1, 52/3–4, 6).

[10] A social fact "is general over the whole of a given society, while having an existence of its own, independent of its individual manifestations" (RSM: 59/14); see also RSM: 51/4.

are not material things – their reality is *sui generis* – but it is far from clear what he means in saying that, despite this ontological difference, social facts are *at least as real* as material things. One way he expands on this claim is by saying that something's being real implies that it cannot be immediately known to consciousness – as mental states might be known through mere introspection, for example – or altered merely by an individual act of will: "A thing is every object of consciousness that is not naturally co-penetrable (*compénétrable*) to intelligence" or "everything the mind (*l'esprit*) cannot understand except by going outside itself through observation and experimentation" (RSM: 35–6/xii–xiii). In both cases his claim is that what makes social facts real is that *institutional reality imposes limits on individual subjects, offering resistance to their efforts to know or alter it* (RSM: 70/29). Resistance, however, is a concept borrowed from physical reality; when applied to institutions, it can have only a metaphorical meaning that must be spelled out if we are to grasp the specific character of social reality. If we restrict ourselves to the case of knowledge, then Durkheim's point must be similar to Searle's when he claims that social reality, while not ontologically objective, is nevertheless a source of epistemically objective truths that, rather than reporting merely subjective states, make claims about how the (social) world is constituted that are subject to objectively valid criteria for rational belief (CSR: 7–9). We might be tempted to express this point on Durkheim's behalf by saying that what is thinglike in the sense of "real" places rational *constraints* on our cognition and will, but, as we will see below, this statement, though true, is potentially misleading since Durkheim tends to use this term to indicate a different sense in which social facts are like things.

Another feature of the thinglike nature of social facts for Durkheim is what he alternatively calls "constraint" and "coercion" (RSM: 50/4). At issue here are constraints that social facts place on the wills of social members, but not in the sense noted above, where social facts resist efforts to alter them. Durkheim often gives the impression that social facts possess causal powers that affect the behavior of social members.[11] If this were true, then social facts would indeed have a reality similar to that of material things. But no sympathetic reader who has grasped the normative

[11] "There are certain currents of opinion that push us with unequal intensity, for example, to marry ... Such currents are plainly social facts" (RSM: 55/9). See also RSM: 52–3/6–7. Social currents that produce collective emotional reactions in large assemblies (if there are such things) might be construed as causally determining certain forms of behavior, but they surely do not mechanically determine some of us to marry and others not.

character of institution-defining rules for him will saddle him with the view that such rules are efficient causes of behavioral regularities observed in human society (IM, 90). Even if we reject this possibility, there remain three further forms of constraint or coercion that appear in Durkheim's texts[12] (without being clearly distinguished) – although, as I will argue, neither is appropriately called "coercion." In the first place, social facts, or institutions, place moral "pressure" (RSM: 44/xxi) on their members insofar as they impose obligations on them, the nonfulfillment of which results in punishments or informal sanctions on the part of society. Yet, as can readily be seen when Durkheim describes the phenomenon further, "coercion" is a misleading term: moral or normative pressure is not the same thing as putting a gun to someone's head and telling him what he must do (and neither phenomenon involves the causal determination of behavior).

Even though the normative pressure characteristic of institutions includes a threat of punishment or sanction, there would be no social reality in Durkheim's (or Searle's) sense if that threat were the principal reason individuals complied with society's demands. In the case of genuinely social facts, compliance is typically – though, of course, not always – motivated by an acceptance of the authority of institutions' rules rather than by fear of the consequences of noncompliance: "What is distinctive to social constraint is that it is due … to the prestige with which certain representations [or rules] are invested" (RSM: 44/xxi). In other words, *institutions constrain individuals' actions by virtue of a normative authority collectively bestowed on them by their members*. Hence, the normative authority of social rules represents a second respect in which institutions constrain individuals' wills – not this time through moral pressure from without but in the form of rules that constrain, as Durkheim sees it, our natural egoism, by virtue of the moral authority we take them to have.

At the same time, it is important not to lose sight of the fact that a sense of obligation is only one aspect of what for Durkheim motivates social members to comply with institutional rules:

> The coercive power we attribute to the social fact is so far from being the whole of its power that it can just as much present the opposite characteristic. For at the same time that institutions impose on us, we are attached to them; they obligate us, and we love them; they constrain us, and we find satisfaction (*nous trouvons notre compte*) in their functioning and in this very constraint. This antithesis is what moral philosophers have often pointed

[12] For more on constraint in Durkheim, see Giddens 1984: 169–79.

> to by the two ideas of duty and the good, which express two different but equally real aspects of moral life. There is perhaps no collective practice that does not exert on us this dual action, which is contradictory only in appearance (RSM: 47n4/xx–xxin2).

Although it still makes sense in light of this point to speak of institutional rules as constraining us – they prohibit us from doing certain things we might otherwise be inclined to do – "coercion" is clearly inadequate to capture the dual character of what moves individuals to comply with institutional rules, one part of which is described as attachment and even love (ME: 100, 105–6/xxx–xx),[13] echoing Hegel's claim that institutions, despite being the source of obligations, are not experienced by their members as *foreign* powers (PhR: §147). This point is especially important to Durkheim's and Hegel's accounts of healthy social life: institutions that work only by placing burdensome obligations on their members are bound to lack the spontaneity and widespread acceptance required not only for their efficient functioning but also for their members' subjective satisfaction and freedom.

The simultaneously internal and external character of institutional rules – the fact that despite being "in us," they are external authorities that individuals do not themselves create – is expressed in the following passage:

> When I fulfill my task as brother, husband, or citizen, when I carry out the commitments I have acquired through contract, I discharge duties defined outside me and my acts, in law and mores. Even if those duties accord with my own sentiments, the reality of which I feel internally, their reality does not cease being objective. For it is not I who made them facts; I received them through socialization (*l'éducation*). Moreover, it happens sometimes that we do not know the obligations that fall to us and we consult the Code and its authorized interpreters to find out what they are! (RSM: 50/3–4)

The third respect in which Durkheim takes institutions to coerce or constrain us concerns not the obligations they impose on us once we are within them but, more fundamentally, the practical necessity of participating in them at all. Refusing participation in institutions *tout court* is not an option for human agents. With respect to society's fundamental institutions, we have no real choice whether to participate in them or not and, so, no real choice as to whether we are bound by their constitutive rules: "I am not obligated to speak French with my compatriots or

[13] See also DLS: lii/xxx, discussed in Chapter 9.

to use legal currency; but it is impossible for me to do otherwise" (RSM: 51/5). As Descombes and Searle emphasize more than Durkheim, institutional rules constrain but also enable. That is, institutions greatly expand our possibilities for acting and for developing our individuality, and they do this at such a fundamental level that we barely notice it and that, even if we did, we could not withhold our participation. No one is forced to follow the grammatical rules of one's mother tongue or the rules of private property, but if one does not, one will be, in the first case, mute and mindless and, in the second, unable, in modern society, to satisfy one's needs. As Descombes emphasizes, following Hegel, institutions are sites of freedom not because we choose to participate in them but because they expand our capacity for action and even make our agency and individuality themselves possible: *we are constrained to participate in social institutions because opting out of them altogether is practically impossible* – in some cases (language) because they are conditions of belief and action in general and, in others (money) because our world is contingently such that satisfying basic needs is in effect impossible outside them.[14]

In sum, if we restrict ourselves to those aspects of Durkheim's view of the thinglike, or objective, nature of social reality that, in essential agreement with Hegel, Searle, and Descombes, we should want to retain in our social ontology, we are left with five theses: First, institutions temporally precede and survive the individuals who live within them; second, institutions have an enduring existence independent of any particular instance of the practices they make possible; third, institutions impose limits on individuals, offering resistance to their efforts to know or alter them (where "resistance" to knowing refers to social reality's being the source of objective truths about the social world that rationally constrain our beliefs); fourth, institutions constrain individuals' actions not through physical constraint but both by the moral pressure exerted by the threat of sanctions in the case of noncompliance and by virtue of a normative authority (in the broadest sense) collectively bestowed on them by their members; and finally, we are constrained to participate in social institutions because opting out of them altogether is a practical impossibility.

[14] The fact that there are institutions we cannot opt out of in no way conflicts with the claim, discussed in Chapter 6, that members of institutions *accept* and *follow* the rules governing them.

Descombes's Conception of Objective Spirit

All of the respects mentioned above in which Durkheim takes social facts to be thinglike find rough equivalents in Descombes's account of social reality as objective spirit.[15] The most important difference between the two views has to do with how Descombes connects the objective character of social reality with its quality as mind or spirit. In a nutshell, Descombes's central insight is that social institutions furnish the conditions that make objective *meaning* – and therefore genuine thought and action – possible. In fact, a less worked-out version of the idea that social institutions make meaningful phenomena possible – that they are, in other words, world-constituting – is probably implicit already in Durkheim's account of social reality. Descombes goes beyond Durkheim by articulating this thesis more precisely, defending it more convincingly and explicitly expanding it into the more sweeping claim that there can be no meaning at all outside institutions (where languages and the "forms of life" [IM, 90] they are inseparable from count as institutional in the relevant sense).[16]

Hence, even here Descombes might be understood as expanding on a claim repeated throughout Durkheim's texts but not sufficiently worked out, namely, that social institutions are supraindividual "ways of *acting and thinking*." That institutions are ways of acting is a less surprising claim than that they are ways of thinking. For one version of Rousseau's thesis concerning the artificiality of social institutions appears to assert precisely this in claiming that the continued existence of institutions depends on our ongoing participation in them: an institution ceases to be when its members stop carrying out the activities that make it what it is. There is no reason to think that Durkheim or Descombes would disagree with Rousseau's claim,[17] but it is not what Durkheim means in saying that institutions are ways of acting. His main point is not simply that institutions are constituted by human activity but that they are *ways* of acting, where "ways" – a translation of *façons*, *manières*, and *types* (RSM: 40, 45, 51/xvii, xxii, 4) – denotes something more abstract than the multitude of specific activities on which the existence of any institution depends. This more

[15] My discussion of Descombes is indebted to conversations with Richard Moran (who, however, bears no responsibility for any distortions my attempt to join Descombes's and Hegel's projects might result in).

[16] Even this general claim is implicit in Durkheim, insofar as he includes "the system of signs I employ in order to express my thoughts" among the institutions that have an objective existence prior to individuals (RSM: 51/4).

[17] In fact, Durkheim endorses it (RSM: 62/18).

abstract aspect of institutions is something like a *pattern* or *form* of acting that has no reality in the absence of specific actions that manifest it but that also is something more than the set of all its particular manifestations. Moreover, the pattern or form of acting that a social institution consists in for Durkheim (and for Searle and Descombes) is not like the pattern of foot traffic in a busy subway tunnel – where a pattern emerges out of thousands of independent decisions of individuals as to how to arrive at their own goal most efficiently – but rather a normative[18] rule that prescribes to the members of institutions what, in the broadest sense, they ought to do.

When we interpret "ways of acting" in this way, it becomes clearer how for Durkheim institutions are also ways of thinking. For, as described above, the rules defining social institutions tell individuals how to conceive of what they do, as well as, ultimately, how to conceive of themselves. There is no way of acting that is not also a way of thinking – and even, as Durkheim sometimes adds, a way of feeling (RSM: 51/4). What Descombes adds to Durkheim's point is that institutions – recall that for both language is an institution – are not only ways of thinking but also the very conditions of thought, if "thinking" is taken to be subject to intersubjectively valid norms. Expressed in Hegelian terms, Descombes's thesis is not merely that there is a realm of objective spirit – of externally embodied mindedness – but also that this domain of reality is prior to (a condition of the possibility of) the meaning, and hence the valuing, essential to subjective mind: "The subject, in order to have a mind, must be situated within a milieu formed by institutions as providers of meanings that individual subjects can make their own" (IM: 9). This aspect of Descombes's position points to yet a further sense in which objective spirit is objective: the "external" character of institutional rules – the fact that both their content and validity depend on none of us in particular – makes the meanings made possible by institutions objective in the sense of *intersubjectively valid.*[19] Or, to repeat the claim made above, *social institutions furnish the*

[18] Descombes's account of the social is normative in that, like Searle's, it makes normativity essential to its object: we do not have genuinely social relations unless individuals take themselves to be following rules that specify the rights and obligations of their statuses. In this expanded sense the rules of English grammar are normative – they tell us how the language ought to be spoken – but breaking them is not a moral offense. Similarly, an account of gift-giving can be normative without taking a stand on whether that practice is good. Descombes's account of objective spirit is normative in the first sense but not (for the most part) in the second (but see note 8).

[19] "Cooperation among the members of society presupposes the existence of institutions that provide their acts with a common context from which they derive their 'objective' meaning, the meaning these acts have for everyone" (IM: xxiv); see also IM: 293.

conditions that make objective meaning – and therefore genuine thought and action – possible.

The full scope of Descombes's contribution to the Hegel–Durkheim tradition of social philosophy comes into view, however, only when we attend to his account of the *holistic* nature of social reality – to his understanding of how institutions are prior to and make possible the mindedness of individuals.[20] The best way to enter into this aspect of Descombes's position is to begin with the principal target of his critique of Durkheim's social ontology, the latter's conception of *collective*, or *common*, consciousness. According to Descombes, Durkheim failed to explain "how … collective representations could be present within the thoughts of individuals. By referring to a 'collective consciousness' he opened himself up to [the] objection … that he had turned society into a great collective individual with its own states of consciousness separate from the individuals who constitute it" (IM: xix–xx). Although Descombes retains Durkheim's emphasis on the importance of rules – and therefore of supraindividual intentionality[21] – in constituting social reality, he thinks that Durkheim gets the ontology of social reality wrong by conceiving of institutions as grounded in a collective consciousness of their constitutive rules.

Figuring out what Descombes means to reject in Durkheim's position is complicated by the fact that Durkheim characterizes collective consciousness in contradictory ways, suggesting indeed, as Descombes charges, that he is confused about what collective consciousness is supposed to be. Sometimes, especially when using the term "common" rather than "collective," Durkheim speaks of collective consciousness as though it were a mere abstraction – "the ensemble of beliefs and sentiments common to … the members of a given society."[22] If this were all collective consciousness amounted to, it would be ontologically unproblematic – but also "unreal" and inadequate to the theoretical purposes Durkheim wants the concept to serve. At the same time, while insisting that collective consciousness "has no unique organ as a substrate," Durkheim asserts that "it forms a determinate system with a life of its own" and "has specific characteristics that make it a distinct reality" (DLS: 38–9/46). Collective consciousness's status as a distinct reality plays a role in various claims Durkheim makes

[20] Richard Moran makes similar claims in explaining how practices have mindedness, or an intentional structure, not reducible to the conjunction of the individual mental lives of their participants (Moran 2016: 319ff).

[21] In acknowledging the role that impersonal norms, or rules of objective mind, play in institutions, "we exit the subjective perspective but without leaving the realm of intentionality" (IM: 75–6).

[22] This phrase is taken from Lukes 1985: 151–7.

about it: that it possesses a life and development of its own; that it endures, while individuals pass away; and that it causally interacts with other realities (DLS: 42–3/50–1) in that, for example, its reactions both produce the punishment of crime and reinvigorate the collective consciousness and in that its weakening is a secondary cause of a more extensive division of labor in organically structured societies. But if collective consciousness is real, yet not a consciousness of the sort individuals have,[23] what kind of reality does it possess, and in what does its mindedness consist?

Descombes is dissatisfied with both components of the term "collective consciousness." First, he believes that "collective" mystifies the nature of the relation between whole and part – between institutions and the humans who participate in them – that is characteristic of social reality, a relation that, in dialogue with Hegel and French structuralism, he reconceives in accordance with "anthropological holism" (IM: xvi–xxii, 9, 87–90, 249–50). Second, he believes that "consciousness" obscures the nature of the external, objective mindedness in which social reality consists. This point can be understood as a specific version of the claim that the domain of the mental extends beyond that of the conscious. In attempting to articulate the kernel of truth in Durkheim's doctrine of collective consciousness, Descombes retains his predecessor's two Hegelian insights, namely, that social reality depends on a form of objective mind external to the consciousness of social members and that the former is metaphysically prior to the latter. At the same time, he provides an account of objective mind that does not locate it in a collective consciousness and that can explain how objective mind "could be present within the thoughts of individuals."

Like Searle and Durkheim, Descombes locates the basic unit of social reality in *institutions*. Unlike Searle, Descombes finds much (but by no means all) of the inspiration for his account of institutions in his European predecessors – in Hegel and Durkheim obviously, but also in Claude Lévi-Strauss's structural anthropology (the source of Descombes's "anthropological holism") and in the work of Marcel Mauss and Louis Dumont. It is telling that in setting out the main elements of his account of institutions in the preface to *The Institutions of Meaning*, Descombes begins with a quote from Lévi-Strauss: "Institutions … are structures whose whole – in other words, the regulating principle – can be given before the parts"

[23] "The states of the collective consciousness are of a different nature from those of individual consciousness; they are representations of a different kind. The mindedness (*mentalité*) of groups is not that of particular individuals; the former has its own laws" (RSM: 40/xvii).

(IM: xx).[24] The thought here is that the key to conceiving of institutions as external and prior to their individual members is to identify institutions with their regulating principles – with "rule[s] governing the differentiation of the parts according to distinctive oppositions and relations of complementarity within the whole" (IM: xx). Institutions in this sense count as spirit or mind because they are the "established rule[s] in virtue of which our conduct has a given meaning." Moreover, such rules give meaning to what we do and believe by "defining various statuses that we recognize one another as having." Institutions, then, are "'ways of acting and thinking' that individuals, in coming to the world and acting within it, find already established and already defined" (IM: xxix).[25]

Hence, similarly to Searle, Descombes takes the constitutive principles of institutions to be rules that define recognized, differentiated deontic statuses ascribed to members of institutions. Those rules are prior to individual[26] members in two senses that are sometimes not adequately distinguished. The first is a temporal priority referred to in the final citation of the previous paragraph: once institutions acquire a stable existence, they temporally precede future members, who are born into institutions, which for that reason confront them as a kind of second nature. Such institutions are *second* nature because they are the results (or embodiments) of human agency, not creations of nature; they are second *nature* because they appear to individuals who are born into them as merely given, not in the same manner as mountains and rivers, but as the source of rules to which individuals must conform if they are to take their place both within society and the space of reasons that makes meaning in belief and action possible.

The latter idea points to the second, metaphysical sense in which institutions are prior to their members: the very ability of the latter to act and to believe depends on their participation in institutions, of which language is the most fundamental. Human beings can exist as agents and knowers

[24] Although Descombes is inspired by this thought of Lévi-Strauss, he rejects the specific way his predecessor understands it (IM: 251–9).

[25] As noted above, "ways of acting, thinking, and feeling that ... exist outside the consciousness of individuals" is Durkheim's standard way of defining institutions or social facts (RSM: 51/4).

[26] Because of Descombes's extensive discussion of the term "individual," it should be noted that I use the term nontechnically to refer to particular human beings in contrast to its logically technical sense, according to which an individual is an independent, indivisible being that for logical reasons cannot itself be made up of individuals. On this usage, it is more appropriate to call human participants in institutions *members* because the nature of institutions is such that a real whole (a "concrete totality") "is given" before its members, who for that reason are not individuals in the technical sense (IM: 123–56). Hegel, Durkheim, and even Descombes all (also) use "individual" in the nontechnical sense; (e.g., PhR: §145; RSM: 51/4; and IM: xxix, 9, 202).

only insofar as they integrate themselves into institutions larger than their individual will or consciousness and subordinate themselves to institutional rules. One might even say, with Hegel, that *who* those beings are – the specific practical identities they embrace – is dependent on the institutions to which they belong. As Searle recognizes, a father is a father only in the context of a set of normative principles that define what fathers should and should not do and situate that status in relation to complementary statuses such as mother, daughter, son, and so on. Neither of these senses in which institutions are prior to their members implies the false claim that institutions could exist even if the world contained no conscious individuals.

We are now in a position to understand the main thesis of Descombes's anthropological holism – a version of "mental holism" – namely, that "the description of someone's state of mind is not the description of an internal state and that it is not possible to describe the state of mind in which a subject acts, responds, or exists by abstracting that subject from its social and historical world." Or, alternatively: "We must understand something of a being's form of life before we can understand the meaning of its statements" (IM: 86,90). What makes Descombes's holism anthropological is its appeal to the idea of a form of life in a distinctively human sense, which encompasses "a psychological core of needs, desires, and natural reactions as well as a historical core of institutions and customs" (IM: 90). The most important element by far in his treatment of distinctively human forms of life is their institutional character, which, as we saw above, is what makes subjective mind – meaning in thought and action – possible.

The claim that the institutions of social life that make subjective mind possible must in some sense precede the individuals who belong to them commits Descombes to a form of holism, but what distinguishes his from Durkheim's and Searle's and brings it close to Hegel's is that it regards institutions as "concrete totalities" (IM: 143), where "concrete" (Hegel's term) denotes a specific type of systematicity that institutions possess. Their systematic, or holistic, character consists in being structured by principles that define their parts through internal differentiation of the whole. In a concrete totality, "the parts … are only identifiable within the whole so that one must begin with the whole (or with the relations among the parts) rather than with disconnected elements if one wants to describe those parts" (IM: 157). Descombes's point here is similar to Searle's claim that the constitutive rules of institutions make genuinely social reality possible only by assigning deontic statuses to their members. Descombes, however, emphasizes the holistic implications of this point by stressing that institutional statuses are logically *complementary* and *definable only in relation to other statuses* within the same institution. Much of Descombes's view is worked out in conversation with Lévi-Strauss's and Mauss's holistic analyses of gift-giving in non-Western societies, but the same point can be made with respect to the modern institution of private property (as Hegel does in "Abstract Right" [PhR: §§41–6, 71]): my owning something appears at first to be a relation between a subject (me) and a thing, but in fact ownership cannot be grasped in these terms alone. I can own some particular thing only insofar as a generally recognized rule bestows complementary deontic statuses on all of us, in this case, *owner* of *X* and *nonowner* of *X*. The first status gives me the socially recognized right to dispose of *X* as I alone see fit (within certain limits), while the second obligates all others not to interfere with my disposing of *X* in the permitted ways. It is only within this internally differentiated whole, or institution, that I and other participants in the institution can *be* owners of property. Thus, what C. S. Peirce asserts about gifts – "there must be some kind of law before there can be any kind of giving"[27] – holds equally for private property.

The thought that within institutions the whole precedes its parts must be understood as the claim that the various statuses of institutions' members are possible only on the condition that those statuses are defined by an impersonal rule that defines the institution in question. This does not imply that institutions preexist the human beings who belong to them. The claim, rather, is that human beings cannot exist *qua* property owners (or parents or citizens) in the absence of institution-defining rules. One might formulate this point using the much-maligned language of internal relations – "relations that precede and constitute their terms" (IM: 185) – but only if one makes clear that the relations in question are internal to individual members only under certain status descriptions. *Qua* parent, my relation to my children is internal to me, but that relation is not a condition of my existing as a living human being.[28] Drawing on the idea of internal relations, it is sometimes said that if my children die, I cease to be myself (since my relation to my children is internal to me). Such language, however, is misleading because in such a case I do not cease to be a person or a living organism; I cease to exist only *qua* parent because my relation to my children is internal only to that institutionally bestowed status. It is for this reason that a concrete totality cannot be understood as a composition of independently specifiable parts, for its parts are definable only "through the internal differentiation of a totality that is already given" in an institution's regulating principle (IM: xvii). The concrete totalities making up objective

[27] Cited at IM: 241.

[28] "Social relations are … external to pure specimens of the human species but interior to individuals that have been specified … by their status" (IM: 202).

spirit are real, and their mode of existence is, as Durkheim struggled to grasp, *sui generis*. But that mode of existence does not consist in some form of collective consciousness; the reality of concrete totalities, or institutions, "inheres [rather] in their organization" (IM: 143). It may help to formulate the claim that such totalities are real as follows: the regulative principles of institutions are the source of – they are what make possible – certain real properties of their members, properties that, to use Searle's terminology, are not ontologically objective (real independently of any subjectivity whatsoever) but, nevertheless, the correlates of epistemically objective facts (CSR: 8). Such properties are possessed by persons only insofar as they are integrated into a system in accordance with an institutional rule; this makes them differently real from natural properties but no less real nonetheless.

It is now possible to understand why Descombes rejects Durkheim's account of social reality in terms of a *collective* mind: using "collective" to understand objective spirit – especially given the long history of attempts in Western philosophy to make logical sense of "collective individuals" (IM: 124–43) – encourages us to think of institutions as bringing together parts of a whole that can be identified as the parts they are prior to being related to one another within the institution by its regulating principles. In other words, thinking of institutions as collective entities – or of objective spirit as collective mind – obscures the most important structural feature of institutions, namely, that it is only through their defining rules that their parts are first differentiated as such and hence only within that whole that they exist as the institution's parts. What goes by the name of "collective representations" are in truth the rules in accordance with which institutions define their members (IM: 268–9).[29]

[29] The thought that mind exists outside individual consciousnesses might seem foreign to Searle's emphasis on intentionality in explaining mental phenomena. But his position is closer to Hegel's and Descombes's than it first seems. The part of Searle's view that comes closest to Descombes's conception of objective spirit is the collective intentionality he invokes to explain the acceptance of institutions' constitutive rules (Searle 1995: 23–6). Positing collective intentionality implies that there is a realm of the mental that cannot be accounted for by building up from individual conscious states, and this implies a weak holism that bears some similarity to the versions of this embraced by Hegel and Descombes. For Searle, as the member of an orchestra, I can have an intention to play my part in the performance of a symphony only as part of our collective intention to play the symphony together. Collective intentionality, Searle insists, cannot be reduced to "I-intentions" supplemented by beliefs each has about others' beliefs; in genuine instances of acting together, "the individual intentionality that each person has is derived *from* the collective intentionally they share" (CSR: 25). Nevertheless, Searle's insistence that "all my mental life is inside my brain, and all your mental life is inside your brain" (CSR: 25) – reinforced by a cartoon drawing of collective intentionality in which "we intend" appears inside each participant's head (CSR: 26) – speaks against categorizing him as a theorist of objective mind.

The holism embraced by Descombes gives content to Hegel's characterization of objective spirit as an ensemble of relations among subjects that has the structure of "an *I* that is a *we* and a *we* that is an *I*" (PhG: ¶177): the possibility of *my* believing and acting depends on my membership in institutions that make the objective meaning of beliefs and acts possible, where such membership depends on my incorporating myself into a *we* in the sense that I subject my beliefs and actions to the authority of institutional rules that hold "for us." (This is the upshot of Descombes's claim that "the subject, in order to have a mind, must be situated within a milieu formed by institutions as providers of meanings that individual subjects can make their own" [IM: 9].) A further aspect of Hegel's "*I* that is a *we*" finds expression in Descombes's anthropological holism, according to which the statuses accorded to me within institutions can be defined only by regulating principles that govern a *we*. Such principles define complementary statuses that I must accord to others if I am to possess and act in accordance with mine, implying that "who *I* am" in a specific institution is settled only by a rule accepted by *us*. In other words, objective spirit is present for Descombes when a plurality of *I*'s become the particular subjects they are only by being governed by a collectively recognized rule specifying what it means to occupy the particular roles occupied by the respective *I*'s. (Or: in institutions "the parts ... are only identifiable within the whole so that one must begin with the whole ... if one wants to describe those parts" [IM: 157].) This aspect of Descombes's view brings him close to a thought prominent in Hegel's conception of objective spirit, according to which *recognition*[30] – a mutual according of statuses among subjects – is central to the relation among *I*'s and *we* constitutive of objective spirit.

Finally, there is a further respect in which Descombes and Hegel arrive at similar views regarding the structure of objective spirit. The example of gift-giving emphasized by Descombes and his anthropologist forbears might mislead one into thinking that there is a basic asymmetry between the individuals involved in a single instance of institutionally regulated action: in gift-giving it appears that one individual is active while the other is passive. But this is a misdescription of "social action" on Descombes's account. Even though gift-giving appears to be one action carried out by a single agent, the donor, the completion of this one action in fact depends on the donee's agency as well. The point is not that giving a gift imposes an obligation on the donee that must be discharged in a subsequent act

[30] See also Thompson 2008: 195–6.

of giving in which the roles of donor and donee are reversed. The point, rather, is that the giving of a single gift depends on two agents acting in concert. My offering a gift to you does not count as gift-giving unless you take up my initiative in the appropriate way and join me in regarding my behavior as part of the rule-governed institution of gift-giving we both participate in.[31] Hegel expresses this point about the cooperative nature of all expressions of objective spirit in articulating the structure of recognition:

> This movement of self-consciousness in relation to another has been represented as the action of *one* self-consciousness, but this action of the one has itself the double significance of being both its own action and the action of the other … Action by one side would be useless because what is supposed to happen can be brought about only by both. (PhG: ¶182)[32]

Hegel's point is not that my recognizing you is complete, or satisfying, only if you recognize me in return. (He believes this, but it is not what is at issue in the passage cited.) His claim, rather, is that my success in recognizing you (or in giving you a gift) requires you to understand my bodily motions as an instance of a shared, rule-governed practice in which each of us accords a certain deontic status to the other. Objective spirit is possible only as joint, or collective, action.

From Descombes's Objective Mind to Hegel's Living Good

Descombes's account of the nature and ontological priority of institutions helps considerably to clarify the social ontology we should ascribe to Hegel and Durkheim. There is, however, a central element of Hegel's understanding of objective spirit that receives short shrift in Descombes's account and is crucial to understanding Hegel's alternative characterization of the being of social institutions in terms of the living good. That this element is (mostly) missing from Descombes's account is explained by the fact that his aims differ importantly from Hegel's. While *The Institutions of Meaning* is conceived primarily as an intervention in the philosophy of mind and in social ontology, Hegel's theory of ethical life belongs to normative social philosophy, for which questions about the good, just, or rational social order are of central importance. Hegel's characterization of

[31] Hegel makes a similar point in discussing the relations among the wills of persons who are party to a contract (PhR: §76).

[32] Descombes expresses this by saying that a single act of gift-giving "requires two operations (one on the part of the donor and the other on the part of the donee)" (IM: 247).

social reality in terms of the living good indicates that for him, much more so than for Descombes's less ethically inflected conception of objective mind, human social life is intimately bound up with realizing, or at least pursuing, the good.

Before addressing this major difference between Hegel and Descombes, I want to consider three other respects in which their positions diverge or appear to do so. The first almost certainly marks a difference in emphasis rather than substance. Descombes's accentuation of the role played by rules in constituting institutions has the potential to engender confusion over what precisely the being of institutions, or social reality, consists in. The position we should avoid is one that *identifies* institutions with the rules (or organizing principles) that define them, as if institutions were nothing more than, in Peirce's formulation, the *law* that makes giving or other social practices possible.[33] Some of Descombes's statements might suggest this overly idealistic social ontology, according to which institutions are simply identical with their defining rules, are exhausted by their meaning-providing function, or are nothing more than a structure or form of organization.[34]

A similar misunderstanding is possible if one interprets Descombes's claim that in objective spirit the whole is prior to its parts as claiming that *institutions* are prior to their members (under certain descriptions).[35] Strictly speaking, it is the institutional *rule* that is prior to social members, not the institution itself, which must be understood not merely as a rule or structure or form of organization but, more broadly, as the ensemble of social activities, carried out in a spatially and temporally real world, governed by the rule in question. This point comes out more clearly in Hegel's idea of the living good, especially in the thought that institutions constantly produce (and reproduce) something in the world – namely, the good. An institution is not merely a principle that guides social action but

[33] Cited at IM: 241.

[34] In most cases the possibility of misunderstanding is explained by the specific interlocutors Descombes is addressing: contemporary philosophy of mind, with its focus on meaning and the phenomena of subjective spirit; the French structuralists, who focus on institutional structure; or Montesquieu, who emphasizes ways of thinking over ways of acting.

[35] This misunderstanding is encouraged by a mode of speaking, borrowed from Lévi-Strauss, that identifies "the whole" with its "regulating principle" (IM: xx). (Should not the whole be the regulating principle and the actions it determines?) The same misunderstanding is a possible consequence of Searle's talk of the constitutive rules of institutions. Such rules are institution-constituting in the sense that, like the constitutive rules of games, they specify the conditions that make specific actions count as participation in a given institution. They are not constitutive of institutions in the sense that rules exhaust what institutions are (just as there is more to games than their constitutive rules).

the entirety of institutional life itself – the realization of a principle in the concrete actions of social members. Although some of Descombes's formulations might encourage the ontological confusion I want to avoid,[36] in most instances he is careful to endorse the broader account of institutions pleaded for here.[37]

A second respect in which Hegel's position appears to diverge from Descombes's – and actually does – concerns the extent to which institutions bear a resemblance to biological organisms. Descombes broaches this topic in saying that the holism he endorses, where the division of the whole into its parts is accomplished through internal differentiation (the specification of mutually complementary statuses) according to a rule of the whole, bears similarities to the structure of organisms, a fact he takes to explain the persistence of organicist metaphors in social theory (IM: 142). But there is an important respect in which Hegel's account of ethical life makes institutions more like organisms than is implied by Descombes's holism. In a word, Hegel's account requires real *specialization*, including permanent differentiation, among members of institutions, whereas Descombes's does not. When Descombes discusses private property, he emphasizes that the statuses specified by that institution, or any other, must be defined complementarily – in this case, as owner and nonowner. Thus, all interactions within that institution require differentiated statuses, but in this case and many others – language, for example – differentiation does not imply enduring specialization among an institution's members: I am an owner in relation to my property, but at the same time I am a nonowner in relation to yours, and nothing in the constitutive rules of private property requires that I be different from you with respect to our enduring qualities. In this case, all participants in the institution can occupy both defined roles, even simultaneously, and with respect to the particular things owned, all can move in out and of complementary roles with relative ease.

For Hegel, in contrast – as for Durkheim and Plato – an enduring specialization of parts, carrying out specialized functions, is necessary in the specific institutions he focuses on: Families require children and parents,

[36] E.g., "an institution is the established rule in virtue of which our conduct … has a given meaning" (IM: xxix). Perhaps, however, "*established* rule" – the passage in question is not part of the original French text – is meant to signal that institutions are more than mere rules.

[37] In one place he characterizes institutions as "various rules *and practices*," and in another, following Durkheim, as "ways of thinking *and acting* that individuals find established in the world" (IM: xxvi, xxix; emphasis added).

as well as husbands and wives; civil society requires specialized professions that take a lifetime to master; and the state requires a class of trained bureaucrats, distinct from the productive classes, if its machinery is to run effectively. As Descombes's social ontology makes clear, nothing in the nature of institutions *most generally* requires specialization in this sense, even if differentiated, complementary statuses are essential to them. While Descombes's account of institutions as requiring internal differentiation of parts in accordance with a rule governing the whole might help to explain the constant reappearance of organicist metaphors in social theory, it is far from being the whole explanation. As we will see in discussing Hegel's concept of the living good, normative social philosophy, whose main concern is not to elucidate the abstract structure of institutional reality in general but to understand and evaluate real human societies, frequently appeals to organicist metaphors because the central tasks of real social life are so complex and consuming that well-functioning institutions cannot dispense with the kind of specialization Hegel's theory relies on (even if one can reject specific forms of specialization – in the family, for example – that he believes necessary).

A third difference between two positions is also attributable to the fact that Descombes aims to understand the being of institutions in general, whereas Hegel focuses on the normatively most important of them, the three central institutions of modern ethical life. Descombes's critique of traditional versions of the doctrine of internal relations is both compelling and consistent with Hegel's understanding of objective spirit. Yet Hegel's account of ethical life implies a stronger, but still unobjectionable, version of that doctrine than the weak one endorsed by Descombes. The point at issue concerns the claim made by proponents of internal relations that in the case of phenomena for which that doctrine holds, it is impossible to change the relations in question without changing the reality of the related terms, even to the extent that if those relations cease to exist, so, too, do their relata (IM: 187), such that, for example, if a father's children die, he ceases to be (or to be himself) because his relations to his children are internal to what he is. As we saw above, Descombes rightly rejects this way of speaking because it fails to distinguish the man's existence in general from his existence – his institutional *status* – as a father, encouraging the illicit conclusion that the man's very being (in every sense) resides not in himself but in his relations to others. Yet there is a further sense, more robust than the one allowed by Descombes, in which for Hegel a father's relations to his children are, and in a well-functioning family must be, internal, or essential, to what he is, or to his *practical identity*. I explain in the following chapter how such practical identities are to be understood and why Hegel takes them to be indispensable in social life. For now it is sufficient to note that in this context "who I am" is inseparable from "who I take myself to be" and that to say that my relations are essential to my being is to say that they are the source of commitments and life-defining projects without which my agency would lack an organizing, meaning-endowing core. There is nothing metaphysically illicit in this way of construing how my relations to others constitute my identity, and ordinary language reflects this thought, as when one says after the death of a spouse or child, "I'm not myself anymore."

The theme of thickly defined practical identities that are at stake in the central institutions of social life and that provide meaning to individual lives brings us again to the topic of the following chapter, Hegel's characterization of social reality as the living good. Part of what it means for Hegel to conceive of social reality in this way is that in social life the good is produced and reproduced in the world or, more generally, that ethical values acquire a real existence in an objectively constituted world. This claim marks the fourth respect in which Hegel's conception of objective spirit goes beyond Descombes's. Interestingly, though, a version of this point finds expression in one way in which Durkheim articulates his claim that social institutions are externally existing ways of thinking and feeling. Sometimes Durkheim formulates this claim in a manner reminiscent of Hegel's thesis that the norms of ethical life are not mere ideals or empty oughts but instead living principles that animate real social life and therefore have a functioning existence in the world external to subjective consciousness:

> What is given to us [sociologists] is not the idea of [economic] value humans make for themselves, for that is inaccessible, but rather the values actually exchanged in economic transactions. [Likewise,] it is not one or another conception of the moral ideal [that is given to us] but rather the totality of rules that in fact determine behavior.... One cannot reach [moral and other ideas] directly but only through the phenomenal reality that expresses them. (RSM: 69–70/27–8)

The idea that phenomenal reality – systems of exchange and social life more generally – *expresses values*, moral as well as economic, is crucial to any Hegelian conception of social reality as objective spirit. Socially efficacious ethical ideals, on this view, are not merely subjectively espoused values that reside originally in the consciousness of individuals and are then made real only when, and if, they decide to act on them. Rather, such ideals are inscribed in the social life itself – in the supraindividually authorized rules of its institutions – into which individuals are born and within which they are educated. The ideals of social life are realized in the world through the enduring practices of their members, the characteristic mode of which is *reproduction of the already existing* rather than creation, fiat, or subjective choice. For Durkheim the rules of social life not only prescribe to participants what they must do (including when merely consulting one's own conscience[38] yields no determinate prescription); those rules also infuse reality with ethical values insofar as the functioning of institutions is inseparable from the value-laden rules that govern them. *Institutions*, in other words, *objectively embody, or express, ethical values*.[39]

Searle comes close to expressing this Hegelian point when he says that, as for all institutions, saying what a restaurant is implies something about what a good (well-functioning) restaurant is (CSR: 3–4).[40] The point is better illustrated, however, in Hegel's rich accounts of the institutions of ethical life, as well as in his account of the bondsman–lord relation in the *Phenomenology*, which I discuss in the following chapter. In the modern family, for example, the informal but unmistakably operative rules governing the interactions of parents and children express ethical values: that the needs of children must figure in their parents' decisions gives expression to the moral principle that every individual has interests that cannot be arbitrarily overridden; that children must be brought up so as to be able, as adults, to separate themselves from their parents and to live lives of their own choosing gives expression to the ideal of personhood; and that parents must discipline (and children must obey) gives expression to the ideals of self-restraint and self-mastery. Moreover, some rules that express the values intrinsic to family life are codified in law and enforceable by the state. (This is clearly seen by Durkheim; it undergirds his method in *The Social Division of Labor* of looking to legal codes in order to understand diverse forms of social solidarity.) An example of value-expressive law, for Hegel, is found in the legally enforceable requirement that when the father – the family's head and representative in the economic sphere – dies before his wife and children, the family's wealth falls to them. Such a law expresses

[38] Durkheim's *conscience* can be translated either as "consciousness" or "conscience," and he exploits this ambiguity.

[39] This is the idea behind Charles Taylor's claim that "the meanings and norms implicit in … practices are not just in the minds of the actors but are out there in the practices themselves" (Taylor 2010: 38).

[40] Thompson also argues that the authority of rules governing social practices rests ultimately on some standard of the good (Thompson 2008: 196–8).

the ethical principle that, because a defining feature of the family is that its members live a common life (rather than relating to one another as independent individuals), family wealth is "property held in common," in which each member, daughters as much as sons, has a "rightful share" that cannot be abrogated by the capricious (unethical) choice of the father to dispose otherwise of the family's wealth (PhR: §§170–1, 178–80).

For Hegel the reproduction at issue in social institutions goes beyond the value-expressive reproduction of customs, mores, and practices in general; it includes as well the material reproduction of society, most centrally in the institutions of family, economy, and state (one of whose functions is to ensure the health of the family and economy). This means that it is not merely institutional life of any sort that expresses values but material practices, too (those that make our continued biological existence possible). For this reason, one can say that in a rationally organized society, nature, in the form of animal life, becomes spiritual (and hence is no longer merely natural or animal life). This thought plays a major role in Hegel's conception of social life as the living good.

CHAPTER 12

Hegelian Social Ontology II: The Living Good

When Hegel sets out his account of rational social life, or *Sittlichkeit*, in the *Philosophy of Right*, he begins by identifying it as "the living good" (*das lebendige Gute*) (PhR: §142). The standard English translation of *Sittlichkeit*, "ethical life," reflects this definition: most fundamentally, social life consists in processes of life infused with ethical content. Although human society is an ethical phenomenon for Hegel, the term "ethical life" signals that his account of the rational society has ambitions beyond those of moral philosophy. Calling rational social life "the living good" expresses an ontological thesis about the kind of being human societies possess: like Durkheim, Hegel takes human societies to be normatively and functionally constituted, *living* beings. For Hegel, societies are normatively constituted not only because their functioning requires social members to act – to a large extent, habitually – in accordance with norms they take to be authoritative (because productive of the good) but also because when they function properly, social institutions *realize* the good.

Moreover, human societies realize the good similarly to how living organisms achieve their vital ends, namely, by means of organized subsystems, carrying out specialized and coordinated functions that enable the organism to be self-maintaining and self-reproducing (PhN: §§350–3). In distinction to living organisms, however, the living good realizes itself in human societies through the awareness, will, and actions of its members, human individuals (PhR: §142); in other words, the living good that animates rational social life and is realized within it is a *spiritual* living good. One could express this in Hegel's language by saying that spirit in the domain of the social is *aufgehobenes Leben*, life that is both transcended and preserved within the spiritual. In the first two sections of this chapter I explore this idea by discussing, first, how social life is like biological life (or how biological life is preserved within human society) and, second, how the spiritual life that characterizes human societies goes beyond merely biological life. In the third section I examine the conception of the good at issue in Hegel's description of ethical life as the living good, and in the final section I look at one especially important component of the human good, recognition.

Biological Life

Hegel's concept of life is complex and obscure, and it plays various important roles in his philosophy as a whole.[1] Here I restrict myself to the aspects of his account most relevant to social philosophy, examining several respects in which human societies preserve, or retain features of, the phenomena of life. They do so in two quite different senses: social life both *incorporates* and *mirrors* (or possesses the same structure as) the activities of biological life. The first of these points is the easier to grasp. It consists in the familiar claim that material reproduction – producing the human bodies and necessities of life required if society is to survive over generations – is an indispensable task of human social life. Like Plato and Durkheim, Hegel takes social life to be biologically necessary for humans, just as it is for other animal species: nature imposes biological needs on living creatures, and for humans, just as for ants and bees, satisfying those needs requires cooperating with fellow members of the species. This basic anthropological claim should be uncontroversial, but it often evokes resistance, some of it due to our tendency to rebel at the thought that we are not fully self-determined – that something other than our own doings might place constraints on who we are and what we can become. Some of this resistance, however, has a less prideful source that dissipates once we get clear on the relative weakness, in two respects, of the resisted claim. First, saying that social life is biologically necessary does not imply that material reproduction exhausts, or even accounts for the largest share of, the meaning and importance such life has for us. The fact that biology requires us to cooperate regularly with others, and to spend large parts of our waking life doing so, does not entail that survival and reproduction of the species are the principal ends of social life. Second, nature's placing constraints how we live does not mean that it *determines* that life, dictating specific acts or forms of organization. Hegel, like Rousseau, is well aware that there are uncountable ways in which humans can satisfy their needs for nutrition, shelter, sleep, and sex. Within material reproduction there

[1] My understanding of Hegel's conception of life is indebted to Khurana 2013: 155–93. The best account of life as it relates to social pathology is Särkelä 2018: 220–35. See also Ng 2020, Hahn 2007, and Testa 2020.

is ample room for human self-determination, even if it is always subject to very general constraints that are not themselves self-imposed.

The second sense in which Hegel takes human societies to preserve elements of animal life – in *mirroring* its biological prototype – is more difficult to explain, although some of the ideas relevant to this claim were broached in Chapter 3 in Marx's account of the lifelike nature of capital's circulation. Hegel's claim is that human society is to be understood *on the model of life*, or that the two share a basic structure that nonliving (and nonspiritual) beings lack. (To be clear: all spiritual beings are also living beings, but the converse is not the case.) More specifically, his claim is that human society is purposively organized, carrying out specialized and coordinated functions that enable it to maintain and reproduce itself as the kind of being it is: the rational society "exists as living spirit only as an organized whole, differentiated into the particular functions (*Wirksamkeiten*) that … continually produce it as their result" (PhM: §539).[2] Hence, the basic feature of life Hegel exploits in conceiving of human societies is its functional character. In his philosophy of nature, when articulating what it is to be an animal organism, he appeals to Aristotle's characterization of the living as something that "is to be regarded as acting [or producing effects] in accordance with purposes [or ends] (*Zwecke*)," although it "does not know its purpose as a purpose" – which is why both thinkers describe the behavior of nonhuman animals as "unconscious acting in terms of purposes" (PhN: §360). For Hegel, both societies and living things are defined by their functions in the sense that one cannot understand what they are or why they are constituted as they are without referring to the ends their features and activities serve.[3]

The functional character of life can be illustrated by looking first at lower-level purposive behaviors of organisms. We can ask, for example, "why do certain animals salivate?" (or "why do they have salivary glands?"), which in the domain of life is to ask, "what function does salivation serve?"[4] In the

[2] "The true actuality of freedom is the *organism*, … its differentiation into … abstract … tasks so that out of these … determinate labors and interests the universal interest and work results" (VPR: 269). Also: "The living self-production of spiritual substance consists in its organic *activity*: … the articulation and division of its general business and power into … different powers and businesses" (VPR: 150).

[3] An articulation of a neo-Aristotelian conception of life that also illuminates Hegel's can be found in Thompson 2008.

[4] Biologists might also ask, "By what causal history did salivary glands come to be?" As I argued in previous chapters, this question is less relevant for social theory, but even in biology, determining origin without determining function falls short of explaining completely, in this case, what salivary glands are.

case of living beings, the answer to this question – "salivation aids the digestion of food" – elicits a further question, about the function of digestion, and so on, until one sees how the original behavior serves an end of the organism as a whole. What brings this series of questions to a close is the discovery of a characteristic end of the organism, one that defines it as the kind of being it is. The ends characteristic of life in general for Hegel are self-maintenance and the reproduction of the species (PhN: §§350–3), although for any given type of organism, its characteristic ends – or the form of life characteristic of it – will be more specifically defined than the ends of life in general. This implies – this is what it is to say that life is purposively organized – that the various specialized functions of living beings work together to achieve those beings' vital ends (PhN: §354): the function of salivation is to aid digestion, and digestion's function is to enable the organism to maintain itself, which typically serves the end of reproducing the species.

This points to a further aspect of life's purposive organization: once the features and behaviors of a living being have been shown to serve one of its vital ends, the call for further explanation ends. It would reveal a failure to grasp what it is to understand a living being if one responded, "I see that salivation enables the porcupine to digest its food, which in turn enables it to remain alive, but what's the point of that?" This is part of what Hegel means when he says (taking himself to be following Kant) that a living being is not merely purposively organized – this is also true of certain non-organic artifacts, such as a watch – but also an end in itself (*Selbstzweck*) (PhN: §360; PhM: §423A) (Ng 2017: 272–3), or something that works for the sake of ends that are not themselves instrumental to achieving ends beyond the maintenance and reproduction of life itself. The ultimate characteristic ends of a living being are not external but come, as it were, from that being's form of life itself, which is why Hegel regards the activities of both biological and social "organisms" as self-determined (determined only by the nature of the organism itself – its characteristic ends – that carries out those activities).

Hegel's description of human society as an organized (and self-organizing) whole that reproduces itself by means of differentiated functions is illustrated by his well-known doctrine that a rational feature of modern European society is its division into three semi-autonomous, coordinated spheres: the nuclear family, civil society (the institutions bound up with a rational market economy), and the constitutional state. Regarding these spheres as relatively autonomous subsystems with distinct functions implies that it is meaningful to make the family, the economy, and the state into objects of study in their own right and to ask about each, "How does this sphere carry out its distinctive functions? By what mechanisms does it accomplish its part of 'the general business' [VPR: 150] of society?" If we restrict ourselves at this juncture to considering only the living (nonspiritual) aspects of human social life, Hegel's point is easily grasped. Each of the spheres carries out a distinct function necessary for the material reproduction of society, and this is one major reason Hegel accords special importance to them: the family furnishes the next generation of human individuals; civil society supplies the material goods needed for the sustainment of life; and the state, while less directly tied to material reproduction, shores up the two "lower" spheres and ensures that neither functions at the other's expense. Of course, once we bring the spiritual aspects of human social life into the picture, each sphere can also be seen to carry out specialized spiritual tasks, such as realizing different forms of practical freedom and producing the forms of subjectivity, or self-conceptions, appropriate to each. Moreover – this point cannot be repeated too frequently – society's material and spiritual tasks are not carried out separately. The activities that make up life in civil society, for example, possess both material and spiritual significance – or, more accurately, their spiritual character consists precisely in their integrating two general types of ends within a single activity, making processes of life at the same time ways of realizing freedom.

Hegel's conception of an organized whole, appropriated from Kant's account of a self-organizing being, or *Naturzweck*, implies not only functional differentiation but also organization in accordance with an overarching telos. Thus, apprehending something as an organized whole includes both ascribing an essential end, or telos, to it and grasping how the constitution of its parts is determined by what it requires in order to achieve its defining end. Formulated in Kant's terms, the thing's end "determines a priori everything it is to contain" (CJ: §65). In biological organisms, this telos is the reproduction of the species to which it belongs (including, perhaps nonderivatively, the maintenance of its own individual life), and its organization takes the form of a network of functional subsystems, each of which operates with a degree of autonomy,[5] even though ultimately subordinated to the end of the whole and dependent on its proper functioning (PhR: §270A). This means that the relation between organism and its specialized parts is one of mutual dependence. Not only does the functioning of the whole depend on each component carrying out its specific

[5] "The different parts within the state [or social whole] must exist as members with their own distinctive organization, which are autonomous (*selbständig*) in themselves and bring forth, reproduce the whole" (VPR: 151).

task, but the parts, too, depend on the whole – or, more accurately, on being united with the others so as to constitute a properly functioning whole – in order to realize their distinctive ends. This mutual dependence between parts and whole – the "interpenetrating unity" of universality and particularity characteristic of organisms (PhR: §258A) – implies that both biological organisms and rational societies achieve their defining ends not by squelching diversity but by fostering it and by bringing its diverse elements into a purposeful arrangement, thereby preserving the qualitative richness that difference implies. This aspect of organic structure, of great importance to Hegel's conception of the relation between individuals and society, is for him another feature of a *Selbstzweck*: "Life must be grasped as a *Selbstzweck*, as an end that has its means within itself, as a totality in which everything that is distinguished [from the whole and from other parts] is at once both end and means" (PhM: §423A).

There are two related aspects of Hegel's conception of rationally organized wholes that should give us pause insofar as it is supposed to apply not only to biological organisms but to human societies as well. One is the idea of *perfect* teleological organization, where every detail of a being contributes to the whole's functioning and precisely complements its other parts. This idea belongs to Kant's conception of a *Naturzweck* – an Idea of reason never perfectly instantiated in experience – and it may well be approximated in real living organisms (despite vestigial structures, now explained by evolutionary theory). Regardless of how perfect Hegel thought the organization of a rational society could be, contemporary social theorists do best to follow Durkheim's lead on this issue; he recognized that human societies, though functionally constituted, are less perfectly organized than living organisms (RSM: 120/91). Closely related is the further idea that the diverse parts of an organized whole ultimately serve a single overarching telos. I have already indicated that even in the biological case, organisms' behavior might not serve just one defining end, insofar as the end of species reproduction can be pulled apart from that of a single organism's self-maintenance.[6] In the case of human societies it seems even less plausible that the various functions of their parts could be explained by referring to just one end of the whole. On this point, however, it is worth remembering that the single telos Hegel ascribes to rational social life, conceived of as the living good, is a very expansive conception of the *good* that includes various social

[6] In most species, organisms continue to preserve their own lives even when they can no longer contribute to reproducing the species.

goods – the well-being of individuals and their freedom of various types – that non-Hegelians might regard as distinct ends of social life.

Thus far I have talked about individual living organisms, but Hegel usually conceives of life more generally, as the vital activity of the *species* (EL: §220),[7] which itself consists in nothing beyond the vital activity of multitudes of living individuals (PhM: §423A). Underlying this shift in focus from the individual to the species is an ontological thesis concerning the (existential and conceptual) priority of the latter over the former (PhN: §§366–7) that echoes the priority thesis examined in the preceding chapter's account of objective spirit with respect to institutions and their members. One feature of living beings, in distinction to rocks and raindrops, is that a relation of dependence holds between them and their species. Living individuals, according to this thesis, are not independent (*selbständige*) entities because their being what they are depends on something greater than themselves alone: the species to which they belong.[8] This can be interpreted existentially, meaning that living beings are the biological offspring of other living beings, which in turn are the offspring of others, but Hegel's deeper claim concerns the conceptual dependence of individual organisms on their species. Most generally, this means that explanations of the makeup and behavior of individual organisms rest on judgments about the activity appropriate to their species' form of life, where this involves attributing a function to that activity connected to the aims of life.[9] In other words, characteristic behaviors of living beings "refer to" their species and can be understood only in that relation. A clear example of this is when the sexual drive impels living beings to behave in ways that serve no end from the point of view of their own individual vital needs. Living beings characteristically engage in behavior that serves only universal ends,[10] which means that we can understand such beings only by situating their behavior in relation to the species' ends as a whole (the reproduction of the species, as opposed to individual self-maintenance). Living beings are what they are only insofar as they participate in the life of their species.[11]

[7] At its most comprehensive, life includes the activity of all species, since the vital activity of one typically requires relating to (e.g., eating) members of others (PhN: §251).

[8] The species' existence also depends on that of its individual members since it does not exist beyond them and their activities.

[9] Thompson 2008: 24–30 and ch. 4.

[10] Pleasure or release of sexual tension is not an end in this (Kantian) sense; an end is a state of affairs brought about in the world.

[11] Hegel makes a similar claim about human individuals and their social life, but the primary analogy his social ontology relies on holds between *society* and the life of a biological *species*.

Before examining how spirit – and, so, social life – goes beyond the biological, we must note a further respect in which Hegel takes society to mirror the structure of animal life. In characterizing living beings he makes the following puzzling remark: "A being capable of holding the contradiction of itself within itself and of *enduring* that contradiction is a *subject*; this constitutes the subject's *infinitude*" (PhN: §359; Hegel's emphasis). In the *Phenomenology of Spirit*, a subject is further characterized as "pure, *simple negativity*," as "the bifurcation (*Entzweiung*) of the simple,"[12] and as a doubling that posits difference (or opposition) internal to itself and then "negates" this difference by restoring identity between the opposites it has posited (PhG: ¶18; Hegel's emphasis). Thus, the nature of a subject, according to these passages, is to engage in a distinctive activity that consists in dividing itself into two and then negating this division by unifying what it is has torn apart, albeit in a way that both negates and preserves – thus enabling the subject to "endure" – that internal division. That this defining activity of subjectivity is crucial to understanding both life and spirit is evident in Hegel's claim that it is what makes spirit a "living substance" rather than an inert, lifeless one (PhG: ¶18); less clear, unfortunately, is what all of this means.

From our discussion of Marx in Chapter 2 we are already familiar with some aspects of Hegel's thesis that living beings, and life in general, are "subjects" (PhN: §337) and that what makes them subjects is their ability to endure internal contradiction.[13] Various features of living beings inspire Hegel to make this odd-sounding claim, but I restrict myself here to those most relevant to his understanding of human society.[14] The main idea is that a living being does not merely exist, inertly, like a stone; it is not, as Hegel puts it, a mere *Seiendes*, something that merely *is* (PhN: §352). Life's mode of being, rather, is like that of a subject – recall Fichte's claim that a subject has no being beyond its own self-positing activity (Neuhouser 1990: 102–16) – in that it exists only as a *sich Reproduzierendes*, only insofar as it constantly reproduces itself, including reproducing its characteristic form (PhN: §353; Rand 2015: 74–6): "The living being … exists only by making itself into what it is" (PhN: §352). This, too, is part of what Hegel

[12] For the role of *Entzweiung* in life and self-consciousness, see Ng 2020: 72–4, 98–9, 103–4, 106–7, 212–2, 268–9.

[13] Recall Marx's claim that the circulation of capital is a "subjective" activity because capital ceaselessly reproduces itself by positing and negating contradictions internal to itself (taking the form of money, then of a commodity, then of money again, and so on) and only thereby constitutes itself as a single process.

[14] For more detail, see Khurana 2017: §§65–8.

means in characterizing life as a *Selbstzweck*. For the point that a living being exists only insofar as it constantly (re)makes itself into what it is can also be expressed by saying that it is an "antecedent end (*Zweck*) that is itself only the result" of its own life activity (PhN: §352).

Hegel's thesis that life exhibits the structure of subjectivity also includes a claim about *how* a living being reproduces itself, namely, by constantly relating to what it treats as *not* itself[15] – air, water, other organic material – or, more precisely, by assimilating what is foreign to it, not by simply ingesting its "other" but by transforming it so that it can then be subordinated to the organism's vital ends. A living being is "infinite," and like a subject, because it relates to itself – it maintains its own living form – only by relating to what it is not and in doing so makes what was other into itself: "Animal subjectivity is this: in its corporeality and its being affected by an external world, maintaining itself and remaining with itself (*bei sich selbst*)" (PhN: §350A). One could say that such a being endures internal contradiction (the contradiction between itself and what it is not), but it might be more perspicuous to say that it feeds on – lives off of – a certain way of negotiating this contradiction. The living being "*is* and *maintains itself*" (PhN: §352; Hegel's emphasis) as what it is – a purposively organized being "set up" to maintain itself and reproduce the species – only by interacting with what it is not, and it does so fluidly, that is, constantly and without losing its identity (without flipping back and forth between states of being itself and being its opposite).

As we should expect, Hegel ascribes this subjective structure not only to individual living beings but also to life more generally: a biological species maintains itself by breaking itself up into individual living beings, each distinct from its species, even if at the same time it is identical to its species in the sense that its form of life is that of the species. It is these individual living beings that, in striving to live and to reproduce, and ultimately in dying, carry out the vital functions on which the continuation of the species depends. Only by positing differences internal to itself (in the form of individual members of the species) (PhN: §371) and then abolishing those differences does the species as a whole reproduce itself as what it is. Individual living beings are nonidentical with their species not only because the two can be distinguished – a single amoeba is not the same thing as the species of amoebae – but also because they have ends not

[15] One basic feature of organisms is the capacity to distinguish inner from outer (PhN: §§357, 359; Pinkard 2012: 24).

fully identical to those of their species. Living organisms, in other words, have a degree of individual integrity within their species' life insofar as one organism's survival is to a large extent independent of others' and any individual is capable of turning against the species, or "disowning its … continuity with it" (PhG: ¶171).[16] (Another example of how a living species posits internal differences, clearly relevant to Hegel's social theory, is that it produces individuals of different, specialized sexes [EL: §220].) A species' abolishing of difference – its negotiating contradiction in the manner of a subject – consists in asserting its identity with the individual living beings that compose it by putting their activities (of self-maintenance, reproduction, and ultimately, dying) in the service of its own larger end, the reproduction of the species.

A further aspect of Hegel's thesis that life posits differences internal to itself and then endures, or negotiates, those contradictions is the claim that life's characteristic activities are *processes* (PhN: §§215, 337, 367, 369). (This point, recall, is appropriated by Marx in his claim that the countless instances of $M-C$ and $C-M$ in the circulation of capital constitute a single, "subjective" process.) A process in the sense intended here is a temporally extended, self-generated series of changes (where changes are "negations" of what previously was) in which different ("contradictory") states succeed one another. Successive states acquire a coherence and unity – become *stages* of a process – by virtue of the end the series as a whole serves, which in the case of life is the reproduction of the species. The reproductive cycle of a fruit tree, for example, involves the sprouting of leaves, the blossoming of flowers, the transformation of the latter into fruit, the production of seeds that will begin the next generation, and so on. Again, this feature of life applies to both individual organisms and the species as a whole. Not surprisingly, Hegel will also claim that *social* practices are processual in this sense, even if in this case (where life is spiritual), in order for the changes in question to constitute a process, they must be interpreted as such by the agents of those changes, namely, self-conscious human individuals.

Spiritual Life

This thought leads naturally into the topic of how social life, as a form of spirit, differs from merely biological life – how, in other words, it not only preserves but also negates (goes beyond) mere life, imbuing processes

[16] Sows, for example, sometimes kill their young after birth; similar phenomena occur in other species, too.

of human life with spiritual significance. The most important respect in which spirit is more than mere life is easily stated: although a biological species maintains itself by positing differences internal to itself and then abolishing them, as well as by assimilating its external "other" to itself and making the latter serve its own purposes, mere life, unlike spirit, is unconscious of itself and its characteristic activity (PhM: §§381A, 386A). Expressed in Hegelian language, life can exist as what it in truth is (a unified process) only for another – its unity can be apprehended only by a conscious being – whereas spirit exists "for itself," which implies that members of human societies are conscious of the ends their social activities serve and that unify their various actions over time into a practice. Since a merely living being passes through its reproductive cycle without awareness of the end that unites its diverse states into a process, the latter differs from social processes, or *practices*,[17] in which the activities constituting them are mediated by an understanding of their governing ends. Spirit, then, could be defined as *self-conscious life* (PhM: §385A),[18] and spiritual beings, one might say, are self-conscious animals (Brandom 2007: 127–8).

This definition encourages us to ask: What becomes of life and its essential functions when self-consciousness is added to them?[19] There is nothing wrong with posing the question in this way as long we take care to avoid a misunderstanding the word "add" might encourage.[20] Spirit is not to be understood as self-consciousness *alongside* life, or as the additive result of "mixing" the two together. Rather, objective spirit is *spiritual life*, or processes of life carried out in the mode of – and therefore thoroughly permeated and, so, transformed by – self-consciousness. It is not possible to explore here all of what self-consciousness includes for Hegel, nor all the complications that arise when it is introduced into the living.[21] Instead, I note four of the most important implications of joining self-consciousness and life in the idea of spirit.

First, the telos of life is transformed when living processes are imbued with self-consciousness. This is because of the connection between

[17] For an account of social practices that bears similarities to Hegel's, see Thompson 2008: 159–201. A comprehensive discussion of the concept of a practice can be found in Stahl 2013: 260–75.

[18] See Ng 2017: 286–7 and Menke 2018: 82–116.

[19] For more on the relation between self-consciousness and life, see Ng 2020: 101–12 and *passim*; and Ng 2015: 393–404. Alznauer 2016a: 196–211 examines the normativity appropriate to spiritual phenomena and compares it to that of merely living beings.

[20] See Khurana 2017: §74; Boyle 2016: 527–55; and Haase 2013: 1–2, 91–106. For a provocative take on what, for German idealism, self-consciousness "adds" to life, see the essays in Kern and Kietzmann 2017.

[21] For a treatment of this topic with respect to freedom, see Pippin 2008: 36–64.

self-consciousness and free agency. In spiritual beings, life's supreme end, biological reproduction, expands into the end of freedom (where freedom requires self-consciousness of various kinds, to be discussed below): "The *essence* of sprit is *freedom*" (PhM: §382), not merely biological reproduction. This does not mean that the end of biological reproduction is replaced by freedom in spiritual beings; rather, their supreme end becomes the reproduction of life in accordance with the (demanding) requirements of freedom. Spiritual life is life "infused with" freedom; it consists in life processes through which – and *in which* – its participants achieve the complex panoply of freedoms made possible by self-consciousness.[22]

The second implication of the self-conscious character of human social life is expressed by Hegel as follows: "Ethical life is the *idea of freedom*, which, as the living good, becomes actual in self-consciousness – in its knowing, willing, and acting" (PhR: §142). Taken by itself, this is merely another way of saying that human society achieves its defining end, the free reproduction of life, in the mode of self-consciousness. If we add to this, however, that society possesses no organ of consciousness of its own, independent of those of its members, then it follows that social life accomplishes its ends only in and through the self-conscious knowing, willing, and acting of its members, who participate in it as individuated bearers of consciousness. The first two respects in which spiritual life differs from, while also remaining continuous with, mere animal life could be summarized by saying that spiritual beings reproduce themselves as living beings in the mode of self-conscious freedom, where the latter describes both the end of spiritual activity and the manner in which it is carried out. (Durkheim, Searle, and Descombes endorse a version of this point insofar as they regard the following of rules by self-conscious subjects as intrinsic to social reality.)

As Hegel and Durkheim recognize, that the social world is composed of parts that are individualized units of subjectivity marks an important difference between social and biological "organisms," for it means that members of society are themselves spiritual beings (PhR: §264). Thus, it is not merely society as a whole, through some mysterious collective consciousness, that is self-aware. Its parts, individual humans, also

[22] Regarding "through" and "in": life processes make freedom possible (having enough to eat is a condition of being free), but in a rational society they also are the sites of freedom's realization (participating in family life is itself an expression of freedom). For the latter doctrine, see Neuhouser 2003: 33, 82–4.

possess self-consciousness, such that the content and adequacy of society's self-consciousness depends directly on that of its individual members. This has implications not only for understanding how social life accomplishes its various tasks – via processes that operate through the subjectivities of its participants – but also for the normative standards appropriate to social life: if individuals are spiritual beings, then the tasks of social life must be carried out in a way that does justice to their spiritual nature. In other words, what is at stake in rational social life is the welfare and freedom – the good – of individuals (PhR: §264) and not merely (if this idea can be made sense of) the welfare and freedom of some collective whole. This in turn means that, like Durkheim, Hegel is not a normative holist in the strongest sense, holding that the social whole has interests of its own not traceable back to the good of its individual members. For both, members of human society have a greater integrity and independence than the model of the biological organism suggests: individual social members, while dependent on their species and its life processes, are fundamentally different from the mitochondria or food vacuoles of amoebae.

The third respect in which self-conscious life differs from mere life follows from the second: processes of spiritual life are carried out with awareness on the part of those who do so not merely that such processes are taking place but also what their function, or point, is. Social members' activities not only have a point; they also have a point *for* those members (even if there is also room for them to have only partial or mistaken understandings of what they do and why). In other words, processes of spiritual life have a *meaning* and not merely a function; expressed in Hegelian terms, their function resides in the domain of the "for itself" rather than, as in the case of nonhuman life, in that of the "in itself." An ontological implication of this is that self-conscious (or potentially self-conscious) meaning is intrinsic to social reality in a way it is not in nonspiritual life. In contrast to the behavior of fish, one cannot capture what humans do within social life without referring to the (more or less explicitly) conscious meanings it has for them. This means that understanding social reality must be a hermeneutic enterprise, inseparable from the interpretation of human meaning; in the case of social life, the conscious purposes of its participants are part of what the thing is. Or to put the point in another way, there is no fact of the matter about what it is that social members do that is independent of their view of what they do. Moreover, the discussion of objective spirit in the previous chapter implies that the meanings that distinguish the doings of spiritual beings from those of mere animals are always *social* meanings, not merely because the activities they interpret are ones we carry out together but, more fundamentally, because ascribing meaning of any sort requires appropriating shared meanings made possible by subjecting ourselves to jointly accepted rules and norms. Spiritual being is intrinsically social being because the meanings constitutive of the former are possible only intersubjectively.

What we have seen thus far about the living good enables us to say more about the meaning social members' activities have for them. As we know, the characteristic activities of human social members serve vital functions (of both society and the individuals who compose it). Raising children (in the family) and producing commodities (in civil society) are two such activities, both of which satisfy needs humans have by virtue of being alive. Such activities serve vital functions, but because they are activities of spiritual beings, they are also carried out with an awareness of them as having an end or serving a function. Moreover – and, again, because they are activities of spiritual beings – their point is not reducible to the mere reproduction of life; as spiritual activities, they aim also at freedom. This means that for members of human societies, more is at issue in raising children and producing goods than merely reproducing life; rather, they are ways of reproducing life that are at the same time taken to promote (or express) freedom, broadly conceived. One might say, then, that the characteristic activities of spiritual beings strive to unite the ends of life with the loftier end of freedom. Those activities are processes of life that at the same time aim at realizing the freedom of those who carry them out.

Finally, self-consciousness brings with it the possibility of a type of *Entzweiung*, or internal division, as well as a way of enduring such internal opposition, that is absent in merely living beings. Whereas mere life cannot know that it both is and is not identical with what is distinct from itself – only an external being endowed with consciousness can see this – spirit is aware of itself as both identical and not identical with what it posits as its other. Spiritual beings, in other words, relate consciously to what they are not, and this expands the potential richness of their relations to self and other. (It also opens up the possibility of taking an evaluative, potentially critical stance to the form of life one inhabits.[23]) The most significant respect in which these relations are enriched derives from the fact that the principal other that a self-conscious subject both distinguishes itself from and identifies with is life itself, understood as the domain of (mere) life generally. Hence, the fundamental contradiction spiritual beings must

[23] If taking an evaluative stance to one's form of life is intrinsic to being a living, spiritual being, then the failure to do so might also be construed as a social pathology; Heisenberg 2019: 15–16.

negotiate is the opposition they posit – the distinction they themselves draw – between self-consciousness, with its essential characteristic, freedom, and mere life, which is taken to be submerged in nature and hence unfree. This means – indicating perhaps the highest sense in which spirit is *aufgehobenes Leben* – that it is characteristic of spiritual beings to be conscious of themselves as part of life and at the same time as something more than life,[24] which is to say (because of their potential to be free) as *higher* than unself-conscious life.

Implicit in this claim about the form *Entzweiung* takes in spiritual beings is that they have at their disposal a conception of their own ("essential") nature, a conception of life, and a conception of their own ambiguous relation to life. Spirituality involves a form of divided consciousness in which what from one perspective appears as alien to spiritual being (our animal creatureliness, our embeddedness in life) is, from another, grasped as essential to one's spiritual nature as self-conscious and free. Spiritual, and therefore social, life is characterized, then, by activities that "mediate," or negotiate, the opposition between free self-consciousness and mere life. To say that a spiritual being has a conception of its own essence is to say that it acts – this is part of what its freedom (and self-consciousness) consists in – in accordance with a self-conception, which has as its most general content precisely what such a being takes to distinguish itself essentially from merely living beings, namely, freedom. This idea is familiar from Kant's moral philosophy, according to which acting freely in the fullest sense of the term is acting in accordance with one's conception of oneself as an autonomous, or self-legislating, agent. For both thinkers, spiritual beings conceive of themselves as free – they regard freedom as their essence and highest value – and they realize their freedom when they successfully act in accordance with that conception. A self-conception of this sort is sometimes called a practical identity (Korsgaard 1996: 100–7, 128; Brandom 2007: 128–9). It is a picture of what is essential to oneself as an agent – or, more precisely, of what one takes to be essential to oneself as a practical being. Such identities are practical because they place normative constraints on what one does: identifying myself as a parent implies that I take myself to be committed to refraining from acting in ways incompatible with being a (good) parent. Hence, to attribute self-consciousness to human beings is to attribute to them a will, for the commitments that self-consciousness makes possible provides such beings

[24] "In *spirit*, life appears partly as its opposite, partly as posited as one with it, where this unity is produced purely through spirit itself" (Hegel [1812] 1969: 762).

with internal standards they take to possess normative authority and that, by instilling a hierarchy among their desires, provides a shape to who they are and what they do.

To act in accordance with a self-conception is to act freely in a formal sense because it is to determine what one does (and who one is) in accordance with a normative standard that, because self-endorsed, comes from oneself. For Hegel, however, the self-conceptions in question also have freedom as their content. This aspect of his view is complex – due to the complexity of his complete account of what freedom in the modern world consists in[25] – but here it is sufficient to note that even particular identities such as parent or nurse can count for Hegel as conceptions of freedom, that is, as conceptions of what one's own particular free life consists in and how acting on the identities bound up with it contributes to realizing freedom in general, that is, others' freedom and not merely one's own. Because of the tight connection Hegel draws between both freedom and the good, on the one hand, and the good and (rational) particular identities, on the other, it might be more perspicuous to say that freedom for spiritual beings consists in acting in accordance with one's own (rationally justifiable) conception of the good.

The Good

If there is an important difference between Hegel's and Durkheim's social ontologies, it resides less in Hegel's conception of human society as *living* than in his characterization of it as the living *good*.[26] The difficult question is in what precisely this difference consists. For although Durkheim rarely invokes the idea of the good in explaining social reality, he believes it is characteristic of modern societies that they realize both of the basic components in terms of which Hegel defines the good: the freedom and well-being of social members.

Before examining the ethical conception of the good at issue in Hegel's account of social reality, it is worth noting that animal life, too, exemplifies the living good insofar as organisms are naturally set up so as to realize both their good and that of their species. Even if only spiritual beings can bring about what is ethically good, living organisms realize the good in

[25] Members of a rational society are practically free in three senses: as persons, moral subjects, and members of ethical life (Neuhouser 2003: 17–27).

[26] The *living* good is, moreover, the *self-sufficient* good – the good that produces and maintains itself – and this makes it a *Selbstzweck* (Pinkard 2012: ch. 3).

a more limited sense: their instincts (as Rousseau famously notes[27]) lead them to seek what is good for them *qua* living beings, namely, survival, reproduction, and freedom from unnecessary suffering. Being constituted so as to achieve one's good naturally (or without external help) is a feature shared by both animal and spiritual life, and it represents another sense in which the two are similarly structured.

Hegel begins to articulate his concept of the ethical good in "Morality," the section of the *Philosophy of Right* immediately preceding his account of ethical life. Here he characterizes the good as a unity of freedom and human well-being (*Wohl*) (PhR: §§129–30), where unity consists not in mere compatibility or co-existence but in a mutual dependence that is both existential and conceptual.[28] With regard to existential dependence, freedom is realized in the world only insofar as self-conscious human beings aim in their practices at realizing the good – hence only insofar as their aims include realizing their own and others' well-being; and, conversely, well-being is achieved only insofar as individuals are motivated by the ideal of freedom. The good, then, requires not only that freedom not be achieved at the expense of well-being (and vice versa) but, more robustly, that each be realized through the other.

Beyond this, freedom and well-being, when united so as to constitute the good, are conceptually interdependent in the sense that the requirements of freedom place constraints on what genuinely good well-being consists in, just as the requirements of human well-being place constraints on what qualifies as genuine freedom. Well-being in general counts as good only to the extent that it is compatible with the realized freedom of all, and the "true" conception of freedom (as a component of the good) is subject to realizability constraints deriving from the nature of human well-being. The latter claim means that a fully determinate conception of freedom must take into account whether it can be realized in a self-sustaining manner by socially related individuals who by their nature cannot help but care about their own well-being.

The details of how and why Hegel thinks modern European social institutions realize his demanding concept of the good are too complex to be set out here.[29] But his main idea is that those institutions are rational because, working in concert, they carry out the specialized tasks

[27] This is the thesis concerning the natural goodness of humans (Neuhouser 2015: 37–41, 109–10, 117, 137–9).

[28] For more, including the relation between well-being and happiness, see Heisenberg 2019: 62–78.

[29] I attempt this in Neuhouser 2003: chs. 3, 4, 5, 7.

required for realizing both freedom (in a variety of senses) and human well-being. To mention just one example: the nuclear family produces the next generation of human individuals on which social reproduction depends. It produces those individuals not only biologically but also spiritually, which is to say that it forms them subjectively such that, as adults, they are motivated to create their own families. In addition, children are brought up to be agents who, as adults, are able and motivated to realize their moral agency (as moral subjects who know, endorse, and act in accordance with the good) as well as their personal freedom (as bearers of rights and owners of property). Moreover, the family promotes the well-being of its members, both children and parents, since it is the source of love and emotional attachment – a form of recognition – as well as, for parents, sexual satisfaction. At the same time, family life, despite its obligations and sacrifices, is itself an instance of living freely – an instance of determining the contours of one's life in accordance with one's own understanding of the good – insofar as family members endorse the ends of family life and regard them as good.

Rational social life, as the living good, is a good that perpetually realizes itself in the world, in space and time, in and through the practices of real human beings. Thus, the good is not only self-realizing; it also acquires an extra-mental (objective) existence in social life that, similarly in this respect to the being of ordinary things, persists independently of its presence in or for the consciousness of any single individual. (Of course, the living good cannot exist, as stones do, independently of *any* human consciousness.) This means that the good enters Hegel's social theory not as a subjective conception of the good consisting in my own, perhaps idiosyncratic, ideas regarding the kind of life most suitable to me. If a "conception of the good" plays a role in Hegel's theory, it is not one that exists in social members' minds in the form of self-fashioned life plans they hope to realize. Rather, the good exists in, and hence can be read off of, the objectively knowable features of the society in which it is widely realized. The good at issue here is accessible, at least in part, even to the empirical social scientist who denies the need to enter the "mindsets" of social members in order to understand how their society functions. In this respect, too, social reality is similar to life. For living beings are also constituted such that the ends ascribed to them by the biologist are inscribed within their empirically accessible features and behavior. This is precisely the principle underlying functional morphology, namely, that the functions of a living being's organs and body parts dictate, and so can be discovered through, their form. Applied to social reality, this principle implies that social theory must not only interpret the subjective understandings of social members (more on this below) but also acquire empirical knowledge of a society's structure and actual functioning, as embodied in its laws and institutions.

The good in its ethical guise becomes central to social reality because Hegel's social ontology expands the normative character of social life beyond an acceptance of deontic rules to a form of acting in accordance with the good. Some of Hegel's reasons for characterizing rational social life as the living good have been anticipated in discussions of Rousseau and Searle in Chapter 6. Although neither of these thinkers explicitly makes the concept of the good central to social reality, both are committed to the following two claims: first, that consensus on some aspects of the good is necessary if there is to be collective acceptance of the constitutive rules of institutions because such acceptance relies on a sense of their point or function; and, second, that in the case of fundamental institutions, constitutive rules, inseparable from a conception of the good to be achieved by institutions, tend to be regulative rules that distinguish better from worse executors of social roles, rather than rules that define precisely which actions social members are permitted to take and which are required or forbidden. (Recall that it is constitutive of being a parent that one look after one's children's welfare, but parents can be better or worse at doing so without ceasing to be parents in the sense implicit in the lament, "He was never really a father to his son.") If we take into account that the constitutive rules of these and other institutions prescribe what to do with less determinacy than the constitutive rules of money or games, it becomes clear that the former are rules only in a loose sense. When we have in view institutions such as the family or state, it is more appropriate to speak of guiding, or regulating,[30] principles, or of *constitutive norms* that provide *orientation* for individual action (but usually not determinate prescriptions), where the norms in question refer to the good of human beings rather than specifying only obligations and rights or tracking what is merely efficient or useful.

This implies that the agency of social members is best described not in terms of applying rules but as interpreting the governing principles of an institution in light of their understanding of the good to be achieved in it. For Hegel, participants in social life are self-conscious in that their actions are guided by an understanding of how what they are doing aims at realizing the human good (or, more precisely, their good and that of

[30] "Regulating principles" is used frequently by Descombes (e.g., IM: xx).

their communities), and this makes social life expressive of their shared values. That what is at issue is the *human* good does not mean that Hegel presupposes a determinate picture of what that good consists in at all times and places. No one is more aware than he that conceptions of the good are historically and culturally variant. One reason it nevertheless makes sense to talk of the human good is that the good at issue in the fundamental institutions of social life is the *living* good, incorporating needs that spring from our biological nature. (Another reason, examined below, is that the human good includes two goods distinctive of spiritual beings: freedom and recognition.) Because biological needs can be satisfied in numerous ways, and because conceptions of freedom vary historically, the requirement that the human good both satisfy our biological needs and accommodate our freedom is compatible with the fact that determinate conceptions of it vary with time and place.

Whereas Durkheim, Searle, and Descombes locate the basic building blocks of social reality in what they call institutions, the term "practices" better captures Hegel's principal object of concern in his account of modern social life. In the first place, "practice" – like *Praxis* in German – brings to mind the idea of a lifelong professional activity in which rule-following *per se* is less important than orienting oneself toward the good one's practice aims at achieving.[31] Another advantage of the term derives from the meanings given it by earlier philosophers: *praxis* suggests human activity that is free, self-conscious, rational, ethically significant, carried out for its own sake, and – if we think of Aristotle – dependent on habits acquired through socialization.[32] I suggested above that processes of merely animal life become, in spiritual beings, forms of practice, insofar as the various activities of the latter are given coherence – held together as a single process – by social members' awareness of the good their activities realize and how they do so. Practices, then, are normatively oriented, self-consciously unified activities that, because they are also cooperative, can take place only within a spiritual community in which agents have an allegiance to norms of the good shared by their fellow beings.[33]

[31] For a similar point about practices, see Taylor 2010: 34–5.

[32] See also the discussion of routine in Giddens 1984: xxiii–xxiv, 60–4, 68–73, 111–13.

[33] For Hegel "the ethical social world consists of contexts of action defined, structured, and distinguished by the fact that each realizes a specific good. Such contexts of action … can be called 'practices,' … the fundamental unities of the practical…. To be an end, a means, a reason, an action, … or to be something good, is to have a specific role or status within a practice" (Menke 2018: 28; see also 27–30).

Following Aristotle, Hegel places habit at the center of his account of social practices conceived of as the living good: the ethical, he says, takes the form of mores (*Sitten*), or ethical habits that have become the "second nature" of participants in social practices, which, as "the penetrating soul, meaning, and actuality of their existence," makes spirit "living and present" in the world (PhR: §151).[34] Habits of the relevant sort have a meaning because they are end-directed and because the purposive behavior they make possible is something the human animal must learn to be able to do. Habits are the results of training, and so they are something we (collectively) give to ourselves; in this respect habits have a spiritual, not a purely natural source. The feature of habit most important to Hegel is that it is a species of meaning that penetrates the bodies of the habituated and through their habit-mediated actions becomes living and present in the world. Habit represents "an inner, thoroughly penetrating formation of the corporeal that makes feelings as such, as well the contents of the will and understanding, *incarnated*" (PhM: §410; emphasis added) and hence a potential source of action in the world. Habit is mind that has taken on a bodily existence and body that has acquired the significance of mind.[35]

Another reason habit is important for Hegel is that those acquired through socialization form our desires, dispositions, and values so as to enable us to participate in social life in a way that is largely spontaneous, or that "comes naturally," analogously to how the vital behavior of living beings comes naturally to them. When habituation takes place in the right way – through socialization in rational social institutions – it forms our inclinations so as to accord with our duties such that we do not normally experience our social participation as dictated by external "oughts" but rather as "what we do" and as what we must do in order to be who we take ourselves to be. Even if habituation makes the demands of social life appear to us immediately, under the guise merely of "what we do," they are not experienced as arbitrary or contingent – as nothing more than "what we, in this specific community, happen to do." Rather, the habits on which social life depends are value-infused in that they make a claim, usually implicit, to realize the good, where this claim purports to have more than merely parochial validity. It is a mistake,

[34] That habituation is compatible with – indeed, necessary for – free social life is argued for in Novakovic 2017: 20–68. The literature on habit and second nature in Hegel is vast; see Khurana 2017: §§82–9, 99–101; Testa 2008: 286–307; Testa 2020; and Haase 2017. For the ethically ambiguous character of second nature, see Menke 2018: 119–48.

[35] For the relation between habit and *living* spirit, see Heisenberg 2019: 37, 43–4, 144–53, 161–2.

then, to assume that social life's habitual character excludes the possibility of having reasons for what one habitually does or of stepping momentarily outside those habits to ask whether the life they animate is good or as good as it could be.[36] That human habits need not be "blind" is apparent even in low-level, only faintly ethical habits such as brushing one's teeth at bedtime. It is even more obvious in the endlessly repeated daily routines in which we feed our children or iron our spouse's clothes; demonstrate both concern and professional distance to our students; or take the symptoms of our patients as more than just the whinings of the spoiled. Habits of the right sort are essential to there being a *living* good, but they do not rule out the possibility of *knowing* the practices they make possible as good.

Closely connected to this is that what Hegel distinguishes as fundamental institutions – those most involved in satisfying the material needs of their members, as well as forming them into subjects who are motivated to reproduce those institutions and who have acquired the subjective capacities required by freedom in its various forms[37] – cannot function well unless their participants have deep, identity-constituting commitments to the values that animate those institutions. As noted in Chapter 6, there are several reasons for this. The first is that the goods aimed at in the practices that take place within fundamental institutions are more numerous and complex than those at stake in bowling leagues and cocktail parties. Second, the well-functioning of the family, civil society, and the state requires that their members regard the goods achieved in each to be more important than those achieved in less fundamental institutions. And, finally, the roles internal to the practices of fundamental institutions are more demanding than other social roles with respect to the time and energy that must be devoted to them and the substantial renunciation of purely egoistic aims they require.[38] A corollary of these points is that

[36] On the compatibility of habit and normative reflection, see Novakovic 2017: 20–68.

[37] Subject formation of this sort is essential to society's *spiritual* reproduction.

[38] With respect to the time and energy devoted to them, the family and civil society might seem to differ significantly from the state since Hegel requires only minimal political participation of ordinary citizens. His thought must be, however, that the *identity* associated with the state – being a member of the Prussian or French state, for example – constitutes an important part of how modern individuals conceive of themselves and provides them with the opportunity to realize a type of freedom (in willing, or endorsing, the general will) and form of selfhood (as "citizens") unavailable in other social spheres (Neuhouser 2003: 135–40, 143–4). As many have pointed out, it is hard to understand why, given Hegel's criteria for identifying fundamental institutions, the educational system does not count among them. If educational institutions were included within the domain of the state – Hegel himself sometimes treats them as such ("philosophy with us is not … practiced as a private art, …

the aspects of the human good realized in the fundamental institutions of social life are more central to human satisfaction than those realized in less fundamental institutions. It is the extent to which such practices address our material needs as well as our deepest spiritual needs that makes participation in them of such importance to the flourishing of individuals. The goods realized in those practices are among the weightiest and most complex goods available to human beings.

Recognition, the Living Good, and the Bondsman–Lord Relation

One aspect of the good Hegel takes to be realized in social life is the good of recognition (*Anerkennung*). It is a central claim of his that in performing their social roles, social members secure the esteem of others and the foundations of their self-esteem as well. They achieve, in other words, the recognition of their fellow social members and thereby satisfy an aspiration that Hegel, following Rousseau, regards as fundamental to being a self-conscious subject, namely, the aspiration to "be someone," to count as a being of value for oneself and for others. In Hegel's words, individuals find in social life their "feeling of self" (*Selbstgefühl*) and "dignity" (PhR: §147, 152), their "recognition and honor" (PhM: §527). Recognition is closely connected to the practical identities discussed above. In characterizing love and friendship, two essentially recognitive relations, Hegel says that "one constrains oneself gladly in relation to another but in this constraint is aware of oneself as oneself ... and wins one's feeling of self" (PhR: §7A). Thus, to have a practical identity as a social member is not only to be committed to fulfilling the obligations of one's various roles but also to know one's social commitments as a source of one's sense of self, to find one's identity in them as a self that matters, such that to be deprived of them is to experience oneself as "defective and incomplete" (PhR: §158A).[39] Recognition, one could say, is a "spiritual need" (VPR: 80) that provides "*subjective* satisfaction" to its recipients (PhR: §124; Hegel's emphasis).

but has a public existence ... in service of the state" [PhR: Preface]) – then it, too, would demand large amounts of time and energy of social members and this respect be more like the other two spheres than might initially seem to be the case.

[39] The importance of recognition in rational social life for Hegel is unmistakable: PhR: §§162, 177, 192, 206, 207, 209, 217, 218, 238, 253, 254.

Even if recognition is a spiritual need for humans, its significance extends beyond being part of the well-being individuals must achieve if social life is to realize the good. For recognition is central also to the normative character of social life; as Hegel's talk of "constraining oneself in relation to another" suggests, recognition is essentially practical and normative. When one person recognizes another, she does so in the name of some conception of the value or standing of the other, and she takes this value or standing to place constraints on her own will. When, in Hegel's bondsman–lord dialectic, the bondsman recognizes the lord as the only free subject among the two, he regards the lord's free status as a normative constraint on his actions. Recognizing the lord as free, and therefore sovereign, means accepting the validity of a norm – roughly, "the wishes of a free will count as my commands" – that determines how a being of that sort is to be treated, and this defines the terms of their interaction. In this case, recognition implies that the recognizing subject regards the will of the recognized – the lord's particular desires – as having authority over him. Love counts as recognition for similar reasons: when I love another, I regard him as worthy of my love – and, hence, as in some sense valuable – and I make his desires and interests into constraints (although not absolute constraints) on my actions. For both the bondsman and the lover, the desires and interests of a recognized other count as an "ought"; they have the status of something I am to respect or promote through my actions.

For Hegel, reciprocal recognition is what makes voluntary self-constraint for the sake of others possible without self-sacrifice or the loss of freedom, for in working for others I win my own self, or identity. It is this that enables us to understand a large part of the glue that holds human societies together, or the phenomenon Durkheim calls solidarity. That is, recognition explains how individuals who cannot help but care about their own particular good can be motivated to follow the egoism-constraining norms that make their social interactions collectively beneficial and coordinate them so as to accomplish the business of society. Recognition provides the recognized with a publicly acknowledged, and therefore objective, confirmation of a specific practical identity that is secured by the norms of the institutions within which social life is carried out. Individuals who live within such institutions have publicly recognized identities that give meaning to their lives and shape their practical engagement in an intersubjectively constituted social world. This explains in part why recognition is crucial for Hegel: it enables socially united individuals to enjoy certain goods they could not achieve outside social life.

When recognition is reciprocal, the actions and identities of the individuals involved are guided and defined by norms or values they jointly endorse. In Hegel's conception of the modern family, for example, husband and wife occupy distinct roles, the rights and obligations of which are defined by institution-defining norms accepted by both. In each case, the two spouses win particular identities – as husband and wife, respectively – that are recognized by the other spouse as well as, through the institution of marriage, by society at large. The tight connection between my being recognized by others and my recognizing them is due to the fact that the norms around which the practices of rational social institutions are oriented are such that to abide by them is to recognize the normative status of those who occupy the other roles of the same institutions. To fulfill the role of father as defined by the (rational) family is to act in ways that demonstrate to the other family members a valuing of them in their respective familial roles.

If relations of recognition are to coordinate the qualitatively distinct (specialized) actions of social members such that together they achieve the good of individuals and society, then the specific content of the norms expressed in recognition is important. Not just any set of recognitive statuses, with whatever rights and obligations, will both provide social members with satisfying identities and coordinate their interactions in socially beneficial ways. (Think of the destructive social consequences of the rights and obligations accorded to the role of mafia boss or feudal lord, practical identities that, if not ultimately satisfying by Hegel's lights, have seemed to many worth pursuing.) One requirement of the institutions of ethical life is that the practical identities they make possible provide social members with complementary desires and characteristics. As such, they possess qualitatively distinct but complementary qualities that enable them to fulfill the specific functions – as children and parents, or farmers and tailors – essential to achieving the ends of social life. The desires and characters of such individuals are not similar, as in Durkheim's segmentally structured societies, but complementary: a mother can realize her desires and ends (as a mother) only if her spouse and children have different desires and ends complementary to hers (and vice versa), and the same is true of farmers and tailors. When family members interact as such – when they reciprocally constrain themselves by the desires and needs of fellow family members – they find in the world, not mirror images of themselves, but numerically and qualitatively distinct counterparts whose particular qualities and actions are necessary for being the particular selves they take themselves to be. To experience such complementarity in one's dealings with others is, for Hegel, to be "at home" in one's social world, or "with oneself in its other" (*in seinem Anderen bei sich selbst zu sein*) (EL: §24A; PhR: §7) (Ramos 2009: 15–27; Wood 1990: 45–51), which counts for him as an especially important sort of practical freedom.

Before concluding this chapter, it will be helpful to rehearse some of its points concerning the living good and the centrality of recognition to social life – bringing in as well the idea of an *I* that is a *we* and a *we* that is an *I* – by examining Hegel's simplest example of social reality as depicted in the *Phenomenology*'s treatment of the relation between bondsman and lord.[40] The story of bondsman and lord is introduced in that work as part of a highly reconstructed account of the historical developments that led to the emergence of modern ethical life in which those developments are shown to be necessary stages in the achievement of full spiritual satisfaction, including the realization of freedom. The relation between bondsman and lord is depicted as a rudimentary and ultimately unsatisfying attempt to unite the aims of life and freedom within social life, and as such it can be viewed not only as a forerunner of the rational society described in the *Philosophy of Right* but also, as we will see in Chapter 13, as a depiction of social pathology. Ultimately, the superiority of modern society reveals itself in the ways it successfully responds to the spiritual failures encountered by its predecessors: grasping why the society of bondsman and lord fails to unite life and freedom in a satisfying way is meant to show the necessity of a different form of social organization – necessary if freedom is to be realized and to be united with life – that responds to the specific ways in which that more primitive society failed in its spiritual task of uniting those two poles of human existence. According to the account given in the *Phenomenology*, modern society is rational because it can be understood as having "learned" (Jaeggi 2019: 221–36) from the failures of the past so as to accomplish more adequately the spiritual task of uniting freedom and life or, what amounts to the same, realizing the human good.

The tale of bondsman and lord is also important as a contribution to social ontology since, as noted above, its appearance in the *Phenomenology* coincides with the emergence of spirit in that text. In short, this tale is relevant to social ontology because it sets out the basic elements and structure of any set of relations among humans that can count as a society, and the relative simplicity of those relations in the case of bondsman and lord brings into relief the kind of thing human society is for Hegel. If spiritual life is most fundamentally about satisfying the aims of both freedom and life, then the tale of bondsman and lord must include not only relations

[40] The literature on this topic is vast; see Neuhouser 2008: 37–54.

among human subjects but also the relations those subjects have to nature (since only through these can they satisfy the ends of life). Simply put, human society is understood as constituted by cooperative practices among biologically needy subjects that the latter regulate in accordance with shared ethical norms grounded in relations of reciprocal recognition and the requirements of freedom. The ethical ideal underlying Hegel's conception of objective spirit, then, is that of a community of humans who are tied both to nature and to one another such that both sorts of ties serve the reproduction of life as well as give expression to and realize the freedom and recognized "self-standingness" – the integrity and noninstrumental worth – of all individuals involved (PhG: ¶177).

Perhaps the easiest point to see about bondsman and lord is that the activities of each serve the material ends of both: the lord commands, the bondsman labors, and together the complementary activities produce the goods needed to reproduce life (PhM: §434). At the same time, these material activities have a significance in relation to freedom. The lord's commanding and the bondsman's obeying are grounded in and expressive of a shared conception of what it is to be free; of the superior value free activities have over those that serve only biological ends; and of who among the two possesses normative authority based on having proved himself to be free ("above mere life") in the struggle unto death. The future bondsman may blink in the decisive moment of that struggle and choose his life over freedom, but his doing so does not signal an abandonment of the basic normative ideal he held before the struggle with the future lord.[41] That the bondsman continues to subscribe to the higher normative authority of freedom, and to the same conception of freedom he had before the struggle, is revealed by the fact that for both him and lord, prestige and authority (recognition) attach to the individual who has expressed in his actions the superiority of freedom over life. This means that the bondsman's obedience and labor for the lord is not merely prudential – because refusing to do so would result in his death – but also ethical: he recognizes his lord's commands as having legitimate authority precisely because the lord is the one who, in risking his life to the end, has given expression to their shared values, which accord a higher value to freedom than to remaining alive.[42] The social relation between bondsman and

[41] More precisely, the future bondsman embraces a new principle – "that life is as essential to him as pure self-consciousness" (PhG: ¶189) – and there is even something progressive in this stance, for life *is* essential to spirit, even if freedom remains the higher value.

[42] Many interpretations of the bondsman–lord relation overlook its ethical character. Despite that relation's ethical deficiencies, the bondsman's obedience and labor are expressions of his allegiance, however perverse in its consequences, to norms that follow from a shared conception of freedom.

lord counts as spiritual, then, because it imbues jointly undertaken activities of life with the value of freedom.

It is often thought that recognition between bondsman and lord goes only in one direction, but this is a serious misreading that obscures Hegel's claim that reciprocal recognition, in some form, is essential to any genuinely social life. The recognition between bondsman and lord is asymmetric and unequal – this is part of what makes their relation deficient – but it is nevertheless reciprocal, even if ultimately unsatisfying, in that each member of the dyad receives some form of recognition from the other. That the bondsman recognizes the lord as the being who proved that he values his freedom above his life is clear, but if this is to be an instance of recognition, there must also be some respect in which the lord recognizes the bondsman.[43] Hegel's thought is that if the bondsman's recognition of the lord is to count as such for the latter – if it is something the lord seeks out and therefore values – then the lord implicitly recognizes the bondsman as an instance of the type of free being that is capable of recognizing another and whose recognition is therefore worth seeking. Regardless of the asymmetry between the recognized identities of bondsman and lord, the latter, in seeking the former's recognition, must attribute to him, at the very least, the capacity to recognize; he must, in other words, implicitly recognize the bondsman as belonging to the class of freely valuing subjects. This thin species of recognition afforded to the bondsman is reflected in the fact that the lord seeks recognition not from any entity whatever – not from a stone or a dog – but only from a being he takes to possess the freedom required to affirm his value (because, as Rousseau saw, valuing presupposes the freedom to judge that something is good).

Thus, the bondsman and lord exemplify a spiritual relationship in which the two *I*'s, though differently situated, constitute a *we*. To say that an *I* is a *we* means, in general, that an individual's being what she is (*qua* social member) depends on her being related to others within the normative context of an institution. We have already seen some of what this idea implies in our discussion of Descombes: first, that my capacity to believe and act depends on my participation in certain institutions, governed by norms that hold for us, that make the objective meaning of my beliefs and acts possible; and, second, that the statuses accorded to me in social

[43] This is the point of the complex account of the dynamics of recognition in general at PhG: ¶¶178–85.

life – those that constitute what *I* am within society – are defined by principles governing a *we* that specify complementary statuses that must be accorded to others if I am to possesses mine.

Hegel's account of bondsman and lord illustrates further respects in which social life involves *I*'s that are a *we* that play no obvious role for Descombes. One of these concerns the complementary roles the two figures play in the material reproduction of life. Bondsman and lord are *I*'s that make up a *we* because neither of their life-related activities is complete apart from the other's: neither commanding by itself nor laboring without direction can satisfy material needs. Each activity is intelligible only insofar as it is integrated, as a differentiated part, into the larger project of the material reproduction of society (which is to say, with Durkheim, that bondsman and lord inhabit specialized roles in a rudimentary form of the division of labor). Beyond this, bondsman and lord constitute a *we* in a more spiritual sense insofar as they are bound by a shared understanding of what is of ultimate value (freedom, however primitively conceived), and this shared commitment determines the relations of normative authority, including relations of recognition, between them. Their shared conception of freedom determines who is entitled to set the terms of cooperation; who must labor and obey in accordance with those terms; and who is recognized as the being of superior value.

When Hegel expands (slightly) on the expression "an *I* that is a *we* and a *we* that is an *I*," he emphasizes that the *we* in question involves the *unity*, or oneness, (*Einheit*) of the *I*'s (PhG: ¶177), but as we know by now this unity is neither numerical nor qualitative *identity*. Rather, the unity in question is comparable to that possessed by the interlocking pieces of a jigsaw puzzle, where the particular character of one piece demands complementary qualities in other pieces, such that, when joined appropriately, they constitute a whole greater than the sum of its parts. Members of such a whole are capable of achieving larger ("universal") ends that each of its members endorses but that are unattainable to them as unrelated individuals. Thus, the members of rational ethical life realize particular identities not common to all only by taking part in collective projects guided by ends or values they share with other members. This sharing of final ends represents a unity among social members that is also bound up with a distinct form of recognition: such members have a commitment to the same final ends, and this implies a confirmation by others of the value of one's own highest ends and values. In rational institutions it is not only my particular identities, as father or worker, that I find affirmed by others but also my competence as a moral subject to judge reliably about the good.

The expression "an *I* that is a *we* and a *we* that is an *I*" can be misleading, however, if it is taken to capture the entire structure of spirit rather than simply its intersubjective character. What is missing from this expression is the relation spiritual subjects have to nature (or to "things" or "Being") that both embody and condition the relations among *I*'s and *we*. For, expressed in Hegelian jargon, a subject's relation to other subjects is mediated by their relations to nature, and their relations to nature are mediated by their recognitive relations to other subjects. The second half of this claim means that the role each plays in the collective project of appropriating nature through productive activity in order to reproduce life depends on and gives expression to relations of recognition between the subjects involved: the bondsman works only because he recognizes the lord as free (as bound by no authority outside himself), and the lord commands with the expectation of being obeyed only because he takes the bondsman to lack the spiritual character on which normative authority (the right to have one's own will normatively constrain others') depends. Conversely, these recognitive relations are mediated by the different relation each has to nature. What gives the lord normative authority in the eyes of both is that he alone risked his life to the end in the struggle unto death in order to "prove" his sovereign status, whereas the future bondsman "gave in" in that struggle, choosing life over freedom. Spirit, then, is a configuration of human life in which subjectivity, with its constitutive aspiration to freedom, and nature are intertwined. It is the world itself, in other words, not merely conscious relations among subjects in which spirit is realized, which explains why Hegel says, in language anticipating Durkheim's, that the spiritual unity characteristic of spirit exists in the world, in the form of *thinghood* (PhG: ¶350).

The interpenetrating unity of nature and subjective consciousness in which spirit consists brings us to a final aspect of the living good I have not yet discussed in conjunction with bondsman and lord, namely, the internal division (*Entzweiung*), and the overcoming or negotiation of such division, characteristic of spiritual life. As noted above, nature and subjective consciousness make up the two poles of the fundamental internal division to be negotiated by spiritual beings, the opposition they themselves posit between their freedom and self-consciousness, on the one hand, and their purportedly unfree immersion in and dependence on the processes of life, on the other. The spiritual task posed by this opposition is to reconcile one's conception of oneself (and real existence) as free with one's understanding of oneself (and real existence) as a living being: How is a spiritual being to negotiate the contradiction involved in being at the same time part of life and above mere life? The aspect of Hegel's answer to this question that is relevant to social theory is that rational social life is made up of real, material processes that mediate this opposition. If this is so, then a society of bondsmen and lords must also exhibit some version of the attempt to negotiate the contradiction between self-consciousness and life. In the following chapter I show that understanding how this attempt fails in the case of bondsman and lord leads us to the richest and most distinctive conception of social pathology to be reconstructed from Hegel's texts.

CHAPTER 13

Hegelian Social Pathology

Although Hegel never offers an explicit account of social pathology – at least not under that name – his social philosophy is fertile ground for theorists who concern themselves with this concept.[1] Demonstrating this, by articulating the various forms social pathology can take from a Hegelian perspective, is the aim of this chapter.[2] Not surprisingly, the link between Hegel's social philosophy and theories of social pathology is the concept of life, which, as we saw in the previous chapter, is fundamental to his social ontology. In treating animal organisms in the *Encyclopedia*, Hegel discusses illness in four substantial paragraphs (PhN: §§371–4), and in "Subjective Spirit" he devotes numerous pages to a discussion of various forms of spiritual (or mental[3]) illness (PhM: §§403–8). Only in one passage of Hegel's early writings is the concept of social illness explicitly, albeit briefly, discussed (Hegel 1975 [1802–03]: 123–4).[4] Here Hegel mentions certain forms social pathology might assume, all of which are consistent with the conception of social pathology I attribute to him here.

Animal Illness

Hegel's discussion of animal organisms makes clear that he regards *dysfunction* as the principal characteristic of illness: "The individual organism …

[1] At one point Hegel appears to rule out the possibility of illness for spiritual beings, but in the same paragraph he calls insanity a "spiritual illness" (*geistige Krankheit*) (PhM: §408A). Moreover, in discussing the state he refers readers to his account of animal illness in the *Encyclopedia*, implying that the state, too, is vulnerable to a version of the same (PhR: §278). That Hegel fails to articulate an account of social pathology is due mainly to his taking the *Philosophy of Right*'s principal aim to lie not in diagnosing social ills but in reconciling us to the present social world by showing it to be already rational in basic outline. I am grateful to Gunnar Hindrichs for making me aware of the complications in Hegel's concept of spiritual illness.

[2] Speaking of the *uninhabitability* of a form of life is another way, compatible with mine, of articulating the concept of social pathology that can be found in Hegel (Pinkard 2012: ch. 4).

[3] Both terms are possible translations of *geistig*.

[4] I am indebted to Brady Bowman for this reference.

finds itself in a condition of illness insofar as one of its systems or organs … asserts itself independently of the others and … works against the activity of the whole, impeding the fluidity of the whole and its [normal] process in all its parts" (PhM: §371).[5] In other words, illness is present when some part of the organism no longer cooperates with other parts so as to enable the whole to achieve its vital ends. Thus, one can also describe health, as Hegel does, using the classical vocabulary of proportion, balance, and harmony (PhM: §371A), as long as we supplement this language with the idea of proper functioning defined in relation to the vital ends and characteristic activities of an organism's species. For Hegel, too, health consists in a harmony of specialized functions, where each part of the organism carries out the functional role it is uniquely suited to play; it is a condition in which "all functions of life are maintained in their ideality" (PhM: §371A) – that is, in the form appropriate to the organism's species, or "concept." In short, animal illness consists in an imbalance among a living being's organs or processes that disturbs its characteristic and vital functions, and health is the absence of such disturbances.[6]

Hegel's texts also suggest a slightly different conception of animal health worthy of mention, given its relevance to a class of pathologies we might also want to attribute to human societies. According to this conception, health is a condition in which "there exists for the organism no inorganic [or foreign] element it cannot overcome" (PhM: §371Z). This definition locates health not in an occurrent state but in an organism's enduring capacity to fight off external threats to its proper functioning. On this definition, a healthy organism is not one that merely happens to avoid dysfunctionality at a given time but one that regularly avoids that condition because it possesses adequate internal resources for warding off potential illness. As the power to overcome whatever might threaten the organism's vital functions, health, on this conception, is a form of *vitality* (*Lebendigkeit*), the organism's ability to react effectively to threats to its vital activities.

Mental Illness as Spiritual Dysfunction

If the concept of illness is relatively clear in the case of animal organisms, it is harder to define when one turns to spiritual beings.[7] The mature Hegel

[5] See also PhR: §278. For an account of illness that emphasizes its character as a loss of *Wirklichkeit*, according to which illness occurs when an organic phenomenon falls back to the less actual (*wirklich*) domain of the chemical, see Särkelä 2018: 245–8.

[6] For more on animal illness, see Khurana 2017: §68. For a different take on the role of dysfunction in illness, see Rand 2015: 68–86.

[7] Contemporary philosophers continue to disagree about how precisely to characterize physiological illness; e.g., Kitcher 1997: 208–10; Boorse 1975a; and essays in Humber 1997. The philosophical

discusses a seemingly infinite number of forms of spiritual illness that fall under the concept of insanity or madness (*Wahnsinn, Verrücktheit*) (PhM: §§408+A). That it takes him so many pages to do so is a sign of the many problems inherent in applying the concept of illness in the spiritual domain, in part because of the difficulties involved in ascribing functions, and therefore dysfunctions, to spiritual processes. These difficulties are due to the more complex – especially, the more robustly reflexive – structure of spirit as compared to that of mere life, which follows from the fact that in spiritual beings life processes are informed, and therefore transformed, by self-consciousness. Yet if spiritual illness is to bear any similarity to animal illness, Hegel's account of the former must rely on some conception of proper spiritual functioning. A clue as to what he takes such functioning to consist in can be gleaned from the fact that the various dysfunctions that appear in his discussion of madness involve some failure of rationality, broadly construed.

On Hegel's most general account, madness consists in a subject's failure to "take possession of," or to appropriate, its various "sensations, representations, cognitions, thoughts, etc." (PhM: §403A). That Hegel conceives of reason as an activity of appropriating what is initially (taken to be) "other" to the thinking and willing subject implies that failures of rationality involve some form of being limited or impeded by one's "other" – or, as Hegel puts it when discussing madness (like Freud after him), of being "subservient to a foreign power" and therefore unfree (PhM: §406A). Thus, dysfunctions in spiritual processes can be described either as failures of rationality or as forms of unfreedom[8] – or, equivalently, as disturbances in the subject's quest to make its "other" its own, another name for which is *alienation*.

The view that spiritual illnesses involve failures of rationality carries implications for both the diagnosis and treatment of such illnesses. It implies, namely, that the diagnostician and curer of both mental and social pathologies must regard the "objects" of its diagnoses and remedies as potentially rational (and therefore free) beings – as *subjects* rather than objects. Hegel goes to great lengths to point out that given the kind of illness madness is, remedies must treat the mad person "humanely," which means, not as a mere animal body but as a being who suffers not from a

literature on illness is even more controversial when addressing human rather than merely animal illness; see Engelhardt 1976: 256–268 and Hesslow 1993.

[8] This may explain why Honneth (2009: 19–42) tends to describe social pathologies as pathologies of reason.

"*loss* of reason" but from a "contradiction in the reason still present" in him (PhM: §408). This does not mean that the mentally ill can be cured with rational arguments but only that their treatment should acknowledge their nature as potentially rational subjects, for example, by avoiding physical force, winning their trust and respect, and attributing to them a modicum of moral sensitivity (PhM: §408A). A further implication seems to be that no successful treatment of spiritual illness, whether mental or social, can come exclusively from outside the afflicted – in the form of a doctor who cures a "patient" – but must instead incorporate in some way the subjectivity of the ill. This point takes on added significance when the illnesses at issue are social pathologies. Since in these illnesses no social member is completely external to the dysfunctions in question, diagnosis and cure cannot take place on any model – such as "healthy doctor treats sick patient" – that demarcates sharply between the agents and recipients of diagnosis and cure.

The "taking possession" that Hegel regards as constitutive of rationality can be spelled out in various ways, corresponding to different types of insanity, but madness's central feature appears to be the subject's failure to bring its inner states into an appropriate relation to an external world shared with other subjects and governed by impersonally valid principles of objectivity. In such a condition, inner states are unbound by normative constraints that distinguish what is merely valid for me – merely part of my inner world – from what is valid for all and therefore part of an objective world: in madness "psyche and world *go their own ways* and become *independent of each other*" (PhM: §406A; Hegel's emphasis).

Hegel also describes madness as a functional *imbalance*, thereby extending the analogy between animal and spiritual illness and further justifying the terminology of pathology: "Illness occurs in *psychic life* (*Seelenleben*) when the organism's merely psychic elements (*das bloß Seelenhafte*) become independent of the power of *spiritual* consciousness and presume to arrogate its function, with the result that spirit loses its rule over the psychic elements that belong to it and, with that, power over itself" (PhM: §406A; Hegel's emphasis). Here, too, the language of alienation and loss of freedom is unmistakable, but this description of mental illness makes clear that the "other" the mad person fails to appropriate is his own psychic life: "sensations, representations, cognitions, thoughts" that are already (but in a rationally deficient mode) his own (PhM: §403). To speak here of functional imbalance is justified because, speaking with Plato, the soul's rational part fails to carry out its proper role – ruling the appropriately subordinate parts – and is, to the detriment of the whole, ruled by them instead. Presumably, the main respect in which spirit's ruling authority is usurped in madness is that its reality-constituting function is taken over by the psyche, or soul,[9] a phenomenon of merely subjective spirit lacking the rational resources to bring inner experience in line with an objective world subject to generally binding norms.

Thinking of madness as a dysfunction makes even more sense if one attends to its practical aspects. The thought is that if the inner life of a mad person lacks the appropriate relations to an external, objective world, it cannot be the basis of genuinely self-determining activity, for the practical involves translating subjective ends into objective states of affairs within a world accessible to all (PhM: §408A). This aspect of madness can be brought into view by considering the person who is not mentally ill:

> The human being of healthy mind and understanding is aware of his actuality (*Wirklichkeit*)—the concrete fulfillment of his individuality—in a self-conscious manner intelligible to the understanding (*verständige*); he is aware of it in the form of his connection to … an external world distinct from him, which he knows just as much as an *internally coherent, intelligibly* (*verständig*) *connected* manifold. In his subjective representations and plans he likewise has in view … the *mediation* of his representations and ends with the universally and internally mediated realm of objectively existing things. This external world extends its fibers into him such that *they* constitute what he actually (*wirklich*) is in the world before him, such that he would himself die off as these externalities disappeared. (PhM: §406; Hegel's emphasis)

This passage addresses both the theoretical and practical aspects of mental health. On the one hand, the person of sound mind knows an external world in which a multitude of causal relations hold among its constituents. On the other hand, it is this relation to an external world that enables the healthy individual to be an effective agent – to carry out her projects with consistency and in conformity with her interests. Moreover, it is only by realizing one's projects and ends that one establishes a practical identity in the world that transcends merely subjective desires and self-conceptions. For the mentally healthy individual it is not only the external world that exists as a coherently connected manifold; by giving reality to her projects and satisfying her interests, she, too, acquires an existence as a coherent and objectively mediated self. It is "this totality of fundamental interests – the … empirical relations an individual has to other humans and to the world in general" – that "make up *that individual's*

[9] For Hegel's conception of the soul, see Sandkaulen 2010: 35–50.

own actuality (*Wirklichkeit*)" and "*concrete* being" (PhM: §406; Hegel's emphasis). Mental illness, then, is a dysfunction in a straightforward sense: it impedes those it afflicts in establishing real, intersubjectively recognized identities as selves who realize their own projects and promote their good.

Social Pathology in Hegel's Early Work

Before addressing the issue of how to attribute functions to spiritual processes in the social domain, it will be helpful to return to the early text of Hegel's, "The Philosophy of Ethical Life," that explicitly invokes the concept of illness (*Krankheit*) in describing defects of social life. This text, written before the *Phenomenology of Spirit*, describes illness in the social domain by appealing to two slightly different conceptions of dysfunction, both of which can be understood as imbalances in which one social institution or sphere extends beyond its proper bounds. In both cases the structural similarity to animal illness as Hegel understands it is patent, and "dysfunction" is a perspicuous term for the defect. One of these conceptions of dysfunction involves what is properly a subordinate part of society arrogating to itself the function of ruling the whole, as when in mental illness the soul, or psyche, claims that role for itself rather than acquiescing to spirit's rule. Hegel's (odd) example of this form of dysfunction in animal organisms is that of a liver that attempts "to make itself into the dominating organ and to force the entire organization to work for it" (Hegel 1975 [1802–03]: 124). Fortunately, this is one instance where dysfunction in the social realm is easier to imagine than in animal life. Hegel's example of the former, expressed in terms developed later in the *Philosophy of Right*, is of one subordinate social sphere, civil society, dominating the others and making itself into the organizing power of society as a whole. What this means concretely is that one way a society can be ill is when the "principle" appropriate to civil society – that of contract – comes to govern all spheres, including the family and state.

This example of social dysfunction is not only intelligible but also of contemporary relevance. The idea on which it rests is that the three main institutions of Western modernity constitute different social spheres in part because each requires a distinct type of relation among social members and because experiencing all three types allows individuals to realize the full panoply of freedoms and forms of selfhood made available by the modern world. Civil society's domination of the whole occurs when the freedom distinctive of contracts is taken to be the ideal appropriate to all spheres of social life and when the relations among social members characteristic of contracts – where sovereign individuals decide what they want independently of their relations to others and come to an agreement regarding their respective property with other, similarly conceived individuals – becomes the model for social relations *tout court*. A society in which relations among family members or citizens were modeled on contracts among mutually indifferent individuals would be one in which both the spiritual benefits available to social members outside the economic domain would go unrealized and those spheres themselves would be unable to function as they should. Treating my spouse as just another contracting party and determining my will in accordance only with what I want for myself, independently of my relations to my spouse and children, would mean both that I would miss out on the distinctive goods afforded by family membership and that my family would not function as it should (because realizing the aims of family life requires lifelong commitments to nonegoistic, though not "selfless," projects).

The second conception of social dysfunction in Hegel's early text is but a slight variation on the first. In this case, one social sphere steps outside its proper boundaries, but rather than becoming the organizing principle of the whole, it invades the rightful territory of, or "colonizes" (Habermas 1987: 232), another sphere. The example given immediately above, if limited to the family rather than applied to society as a whole, illustrates how one sphere, civil society, can colonize another, the family, in ways that lead to dysfunction in the latter. A further example might be an economy structured such that parents are unable both to support their families through work and find the time to raise their children in the ways required by the ideal of freedom. Hegel's own example of this type of pathology involves the state invading the rightful territory of civil society such that "a fully developed police authority completely pervades the individual's existence, destroying civil freedom" (Hegel 1975 [1802–03]: 124). The situation he has in mind is one in which economic activity is too rigorously regulated by laws aimed at achieving the public good (the "principle" of the state), resulting in a loss of the specific kind of freedom and possibilities for self-realization that a market economy affords. Hegel's view, then, is that in a rational society the market must be regulated to some extent – otherwise dire consequences for some are inevitable – but that it is also possible to over-regulate it such that the rational potential of civil society for achieving specific forms of freedom and selfhood goes unrealized.

Material Functions of Social Life and Corresponding Pathologies

These examples are helpful for finding compelling conceptions of social pathology in Hegel's thought, but they do not answer the more fundamental question of how functions are to be ascribed to social institutions in the first place. Both are instances of an institution going beyond its proper sphere and producing dysfunctions in society, but the very idea that an institution has a proper sphere beyond which it ought not to expand presupposes some understanding of the appropriate functions of the spheres referred to. From where, though, do these functional ascriptions come? I address this question here by first recalling the discussion of this topic as it arose in Chapter 8 in explaining Durkheim's method of functional analysis. Much of what was said there in support of his functional analysis applies to Hegel's as well, but the latter also relies on a further principle, grounded in the self-conscious character of human social life, that is less prominent in Durkheim's sociology.

The most significant overlap between the two versions of functional analysis is their underlying principle that material reproduction is fundamental to social life, a *sine qua non* for successful social functioning imposed on humans by their biological nature. It is a mistake to believe, as some (exaggeratedly idealist) Hegelians believe, that biology places no constraints whatsoever on human social life for the reason that spiritual beings, because of their essential reflexivity and appropriative transformation of everything given or immediate, are subject to *no* imperatives imposed from without. This is not, however, Hegel's view, and insisting that it is blinds us to the important truth that Hegel's position has more in common with Marx's materialism than is usually thought.[10] Although our biological nature places no *determinate* imperatives on us – such as, "eat both meat and vegetables!" or "raise children within patriarchal, nuclear families!" – it does impose general constraints on what can count as well-functioning social life. (Anyone who doubts this should consult the North Koreans who survived the famines of 1994–1998, the effects not only of naturally caused floods and droughts but also of economic policies that constrained producers' ability to respond to the crisis. Or, to bring the example closer to home, consult the Californians who have survived years of uncontrolled wildfires produced by socially caused global warming.) Indeed, the purely

[10] Whatever Hegel's idealism consists in, it is not opposed to the modest version of materialism expressed in Marx's claim that "the way in which material life is produced conditions social, political, and spiritual life-process in general" (MER: 4/MEW: XIII.8–9). Marx and Hegel agree that how society reproduces itself materially shapes the consciousness of social members, including how they conceive of themselves. The materialistic aspects of Hegel's social philosophy derive from his view that spirit includes both consciousness and life and that its characteristic activities are at once mental and material. Of course, in other senses of the term Hegel does not share Marx's "materialism," for example, in believing that the state can resolve the conflicts among material interests inherent in civil society.

material constraints on successful social life can be taken as the basis for one relatively simple – but today highly relevant – conception of social pathology made available by Hegel's social philosophy, namely, an inability to satisfy the biological needs of social members that is due in large part to social causes. Examples of such pathologies – materially nonsustainable social practices – might be birthrates too low to reproduce society over generations (where the causes of such rates are social rather than physiological or environmental) and economic practices that, rather than respecting the constraints nature places on our projects, ravage the natural environment and undermine the conditions of human reproduction. Even if Hegel never thematizes this class of social pathologies – he was not in a position to imagine that a society might organize production so as systematically to undermine the conditions of future life – he provides us with the theoretical resources for doing so, insofar as he regards material reproduction as an essential function of human social life.

The claim that material reproduction is an essential function of human society does not imply that every social institution is involved in that task or even that material reproduction is the most important task of any institution. It does, however, furnish Hegel with *one* criterion for distinguishing fundamental from nonfundamental institutions.[11] One reason for this is the absoluteness and urgency, grounded in nature itself, of the requirement that in order to survive, societies must find effective ways of producing both future generations and the material goods that present social members need to go on living.[12] A second reason rests on the claim that institutions involved in the material reproduction of society are more fundamental than others because of the disproportionate amount of time, energy, and commitment those tasks require of their members.[13] Of course, as Rousseau points out in the Second Discourse, the demandingness of the activities involved in material reproduction does not derive from nature alone. In human societies the production of children and of the means for survival are inextricably bound up with spiritual ends: the family aims not only at producing babies but also at forming them spiritually, and

[11] It is, however, only one such criterion. As noted above, another is that membership in each fundamental institution allows individuals to realize the full range of freedoms and forms of selfhood available in modernity; see also Chapter 12.

[12] As noted in Chapter 12, although the state is less directly involved in material reproduction than the other two spheres, it plays a crucial role in supporting and harmonizing them, including insofar as they are sites of material reproduction.

[13] Recall from Chapter 12 that this is also true of the state if educational institutions are included within its domain; it then plays a major role in reproducing society spiritually.

economic production aims at delivering not only a certain number of calories to social members but also food, clothes, and shelter that, judged by historically variant criteria, are *worthy* of human beings. It is only because these spiritual projects are so complex and demanding that the tasks of reproduction occupy so central a place in social life. Social reproduction, never completely free of natural constraints, is always the reproduction of society as the kind of thing it is, which is to say, a form of life in which material ends are imbued through and through with spiritual significance. (Marx gives an interesting twist to this Hegelian idea: if capitalism develops the productive forces to such an extent that material reproduction can be accomplished easily – and if there is no advertising industry to seduce us into desiring always more and "better" – then in a post-capitalist society the spheres Hegel takes to be fundamental to social life might lose their central significance [MER: 440–1/MEGA: XXV.828].)

A further possibility for conceiving of social pathology that depends on a close resemblance between social and animal health mirrors Hegel's characterization of the latter in terms of vitality, or an organism's enduring power to ward off threats to its proper functioning. Although the implications of this aspect of animal health for social pathology are not emphasized by Hegel himself, it is not difficult to extend the idea to the social realm, in which case social pathology would appear as an ossification or paralysis that impedes a society's capacity to respond appropriately both to changes in its external environment – think of climate change today – and to more internal, normative transformations in, for example, our conceptions of what the good of social members requires.[14] One such normative change we are still in the process of accommodating in social life is the rejection of the long-held view, championed by Hegel himself, that women find their fulfillment by restricting themselves to the sphere of the family, that is, to their roles as wives and mothers. In other parts of his system Hegel calls the vitality at issue here "living spirituality," and one pathological counterpart of it is said to be spiritual life that has sunk to the level of "mere habit" (PhM: §258A). A pathological lack of vitality can afflict specific instantiations of institutions, as in marriages we describe as "dead,"[15] but the absence of vitality can also afflict a social institution in general, as evidenced by repeated contemporary (but mostly false) proclamations of "the death of the family."

[14] Other accounts of social vitality are suggested by Särkelä 2018: 159–60 and Heisenberg 2019: 161–2, 168.

[15] The internet is full of sites that offer, e.g., "three sad signs that your marriage is dead" (www.yourtango.com/2015284073/3-sad-signs-your-marriage-dead).

The thought that spiritual beings die spiritually when the processes that sustain them sink to the level of mere habit – even if some forms of habit are also essential to healthy spiritual life (Menke 2018: 127–34) – suggests one general possibility for conceiving of social pathology, namely, when spiritual processes lose their distinctively spiritual features and become indistinguishable from those of their "ontological inferior," mere animal life. This point is expressed well by Arvi Särkelä: "Social pathologies consist in … the reduction of a genuinely social life to the *mere* reproduction of itself"; or, to put the point more generally: "What at the organic level of natural transactions appears as 'healthy' [may] qualify as 'pathological' at the richer level of the social."[16] Since the ways in which spirit differs from mere life (detailed in Chapter 11) are numerous and complex, this general conception of social pathology can take multiple forms, depending on which activities of spirit have degenerated into unself-conscious processes. Särkelä focuses on pathologies of development (or "evolution") in which social members are insufficiently reflective with regard to the ends their practices serve – failing to subject them to critical scrutiny – which leads to stagnation marked by disturbances in the "self-transformative praxis" that is the hallmark of healthy social life (Särkelä 2018: 17).[17]

The thesis that material reproduction is a central function of *social* life, just as it is in animal life, has for both Durkheim and Hegel a weakly a priori character. It is not a priori in the strict Kantian sense because it rests on empirical, albeit very general, claims about human nature and the spiritualized form material needs take in human societies. It is, however, a priori in the following methodological sense: one can think of the social theorist as approaching her object with the assumption that cooperative reproduction of the conditions of life must take place in every human society and then posing two questions about the society at issue: "Where in this society, and in what forms, does material reproduction take place?" and "What are the spiritual implications of – including the spiritual aims implicit in – those forms of material reproduction?" A principle this general leaves room for nearly infinite variation among the particular forms of social life the social theorist might encounter, and only empirical inquiry will put her in a position to make specific claims about how reproduction in any given society takes place.

[16] Särkelä 2018: 321, 193 [emphasis in original]; see also 315, 318.

[17] This form of social pathology is articulated especially by John Dewey. Something similar might be attributable to Hegel if one regards the *Phenomenology*'s method of immanent critique as a model for self-transformative critique in the domain of the social (Särkelä 2018: 28–138).

Spiritual Functioning in Social life and Corresponding Pathologies

Although Hegel's social theory has the resources for viewing failures of material reproduction as forms of social pathology, such failures are far from being paradigm cases of social dysfunction, precisely because they can be diagnosed in abstraction from the spiritual functions of social life. In this context it is worth noting that the weakly a priori principle ascribed to Hegel and Durkheim above has two components, one less controversial than the other: one might accept that the task of material reproduction is central to social life but resist the claim that the activities that serve this end are typically imbued with spiritual significance. Yet this second claim follows from the fundamental (and defensible) thesis about the nature of human, as opposed to purely animal, society under examination since Chapter 6, namely, that social reproduction takes place via normatively governed activities carried out by agents who understand, apply, and recognize the authority of the rules or norms structuring social cooperation. If this is true of human society in general, then its material reproduction cannot be independent of a spiritual reproduction that educates, or forms, its members into agents capable of acting on the basis of reasons whose validity they recognize. This is precisely the thought underlying one guiding assumption of Durkheim's *Division of Labor*, namely, that solidarity, an intrinsically moral phenomenon, is a vital need of human society.

I have argued that Durkheim's ascription of functions to social institutions is guided by broad premises concerning human social life in general but that making functional claims about a specific society depends on empirical investigations that require considerable interpretation before functional ascription can take place. One type of empirical evidence for ascribing functions to institutions that Hegel might be taken to employ along with Durkheim is the real history of social institutions, insofar as it shows a certain social change to respond to a specific problem faced by the society under consideration. In Chapter 8, I suggested that Durkheim's ascription of a moral function to premodern corporations relied on his account of the problems faced in the historical moment prior to the advent of corporations and of how the latter responded to precisely those problems. To be sure, arguments of this sort require that empirical facts be interpreted, and this implicates Durkheim's sociology (and the versions of social theory I defend) in a hermeneutic circle, or holistic "web of belief," that, while ruling out the piece-by-piece verification of specific claims, is nevertheless subject as a whole to standards of rational assessment.

The question is whether this form of historical argument has a role to play in Hegelian social theory and its diagnoses of social pathologies. The answer to this question is complex. On the one hand, the type of argumentation I have attributed to Durkheim is not foreign to Hegel in general. In the *Phenomenology* similar arguments play a central role: what justifies transitions from less to more adequate "forms of consciousness" there is precisely that a new form of consciousness offers a rational solution to the specific problem, or contradiction, encountered by its predecessor. The conception of knowledge represented by "Perception," for example, is shown to be *rationally* superior to its predecessor precisely because it brings together what its predecessor could not, namely, empirical determinacy and the persisting unity of the object of knowledge despite its varying intuitional content. The same type of argument explains, later in the text, the rational superiority of the modern social world over a society structured by the relation between bondsman and lord. On the other hand, the *Phenomenology*'s project is not social theory, and the text concerned with that topic, the *Philosophy of Right*, appears to rely not on historical but on conceptual-dialectical arguments oriented toward finding an adequate conception of practical freedom that does not come into conflict with its expressions in the empirical world. Yet if we take a looser view of the *Philosophy of Right*'s method, it seems likely that its ascription of functions to modern institutions is motivated in part by historical considerations of the sort Durkheim relies on. For example, one of the functions Hegel attributes to the modern nuclear family – the task of bringing up children to be, as adults, self-sufficient providers for themselves and agents motivated to pursue their own individual projects free from the constraints of the premodern extended family – can be viewed as a response to the momentous change from feudalism to a market economy: if modern civil society is to function well, it requires agents who are trained from childhood in ways that enable them as adults to take up the more individualistic roles required by market-regulated production. Looking at how the modern family is in fact structured reveals norms and practices well suited to serving precisely this emergent social need.

For some, attributing empirical arguments of any kind to Hegel will appear misguided. Yet, as Hegel tells us, the conceptual arguments of "speculative philosophy" – of which the *Philosophy of Right* is an instance – cannot get by without attending to the world as it in fact exists. This means that the method of Hegel's social philosophy is more complex than standard interpretations suggest. The complication of interest here is that the *Philosophy of Right* combines a sort of conceptual argumentation (about the nature of freedom or self-determination) with empirical investigation – looking to social institutions as they in fact exist – in order to provide a determinate account of the rational social order, including the appropriate functions of its institutions. It is impossible here to explain in detail how this two-pronged method works, but even a cursory examination of the argument of the *Philosophy of Right* shows it to depend on a thorough familiarity with already existing institutions. If social philosophy's task is "to comprehend *what is*" – to grasp "*the present and the actual*" (PhR: Preface; Hegel's emphasis) – then the philosopher must begin with some empirical knowledge of what exists prior to its being philosophically comprehended. If philosophy must know "what is" in order to go about its business, it cannot be a fully a priori undertaking, providing a determinate account of rational social institutions independently of how they in fact are. Apart from this consideration, it would be absurd to pretend that the *Philosophy of Right*'s accounts of the fundamental institutions of modernity relied only on a priori arguments. There is no sense in which the rich details of those accounts could be "deduced" by purely conceptual argumentation. Hegel acknowledges this in noting that the existence of rational institutions (albeit in an incompletely rational form) must temporally precede philosophical comprehension of them (PhR: §32A). This point is also expressed in his claim that philosophy can accomplish its work only after a certain historical process has taken place – only after "actuality has gone through its formative process and attained its completed state." Social philosophy, according to Hegel, is not the deduction of what should be from concepts of pure reason but rather "*its own time comprehended in thoughts*" (PhR: Preface; Hegel's emphasis).

Comprehension, of course, involves more than empirical apprehension. The latter must be complemented by systematic ("dialectical") reflection on the various components of the concept of freedom and what they imply for the nature of a rational social world. This concept of freedom serves as the criterion that guides the philosopher's *comprehension* of "what is" by distinguishing the "inner pulse," or rational "core," of externally existing institutions from their "multicolored hull," which contains "an infinite wealth of forms, appearances, and configurations" not essential to their rational character (PhR: Preface).[18] This feature of the *Philosophy of Right* points to a respect in which its procedure for attributing functions to institutions differs from Durkheim's: Hegel examines existing institutions with

[18] More needs to be said about this method in order to make it minimally comprehensible; see Neuhouser 2017b: 16–36.

an eye to discovering not only how the society they make up reproduces itself and maintains inner cohesion in doing so, but also how those institutions function to realize their members' freedom. The important role that freedom plays in his functional analysis follows from his view of how spiritual phenomena differ from those of mere life, including that spiritual beings strive not merely to reproduce themselves but also to realize their freedom in doing so – or, more precisely, to reconcile their aspirations to live and to be free within the very same activities. In the context of contemporary social theory this premise needs to be further justified. Before addressing that issue, however, it is worth noting that the realization of freedom also finds a place in Durkheim's account of healthy social functioning: the increasingly individualistic character of modern European societies is said to open up new possibilities for individual self-determination, and when individuals find these possibilities realized, the consensus on which social cohesion relies is promoted.

Justifying Hegel's premise that realizing freedom is a central function of modern institutions requires noting a second respect in which his method differs from Durkheim's: Hegel's attribution of functions to institutions relies on a hermeneutical principle not found (or not prominent) in Durkheim's functional analysis. It, too, can be traced back to Hegel's commitment to treating social life as a spiritual phenomenon, but in this case what he takes from his concept of spirit is less controversial than the premise that realizing freedom is an essential function of social life. The principle in question follows from Hegel's understanding of the self-conscious nature of spiritual beings, more specifically from his claim that human societies carry out their characteristic processes in the mode of self-consciousness. As noted above, this means that the activities of social life are carried out with some awareness on the part of its members of the point of what they are doing. Their social participation has a meaning for them, and although their understanding of what they do may be mistaken in various ways, it is also partially constitutive of what their activity consists in. Parents who have not yet philosophically comprehended family life will not be able to say, "the family's function is to reproduce society while at the same time promoting freedom." If you ask them, however, what ends their parental responsibilities promote – "what is the point of disciplining your children, and why do you avoid physical force in doing so?" – their answers will typically reveal an allegiance to the ideals of self-mastery (a type of self-determination) and bodily integrity, even if they are unable to articulate these points in precisely these terms. Similarly revealing answers will typically be given to questions about why (in civil society) workers should not be forced to work under conditions they find unacceptable or why (in the political realm) the interests of all must be represented in the legislative process.

To be sure, the point of social activities cannot be discovered only by attending to what social members say in response to such questions; as Durkheim sees, the point of such activities is also embodied in the norms and rules "all of us" follow without necessarily being aware of the meanings they bestow on what we do. The methodological point here is that ascribing functions to social institutions requires understanding the implicit and explicit meanings they have for their participants and that the social theorist has at her disposal a wealth of empirical resources for figuring out what those meanings are. This means that social theory, including the diagnosis of social pathology, must be in part an interpretive enterprise, aimed at discerning the conscious and implicit meanings of social activity, and thinking of it as such helps us to understand the considerable power Hegel's account of modern social life possesses even for readers who know nothing about speculative philosophy or dialectical logic. The viability of Hegel's claims regarding the functions of the modern family depends on making sense of empirically accessible details regarding what family members do; what they take their obligations to be; and how they understand the ends that give their activities a meaning. His judgment that the modern family is rational rests on the claims that an inner coherence to these activities can be found in the various ways in which together they serve the intertwined ends of life and freedom. For Hegelian social theory that does not simply point to other parts of Hegel's system in order to support its claims, the plausibility of its functional ascriptions rests on a network of various claims – some less, others more empirical in nature – and on how hermeneutically convincing the Hegelian interpretation of the real phenomena is as a whole.

In Chapter 8, I claimed that Durkheim's functional analyses extend over entire texts and win whatever force they have by bringing various arguments, interpretive suggestions, and analogies together into a compelling picture of how, for example, the division of labor generates social solidarity and why the latter is a vital social need. These arguments move back and forth between broad claims about the nature and needs of society in general and more specific functional claims about the institutions at issue. My account of Hegel's position implies that it moves between "subjective" and "objective" levels or perspectives (PhR: §§144–7). The latter is more or less the perspective of Durkheim's functional analysis, in which the social theorist regards his object of study as a "thing" that "the mind cannot understand except by going outside itself through observation and experimentation" (RSM: 35–6/xii–xiii). In Hegel's case, this perspective includes, along with an empirical familiarity with social reality, general theses regarding the nature of spirit and life, including the fundamental ends at which their activities aim, as well as the idea that spirit, like life, maintains itself as what it is through specialized and coordinated activities.

The subjective perspective, in contrast, regards social life as made up of human practices, or forms of activity informed by the consciousness and will of those who carry them out. Such practices are partially constituted by the beliefs, attitudes, and normative commitments of the members that make that life into an intrinsically meaningful phenomenon. This means that probing and interpreting the self-understandings of participants in social life plays a large role in Hegelian social theory. Regarding social processes as practices is important because it makes it easier to see how they can have a point or function, deriving (in part) from their being carried out by agents who act for the sake of ends. In the standard case, where social theorists themselves belong to the society under examination, this means that comprehending "what is" requires that they regard social reality not as an independently constituted thing but as something to which their own position within social life provides them with a (defeasible) "view from the inside" of the meaning of social phenomena. A social theory that incorporates this perspective will concern itself with judgments regarding "what good our actions are in the service of" or "what we have been up to all along (even if we have not explicitly understood it in that way)."

One more point about ascribing functions to social institutions is worthy of mention because of its connection to understanding and diagnosis in medicine. It is often alleged that theories of social pathology must presuppose a determinate picture of social health – including an account of the proper functions of institutions and practices – in order to be in a position to detect social dysfunction. This charge is often regarded as a decisive critique of such theories since producing a full picture of health in advance of specific diagnoses, difficult enough in the case of physical health, is out of reach in the social domain. It is easy enough to understand the logic behind this allegation, but, like many deliverances of pure reasoning, it is not true to the real phenomena, including diagnostic practices in medicine itself. Fortunately, neither medicine nor social theory relies on a fully articulated conception of health or proper functioning in order to recognize dysfunctions. It is nearly always easier to recognize dysfunctions in a living system than to say exhaustively in advance what proper functioning consists in, and recognizing a specific dysfunction often leads to a better understanding of what an organ's or subsystem's proper function is. To take a specific example: it is not difficult to recognize cancerous growth as dysfunctional – it is a clear physiological example of one organ overstepping its proper boundaries and invading the territories of others – but understanding that the function of T cells is to recognize and fight "foreign" antigens was a consequence of first recognizing and then studying the specific dysfunction of cancerous growth.

The charge that recognizing dysfunction presupposes a positive picture of normal functioning contains a kernel of truth: the two concepts are indeed logically inseparable in the sense that identifying some condition as dysfunctional necessarily implies something about what health consists in. That the two concepts are logically inseparable does not mean, however, that a conception of health is epistemologically prior to the recognition of dysfunction. That medical science proceeds in part along a *via negativa*, where recognizing the dysfunctional precedes knowledge of the healthy, is a widespread view among philosophers of medicine.[19] The same holds for mental health. Freud, for example, begins not with an already formed picture of healthy mental functioning but with recognizable examples of (neurotic) dysfunctions, and only by studying the etiology of such dysfunctions does he arrives at an understanding of healthy development and functioning.

Although all theories of health and illness involve (revisable) judgments about the good in a broad sense, as one ascends the ladder of pathologies from the less spiritual to the more, the conceptions of the good informing one's picture of health become more substantive. In the end, Freud characterizes mental health as the ability to find satisfaction in love and work,[20] where all three concepts – satisfaction, love, and work – imply a substantial, if still general, conception of what a good human life consists in. The same point can be seen in the American Psychological Association's finding that homosexuality is not an illness because it "implies no impairment in judgment, stability, reliability, or general social and vocational capabilities" (Fox 1988: 508–531). The clinical (and overly functional) tone of this pronouncement hides its reliance on a broad conception of the good that differs little from Freud's: as the phrase "social and vocational capabilities" reveals, part of what makes homosexuality nonpathological is that it does

[19] "Disease reveals normal functions to us at the precise moment when it deprives us of their exercise" (Canguilhem 1978: 52).

[20] This points to another convergence in Hegel's and Freud's views: human satisfaction depends on engaging in social activities requiring considerable time, energy, and long-term commitment.

not itself – apart from society's prejudicial treatment of it – impair one's ability to love and work successfully.

A further point is that the conceptions of the good underlying diagnoses of spiritual pathology are historically and socially conditioned. They are, in Hegel's account of *Sittlichkeit*, conceptions of the forms of satisfaction typically available to *modern* (European) individuals, but this feature of them in no way prevents treatment of spiritual illness based on them from being effective, here and now, for individuals who share the spiritual world informed by those conceptions. Moreover, the historically situated character of these conceptions of the good implies that diagnoses of spiritual illness must remain constantly open to revision, as the example of homosexuality attests: what was previously regarded as an illness ceases to look like one once we learn that it implies no spiritual dysfunction, which itself depends on social acceptance of a gender-neutral revision of what the good of marriage or committed partnership consists in. These points constitute another example of the principle that knowledge in a given domain must be appropriate to the kind of thing its objects are; to demand a value-free and ahistorically valid science of mental health is to be confused about what human (spiritual) life consists in – and the same is true of demands for a value-neutral social theory that aspires to prescribe for all times and places.

The idea that science proceeds by a *via negativa* is hardly foreign to Hegel. It is clearly visible in the method of the *Phenomenology*, where progress in specifying what adequate knowledge consists in is always sparked by the discovery of a problem (or "contradiction") in one's current conception of the standards knowledge must meet. On this method, progress can be made only if it is possible to recognize problems and find (local and still not fully adequate) solutions to them by formulating ever more comprehensive concepts of what it is to know, all of which takes places in the absence of a prior comprehensive understanding of what knowledge is. Although this is not the *Philosophy of Right*'s official method, it is hard to imagine that something like it was not part of the original logic of discovery that made that text's systematic account of the rational (modern) society possible. Hegel's familiarity with poverty and the form of alienation described in the text's Preface, including his recognition of them as problems, certainly preceded his mature, systematic treatment of them, and it is plausible that his occupation with them played a substantial role in developing the vision of the healthy society he later came to embrace. Something similar is no doubt true of the path by which Durkheim arrived at his views on the importance of solidarity in modern societies;

indeed, his first claim in the second preface to *The Division of Labor* is simply an acknowledgment of the "juridical and moral anomie in which economic life currently finds itself" (DLS: xxxi–xxxii/ii). Durkheim, too, seems to believe that the way to get started in social theory is to point to real phenomena, the dysfunctional character of which is accessible to a pretheoretical eye.

Despite their methodological reliance on a *via negativa* in diagnosing social pathologies, the positions of both Hegel and Durkheim are complicated by an interesting twist: not everything that appears to the pretheoretical eye as dysfunctional turns out in fact to be so. In other words, initial judgments about what is dysfunctional are themselves subject to revision as our understanding of a society's functioning develops. It is well known that Durkheim does not regard statistically normal levels of divorce, suicide, and crime as signs of social pathology, and something similar is true for Hegel. That is, divorce (in the family), the unlucky economic failures of some (in civil society), and war (between states) are taken by him to be "normal," indeed, *necessary* phenomena that do not indicate social dysfunction. (Hegel does not consider Durkheim's idea of *abnormal levels* of divorce, economic failure, or war, but perhaps these could be considered pathological from a Hegelian perspective, too.) The thought is that once we understand the nature and spiritual point of the family, the market economy, and the nation-state, then divorce, individual economic failures, and war no longer appear as merely contingent (Hardimon 1994: 20, 32, 228–30, 240–1).[21] For example, the important role that feeling plays in marriage[22] in the form of conjugal love is part of what makes that institution distinctly enriching: conjugal love grounds an especially intimate, uniquely satisfying relation to another that has no substitute in civil society or in politics; feeling is what makes marriage an instance of "ethical life in the form of *the natural*" (PhR: §158Z;Hegel's emphasis), a distinctive spiritual intertwining of life and freedom (including recognition in the

[21] For this reason, not all instances of divorce, economic failure, or war would count as pathologies for Hegel; on the contrary, some rates of these phenomena belong, in Durkheim's words, to the "normal" state of society. For Durkheim the same is true of suicide and crime, although unusually high rates of both (measured in comparison to the rates of similar societies) do indicate social pathology.

[22] This might seem to contradict Hegel's claim that the absence of sensuality does not necessarily violate the essence of marriage, but it does not. His view is that the *essence* of marriage resides in the ethical relations of love, trust, and a shared life, and that a marriage that has these without sexual passion still counts as one because it realizes the most important ethical goods that institution offers. Even though sexual desire is a subordinate moment in marriage, a fully actual marriage joins sexual desire with marriage's ethical goods (PhR: §§162–3, including Hegel's handwritten marginal notes).

form of love). At the same time, since feeling is by its very nature "moody, fleeting, [and] volatile" – a "contingent" factor that at any particular time may or may not align with marriage's ethical ends (PhR: §§161A, 164) – it constantly lurks within conjugal union as a potential destroyer of a form of ethical life. According to Hegel, similar points hold for civil society and the state: whereas the "freedom" of the market fosters self-reliance and offers a large arena for individual free choice, its unplanned character means that some participants will inevitably lose out (PhR: §§185, 230); further, the character of nation-states as sovereign individuals, subject to no higher political authority, means that some disagreements among them will be settleable only by war (PhR: §§331, 334). In all these cases the possibility of such "failures" is built into the nature of institutions such that to eliminate the possibility of failure would be to destroy part of what makes those institutions ethically desirable.

Further Implications for Diagnosing Social Pathologies

Once we have considered how Hegel's theory ascribes functions to social institutions, the field of possibilities for diagnosing pathologies expands to include types beyond those considered thus far. For we are now in a position to detect disruptions in *spiritual* functions whose causes are *internal to a specific institution*. The systematic production of poverty in civil society is an example of such a pathology: it is spiritual because its primary danger is not malnutrition or death but exclusion from the spiritual benefits – freedom, recognition, and a subjective attachment to social life – that membership in civil society affords; and it is internal to civil society because its causes – overproduction and bad luck in the market – derive from the very nature of a market economy (PhR: §§241–5). Moreover, such examples make clear that the complexities accompanying the introduction of spirit into the processes of life mean that we should expect to encounter distinctively social forms of pathology that have no strict analogue in the domain of life.

Although one could no doubt adduce many further examples of possible disruptions in spiritual functions whose causes are internal to a specific institution, I restrict myself here to two, both related to specific details of Hegel's account of mental illness discussed above. The first harks back to his claim that the healthy rational subject takes possession of its mental states by subjecting them to intersubjectively valid principles of objectivity, thereby bringing them into relation with an external, objective world. There is a social analogue to this spiritual dysfunction – where psyche and world go their separate ways (PhM: §406A) – that is importantly not a form of mental illness. Drawing on the idea that spiritual beings act with an understanding of what they do and why, the reality-constituting function of rationality goes awry when what social members take themselves to be accomplishing in the world fails to correspond to what they actually do. If a mismatch of this sort has its source in a basic feature of institutions – as in Marx's account of commodity fetishism – then it makes sense to think of that mismatch as a spiritual defect of society itself. Some forms of *ideology*, in other words, are social pathologies, and even though Hegel never uses this term, the phenomenon it names lies at the heart of the *Philosophy of Right*'s account of the alienation faced by modern subjects and its location of the central task of social philosophy in comprehending "what is," itself a spiritual appropriation of the initially foreign.

A second possible social pathology that echoes Hegel's understanding of mental illness can be found in his account of one danger intrinsic to civil society, namely, the possibility that the appetites of participants in a market economy will be "infinitely aroused" by the nearly unlimited possibilities it offers for satisfying ever more specific and complex desires, leading individuals to acquire limitless "needs" that render enduring satisfaction impossible (PhR: §§185, 190, 195, 253). Although Marx recognizes this more clearly than Hegel, the potential for acquiring unlimited desires in civil society is further heightened by the fact that its members' activities and relations to others are mediated by money, a "universal medium of exchange," that measures value only quantitatively (PhR: §204).[23] It is indicative of deeper philosophical differences between the two that Hegel conceives of the problem posed by the dominance of money in civil society slightly differently from Marx. Hegel's worry, taken over from Rousseau, is that individuals in civil society, especially members of the commercial estate, will be tempted to seek *recognition* from others by accumulating wealth, measured in money terms, rather than by pursuing less abstract, "finite" ideals. In contrast to qualitatively defined aims – being a good plumber, doctor, or schoolteacher – measuring one's worth by quantitative standards admits of no natural stopping point in the quest for recognition, for value that can be expressed only in quantitative terms can always be improved on simply by accumulating more. The satisfaction to be had from this way of valuing oneself and of pursuing recognition from others is both qualitatively thin and highly unstable since there is no reason not

[23] Money is said to "reduce" phenomena to a universal measure of value (PhR: §299).

to seek ever larger "quantities" of recognition and hence no reason to place limits on one's quest to acquire more. Moreover, individuals who find their recognition in being good plumbers, doctors, or schoolteachers are not only able to enjoy richer forms of satisfaction; they are also more valuable to society than those whose main goal is to acquire as much as possible. The problem is not so much that satisfaction in *consumption* becomes elusive – Hegel was not familiar with the orgies of consumption we revel in today – as that the stable recognized identities to be had as *producers* in civil society give way to a frenzied, never-resting quest for recognition fueled by a dynamic of "bad infinity."

Hegel's claim that it is only through recognition of the former sort that "the individual gives himself actuality (*Wirklichkeit*)" in civil society (PhR: §207) echoes his characterization of mental health, which involves "awareness of one's actuality (*Wirklichkeit*) – the concrete fulfillment of one's individuality – … in the form of one's connection to … an external world, … which constitutes what he actually (*wirklich*) is in the world before him" (PhM: §406). The point is that mental illness is not the only factor that can impair individuals' ability to establish intersubjectively recognized identities; social institutions, too, can push them in a similar direction. In such cases, the affected individuals are not mentally ill; rather, the dysfunction at issue has its source in a dynamic internal to a specific social institution, which is why it is appropriately thought of as a pathology residing not in individuals but in society itself. Of course, the affected individuals suffer – from the lack of "actual," satisfying identities – but they are not themselves ill because the dynamic responsible for their impoverished state is not intrapsychic but social. That Hegel takes the dysfunction at issue to be social rather than individual is consistent with his claim that it can be remedied by modifications internal to civil society (consistent with the basic structure and raison d'être of a market society). In this case, it is the strengthening of economically based corporations that Hegel regards as the most important remedy. For corporations – labor unions, guilds, and professional associations, in the contemporary world – secure conditions under which their members can achieve socially recognized identities, or their *Standesehre*, as competent executors of a socially useful occupation (PhR: §253).

Pathologies of Bondsman and Lord

In concluding my discussion of Hegelian conceptions of social pathology, I return to Hegel's account and critique of (premodern) forms of social life grounded in the relation of lord and bondsman. Doing so will bring into relief what I take to be his most comprehensive and interesting conception of social pathology – a coming apart of the two principal ends of social life, freedom and material reproduction – which, though virtually ignored by commentators, is his most distinctive, perhaps even (because of its importance for Marx) his most influential contribution to a theory of social pathology.

Hegel's depiction of social relations organized around the categories of bondsman and lord is fertile territory for thinking about social dysfunction.[24] Yet determining what it implies for conceptions of social pathology is more complex than might initially seem. As noted in the previous chapter, material reproduction in such a society takes place through the complementary but hierarchically structured activities of bondsman and lord: one commands, while the other carries out those commands by laboring. There is, however, another dimension along which material reproduction for bondsman and lord is radically asymmetric, namely, the two moments of material reproduction – production and consumption – are themselves torn apart, with each ascribed principally to only one of the relation's opposing poles: the bondsman labors, while the lord "enjoys" the goods produced by that labor. One might think that production and consumption fall to distinct, mutually antagonistic social classes, but in fact each participates in both activities, although on starkly hierarchical terms: the lord participates in production by dictating what is to be produced, and the bondsman consumes, although only enough to reproduce his capacity to labor further. The opposition between the two is perhaps better formulated by situating each in relation to the domains of life and freedom: the lord's will is sovereign for all (and therefore free), and his consumption is free insofar as it is no longer determined by biological need; the bondsman, in contrast, is little more than a beast of burden in his productive activity, and his consumption is limited, in accordance with nature, to what he needs in order to live.

At the same time, as Hegel's account of spirit implies, all aspects of material reproduction are imbued with spiritual significance. As noted above, the asymmetric activities of bondsman and lord are grounded in, and give expression to, shared values that ascribe normative authority only to those who have demonstrated their willingness to renounce life in order to achieve freedom. This stark division of roles is not only grounded in a shared understanding of what is of ultimate value (namely, freedom); it

[24] A different attempt to glean resources for an account of social pathology from the bondsman–lord relation is found in Särkelä 2018: 328–59.

also has spiritual consequences in that one class is permanently absorbed in mere life, and thereby unfree, whereas the other, as far removed from life as possible, enjoys, precisely for that reason, a kind of freedom. The problem is not merely that some are free while others are not but, more fundamentally, that both freedom and life remain empty and unsatisfying. Labor for the necessities of life becomes a site of domination and self-denial rather than freedom, and its counterpart, enjoyment, devolves into luxury consumption that fails ultimately to satisfy. Those who labor satisfy no end of their own other than survival, while those who "enjoy" have a spiritual end in view – realizing freedom – but are moved by a distorted vision of what freedom is, namely, complete independence from the demands of life.

Ultimately – this is what Hegel's idealism in this context amounts to – the spiritual failings of bondsman and lord can be traced back to inadequate self-conceptions, that is, to conceptions of a free subject's relation to life that either disavow that connection or, because of an impoverished understanding of freedom, misconstrue the threat that being subject to the demands of life poses to being free. In other words, bondsman and lord are united by an overly exalted conception of the freedom self-conscious subjects should aspire to. Both take the essential property of freedom to consist in what the *Phenomenology* eventually reveals to be a deficient, one-sided relation to life, in which a subject asserts its status as free by acting as though its being alive placed no constraints on its freedom, thereby demonstrating its elevated standing in relation to the merely living. The inadequacy of this understanding of freedom – and one reason for viewing social enactments of it as pathological – shows itself in the fact that the practices bondsman and lord must engage in in order to assert themselves as what they take themselves to be require an engagement with life processes that makes the attainment of freedom on that conception impossible for both. (This can be understood as a form of ideology where what one in fact does in social life fails to correspond to what one takes oneself to be doing.)

That bondsman and lord subscribe to the same austere conception of freedom explains why the result of their struggle unto death, including the positions they come to occupy in material life, is so asymmetric: if freedom requires absolute independence from every imperative external to one's own will, then avowing one's nature as a living being, as the future bondsman does in the struggle with the future lord, is incompatible with being free. The starkly unequal relation between bondsman and lord is an expression of this shared all-or-nothing view of freedom, conjoined with the fact that, facing the prospect of death, one affirms the value of his life while the other persists in valuing only absolute independence from life's demands. This conception of freedom explains the asymmetric recognitive relation between bondsman and lord, which translates into asymmetric roles in material life. The bondsman's servitude and labor are at once a relating to *life*; a relating to *self* (an expression of his own conception of freedom and attitude to life); and a relating to *another subject* (through relations of recognition that define conditions of authority and subjection). If we recall Hegel's characterization of spirit as "a being capable of … *enduring* [the] contradiction" between freedom and life (PhM: §359A; Hegel's emphasis), it becomes clear that the bondsman's servile labor is *spiritual* activity – the same is true of the lord's enjoyment – because it is a way, however rudimentary, of negotiating the opposition, central to what spirit is, between freedom and life.

If the social pathologies associated with bondsman and lord can ultimately be traced back to inadequate self-conceptions, or impoverished conceptions of freedom, it is tempting to conclude that social pathology is nothing more than false consciousness and that its remedy lies in transforming the consciousness of those afflicted by it. But Hegel's conception of social pathology is more materialistic than this, as it must be, given his view that spirit's activities are at once self-conscious and material. This means that holding together the poles of the opposition between freedom and life is not something that takes place only within consciousness; material practices, too, are part of how spiritual beings mediate, or negotiate, this opposition. If, for Hegel, "social being" does not unidirectionally determine "social consciousness," the latter is nonetheless given determinate expression in and transformed by social practices thoroughly intertwined with the functions of life.[25] Even if there is more to the social than mere engagement with life, it is also true that whatever belongs to the social – relations of recognition, for example – is inextricable from processes of life. This implies that much of what counts as social pathology from a Hegelian perspective must be theorized not simply as false consciousness but, at the same time, as *false material practices*, that is, social practices that embody unsatisfying ways of negotiating the opposition between freedom and life.

[25] Moreover, material practices are essential to spirit's achieving a true self-conception because the only way it can correct its self-conceptions is by discovering how, when enacted in material practices, they produce real relations between self-consciousness and life that belie the conception of freedom those practices were supposed to express.

Where, then, does social pathology enter this picture? It is tempting to locate the pathological character of the bondsman–lord relation in any number of its features: neither party finds genuinely satisfying recognition from its counterpart; their relation manifests itself in domination rather than freedom; and the bondsman is alienated from his true essence (since he is potentially a free subject but regards only his lord as belonging to that exalted category). None of these responses is incorrect, but a more comprehensive (and more functional) description of what has gone wrong is that, taken as a whole, the society to which bondsman and lord belong cannot achieve spiritual satisfaction in the sense that the real conditions of its existence necessarily fail to measure up to the conception of freedom on which it takes itself to be grounded. Its practices aim at realizing freedom under a specific description but end up producing its opposite. When one adds to this picture that realizing freedom entails engaging in the processes of life, one sees that this form of social pathology can also be characterized in terms of society's failure to *negotiate the opposition* between freedom and life in a satisfying way. This failure shows up in the inability of social members to integrate, in both what they do and how they think of themselves, their dependence on the conditions of life with their understanding of freedom. The failure to achieve this integration means that the various ways in which social members participate in life are not, at the same time, expressions of their freedom and that, conversely, the activities they regard as expressions of freedom are not, at the same time, ways of participating in life. Expressed differently, failure here consists in the inability of social members to sustain conceptions of themselves as *free animals*, where the tension between being free and being part of life is (somehow) both present to consciousness and overcome.[26] Spiritual satisfaction, in contrast, requires that life be elevated to freedom and that self-consciousness be filled with the aims of life. On this view, social pathology exists whenever the basic conditions of a society prevent its members – in their self-conceptions, in their recognitive relations to others, and in their material practices – from bringing together their membership in both the realm of freedom and the realm of necessity. Success in doing so – satisfaction in the realm of *objective* spirit[27] – would mean that there is no essential life activity that is not also a site of freedom and no expression of freedom that is not also a material practice.

[26] This ideal could also be ascribed to Nietzsche, perhaps a fruit of reading Hegel carefully.

[27] This claim holds less in the sphere of absolute spirit, even if elements of materiality persist in art and religion.

It is not difficult to recognize in this thought the central idea behind Marx's account of alienated labor: "[T]he [alienated] worker no longer feels himself to be freely active in any but his animal functions – eating, drinking, procreating ..., etc. – and in his human functions he no longer feels himself to be anything but an animal" (MER: 74/MEGA: LX.514–15). For on this account, overcoming alienation depends precisely on humanizing the animal and animalizing the human, or on making the functions of life free and the activities of freedom material. If Hegel is right about the conditions of spiritual satisfaction, however, then Marx is mistaken when he suggests later in his career that true freedom begins only where material production ceases, only where human activity is completely undetermined by nature and no longer undertaken for the sake of satisfying needs and "external purposes" imposed on us by nature (MER: 441/MEGA: XXV.828).[28] Here at least Hegel is more consistently materialist – and less Kantian – than Marx is in those (rare) instances where he posits a sharp opposition between freedom and subjection to the requirements of life.

Thus far I have discussed the ideal of spiritual satisfaction as if it required a remainderless reconciliation between freedom and life in which the tension between distinguishing oneself from life and at the same time identifying with it were completely overcome. This is how Hegel's concept of reconciliation is often interpreted, and not without reason, but his thought offers another possibility as well, one that takes more seriously his claim in the *Phenomenology* regarding the restless, unceasing character of subjective (and therefore spiritual) activity. If, as noted in the previous chapter, subjectivity is nothing but the activity of "self-negation" – dividing oneself in two and then negating this division so as to bring together what one has torn apart in a way that both cancels and preserves that division (PhG: ¶18) – then any final overcoming of the opposition between freedom and life – any ceasing of the first, "dividing" moment – would signify spiritual death.[29] If enduring the contradiction between free subjectivity and life meant eradicating that tension, spirit would become inert and lifeless, and hence not spirit at all. Hegel says precisely this in emphasizing the unceasingly negative character of spiritual activity: "The life of Spirit does not shy away from death and devastation ... but instead tolerates death and preserves itself in it. Spirit wins its truth only by finding itself in its own state of being torn in two.

[28] I am indebted to Rahel Jaeggi for discussion of this point.

[29] Hegel characterizes the ceasing of consciousness's self-negating activity as spiritual death (PhG: ¶32).

This power does not … look away from the negative … but looks it in the face and abides within it" (PhG: ¶32).

The thought that the tension between life and freedom can never be completely overcome in healthy social life gains in plausibility when we take into account that death – the bodily deaths of living individuals – is intrinsic to life and can be "overcome" only if one abandons the limited perspective of the living individual and focuses on its membership in the species, which does not itself (necessarily) die and which exploits the deaths of individuals in rejuvenating itself. In the case of spiritual, living beings, one's own death cannot but be an object of consciousness, for a subject aware of itself as life is also aware of itself as dying; relating to life includes consciously relating to one's death. The necessity of death, however, stands in fundamental conflict with a subject's aspiration to be free because it imposes on living subjects an absolute, immovable natural constraint on what they can do or make of themselves. If the individuals who make up the social realm are self-conscious beings who can be reconciled to their own deaths only by grasping their unity with the self-reproducing species, then one task Hegel must assign to a healthy society is aiding its members in the spiritual process of coming to terms with their own mortality (Heisenberg 2021: 872). Failures to endure the dual attitude to life definitive of spirit must qualify for Hegel as disturbances in the proper "functioning" of a spiritual being. And to the extent that such failures have social causes, they constitute social pathologies.

Reading the *Philosophy of Right* with this thought in mind, one finds that this is indeed a task Hegel assigns to rational institutions, which possess the potential to reconcile their members to death in different ways: raising children within the family, which enables parents to identify with and "live through" the generation that survives them; laboring in civil society, which permanently inscribes one's subjectivity into the objective world; political participation, in which citizens experience themselves as part of a community that will outlive them.[30] This helps us to see why it is far from a perversion of ordinary language when Hegel refers to society as objective *spirit*, for by helping to reconcile us to our death social life acquires a spiritual dimension in a completely ordinary sense of that term.

Although these practices are ways individuals can lessen the constraints their own death imposes on them, it would be folly to think – and Hegel did not – that they can fully remove death's sting. Life and the finitude intrinsic

[30] For more on how all three institutions help to reconcile us to our own death, see Heisenberg 2021: 874–84.

to it can be made more compatible with the subject's aspirations than it is without these practices, but reconciliation here cannot produce differenceless unity. That later, in the domain of absolute spirit, Hegel also calls on art, religion, and philosophy to help with this spiritual task is an indication that healthy social life cannot by itself fully resolve the tension between freedom and life's requirement that individuals die. This suggests that in the social realm the negotiating of contradiction constitutive of spiritual activity is best understood not as seamless reconciliation but as similar to the process psychoanalysts call "working through," where conflicts are made tolerable – compatible with healthy, self-affirming life – rather than eliminated. Regarding the enduring of the contradiction between life and freedom in this way opens up a further possibility for conceiving of social pathology, in which "sick" would not only describe societies whose institutions failed to enable their members to work through their own mortality in the ways available to parents, producers, and citizens, but would also apply to societies that eliminated the felt tension between freedom and life by in some way making the spiritual problem of human mortality invisible to its members. While it is possible to imagine social arrangements that discourage individuals from grappling with their finitude, it is no accident that the likeliest of such deterrents come from the realm of culture. One might think, for example, of religions that sugarcoat death, but social phenomena can make the problem of death invisible as well, such as economic institutions that encourage an all-absorbing get-rich-or-die mentality so burdensome that individuals have no psychological space to concern themselves with existential problems. I raise this issue here, without being able to pursue it further, because it points to questions regarding the relation between culture and society that are important to Hegel and that theorists of social pathology need to address. Hegel's position on this topic is interesting because, on the one hand, he insists on a difference between the social and the cultural – in distinguishing objective and absolute spirit – and, on the other, he implies that healthy social life cannot by itself fully accomplish this spiritual task. The relation between society and culture raises a host of thorny issues. One is whether there is such a thing as *cultural* pathology and if so, in what relation it stands to pathology in the social domain. Addressing these questions would lead us into the thought of Kierkegaard, Nietzsche (Neuhouser 2017a: 355–68), and Freud, all of whom offer pictures of cultural illness that although outside the domain of social theory proper, certainly bear some relation to thinking about what healthy social life would be.[31]

[31] The place to begin in addressing these questions is the work of Max Weber.

Finally, an important aspect of Hegel's tale of bondsman and lord is that, in modeling a form of immanent critique, it enables us to see a way out of the philosophical conundrum encountered by Durkheim's conception of social theory as a science of morality. The problem noted there arises from the fact that his science of morality aspires both to explain changes in a society's ethical norms (as functional responses to crises caused by changes in social structure) and to justify those norms without reducing the point of morality to functionality in a nonmoral sense. Durkheim's conundrum arises because he characterizes the crises that occasion social change in narrowly functional terms and not as also ethical crises. His claim, recall, is that moral transformations occur when changes in the structure of societies make existing forms of solidarity obsolete (DLS: xxvi/xxxviii). This means, however, that new social formations appear as rational solutions to narrowly functional problems but that there is no sense in which the ethical aspects of the new societies appear as *morally* superior to those of the old. They are, instead, adjustments in response to changes in social structure, and their justification is limited to the fact that they are the ethical norms that functionally support the new structure. This is illustrated well by the example of social transformation considered in Chapter 8, where population growth and increasing social density efface the boundaries of existing segmental societies, merging them into larger but still homogenous units, and the ensuing scarcity and weakening of the common consciousness leads to increased specialization among individuals, weakening the old "mechanical" solidarity and making necessary a new, organic form of solidarity, together with its corresponding ethical norms valorizing individual freedom. The point is that nothing in this account shows the new norms to be ethically superior to those they replace.

In this context the theoretical virtue of Hegel's account of bondsman and lord is that it locates the failure of a society based on that relation in a dysfunction that cannot be fully described without referring to what by its own lights – by the ethical ideals governing and expressed in the cooperation between bondsman and lord – must be regarded as an ethical failing. (That the normative standards in terms of which failure is diagnosed are internal to the object judged is what makes evaluation of this sort immanent critique.) That is, the crisis, or contradiction, inherent in such a society is moral-functional; the reason such a society ceases to function is inextricably bound up with its failure to realize its own ethical ideals. As we saw above, the real conditions under which cooperation between bondsman and lord is carried out necessarily fail to measure up to the conception of freedom on which they take their relation to be grounded: neither finds satisfying recognition, and neither is truly free since their relation results in domination for the bondsman and only a hollow freedom for the lord. The dissatisfaction of social members in their aspiration to make freedom real and to be appropriately recognized in doing so is intrinsic to their society's dysfunction.

If social crises are always also ethical crises, then a rational response to them must include ethical innovation that revises and enriches what has shown itself to be an impoverished conception of freedom and of the requirements of satisfying recognition. Moreover, ethical innovation of this sort always implies changes in how social members cooperate in the various tasks of material reproduction; those practices themselves change, as can be seen by comparing what child-rearing in the patriarchal family looks like with the form it takes in a family where parents regard themselves as equals. Hegel's advantage over Durkheim is that when explaining social transformation he is more consistent in regarding morality and social reproduction as inextricable aspects of social life, neither of which can be reduced to the other. By understanding new ethical ideals as part of a rational response to crises that are inseparably functional and moral, he is in a position to regard those ideals as rationally superior to their predecessors and therefore as justified in a more robust sense than is implied in taking them to be determined by the (narrowly) functional requirements of a changed social structure. If, following Hegel, we regard practices that unite the aims of life and freedom as spiritual, then it makes sense to say that he, more than Durkheim, possesses the philosophical resources for claiming that modern societies – roughly, those embodying the ideals of the French Revolution – are not merely continuous with[32] but also spiritually, and not merely functionally, superior to the premodern societies they replace, where this superiority lies in their having responded successfully to spiritual failures encountered by their predecessors.

In this chapter I have laid out resources in Hegel's thought for diagnosing pathology in the social realm. I have argued that for Hegel social pathologies should be understood not only in term of impaired functioning in general or as imbalances among specialized functional spheres but also as ways in which society fails in the spiritual task of enabling its members to relate to life in the mode of freedom, a task that can be characterized as negotiating the opposition between self-conscious freedom

[32] Recall that another oddity of Durkheim's account of social change is that it requires no continuity among the ethical ideals of earlier and later societies: the ideals at work in segmental societies appear to be replaced by completely different ones that valorize individuality.

and life. In addition, I have indicated a variety of specific conceptions of social pathology that can appropriately be described as Hegelian, although I make no claim that my account is exhaustive; there is no reason to think that other conceptions could not be found in the same texts I have relied on here. To summarize, the versions of social pathology I take to be attributable to Hegel include:

- socially caused dysfunction in the material reproduction of society;
- functional imbalance, where dysfunction results from one social sphere transgressing its proper boundaries and interfering with the functioning of another or subjecting the whole to its rule;
- any failure of a specific social sphere to carry out its distinctive function (such as the market-generated poverty internal to civil society);
- an absence of vitality, in which society is impaired in warding off threats to its proper functioning;
- social activities losing their spiritual features and becoming indistinguishable from processes of mere life;
- social conditions impeding the development of practical selfhood due, for example, to inadequate sources of recognition or to the generation of infinite, unsatisfiable desires;
- forms of ideology that involve a mismatch, grounded in social reality, between what individuals do in their social activity and what they take themselves to be doing;
- the failure of social life to bring together the ends of life with those of freedom (including forms of alienation in which social life becomes a mere means for staying alive rather than a site of freedom); and
- socially caused impairments of individuals' ability to reconcile themselves (without illusion) to their own death.

CHAPTER 14

Conclusion: On Social Ontology

A reader who has endured until this point will not be surprised to hear that summarizing this book's conclusions is a difficult task. One reason for this is that in the preceding chapters many possibilities for conceiving of social pathology have emerged. Hopefully this multiplicity reflects the complexity of real social life rather than merely the author's inability to find a single guiding thread among the many ways in which human societies are vulnerable to falling ill.

I have argued that theorists who invoke the concept of social pathology have good reasons for resorting to the language of illness in understanding and criticizing social life, reasons that derive more from their inquiries into real phenomena than from a priori philosophical commitments. The principal advantage of conceiving of society and its problems in terms of health and illness is that doing so brings aspects of social life into view that cannot be grasped by restricting ourselves to the narrower categories of legitimacy, justice, or moral rightness. Thus, rejecting the discourse of social pathology runs the risk of rendering invisible a wide range of phenomena that plague social life.

I have also argued that we can learn important things about social life by attending to past attempts to understand and criticize society using a vocabulary appropriated from medicine. Thus, in addition to asking how we should understand social pathology, I have been guided by a second question: What does its vulnerability to falling ill imply about the kind of thing human society is? Addressing this question takes us into the domain of social ontology, although understood more broadly than is typical in Anglo-American philosophy. One might formulate this difference by saying that Hegelian social ontology incorporates general anthropological assumptions concerning the needs, psychology, and interdependence of human beings and that these assumptions imply, or make plausible, certain normative theses regarding the nature and value of human freedom and well-being – in other words, about the human good. (Recall: that freedom and well-being are constitutive of the human good counts as universally true for this tradition, but this general claim leaves room for great variation in how freedom and well-being are specifically understood in different societies.) That in studying human social life descriptive and normative claims cannot be held strictly apart is a principal thesis of this book.

My position with respect to this second, ontological question is more unitary than might be expected from the multitude of forms of social pathology I adduce in response to the first. I have tried to show that the conceptions of human society espoused by Marx, Plato, Rousseau, Durkheim, and Hegel – as well as John Searle and Vincent Descombes – *converge* (or can, without interpretive violence, be made to converge) into a general conception of the kind of thing human society is. In this conclusion I bring together the various strands of this conception that are scattered throughout the preceding chapters.

All the classical theorists of social pathology focused on here subscribe to some version of the thesis that human societies are to be understood as living beings, or on analogy with phenomena of life: social being is social *life*. This is relevant to whether pathology is a helpful category for social philosophy since life is the domain within which the concept of illness originally resides. My emphasis on the idea of social life means that one should expect from my investigations some account of the respects in which human societies are living beings, or sufficiently like animal organisms (or species), that diagnoses of social illness can enlighten rather than distort our study of them. At the same time, theories of social pathology must avoid reducing human to animal life, which is to say, they must do justice to the ways in which spiritual life differs from the life of nonhuman animals.

The most relevant similarity between animal and social life is that both involve carrying out *functions* of various kinds. If the core idea of social pathology is dysfunctionality, then human societies, if subject to falling ill, must be thought of as carrying out certain functions – processes organized in accordance with a "point" or aim – even if in the case of spiritual beings "function" has a more expansive meaning than in animal biology. Moreover, the functionality of living beings relies on *specialized* and *coordinated* activities of their parts or members, implying that one principal dysfunction living beings can exhibit is imbalance or faulty coordination among their specialized parts. A related feature of processes of life is that they typically take the form of quasi-autonomous *dynamics*, or self-reproducing nexuses of activities that exhibit a characteristic logic or coherence, which means that diagnoses of social pathology rely on a dynamic understanding of social processes and of how they reproduce or transform themselves over time. This feature of life necessitates a kind of explanatory holism in both biology and social theory since it implies that living phenomena cannot be grasped as the merely aggregative results of actions independently undertaken by the parts of the "organism" in question.

Another important aspect of the life-like nature of social life is that reproducing life's material conditions – the bodies of future generations and the goods required for biological survival – is a central and indispensable task of human societies. In this respect social life *is* biological life, even if there is a human tendency to deny our rootedness in nature, perhaps because of the dependence and vulnerability it implies. There is, however, something instructive in the impulse to downplay the material aspects of social life, for, first, material reproduction far from exhausts the significance of social life; and, second, material reproduction in humans is always at the same time spiritual reproduction or activity undertaken in order to satisfy spiritual ends. In human societies, the production of food, shelter, medicine, or clothing is never completely determined merely by what human bodies need in order to survive and reproduce. Rather, such activities are also governed by historically variant conceptions regarding the kind of life it is *appropriate for humans* to have, two components of which are freedom (variously conceived) and recognition from others. This is the idea behind Rousseau's, Marx's, and Hegel's implicit thesis that integrating the ends of life with those of freedom in our social activities is the paramount "function" of human social life and that social pathologies should be understood not only as impaired functioning in general or as imbalances among functional spheres, but also as ways in which society fails in enabling its members to participate in life in the mode of freedom.

Although all living beings are self-organizing, functionally specialized entities, human social life also has features that distinguish it from merely biological life. The core difference is that in the former, processes are carried out self-consciously; human social reality is *self-conscious life*. This means that for humans, social life is mediated by its participants' awareness, potential or actual, of the point of their activities. Even if this awareness is subject to distortion, those activities have a meaning and not merely a function, implying that it is impossible to say fully what humans do in their social activity without referring to the meanings it has for them. In contrast to biological life, social members' consciousness of the point of what they do unifies their activities into unitary, temporally extended processes, which, because they are informed by consciousness, are better thought of as practices than processes.

Closely connected to this is Rousseau's thesis that human society is artificial rather than purely natural. I have interpreted this thesis as claiming that social life is not only meaning-laden; it is also normatively constituted in the sense of being governed by conventions, rules, and norms to which its participants ascribe an authority that bestows obligations on them, even if in healthy social life individuals do not regard those obligations as imposed from without, by a foreign power. Since the acceptance and application of conventions, rules, and norms depends on a sort of freedom – what Kant calls "spontaneity" – the functioning of institutions, as well as their reproduction and transformation, is in principle "up to us," that is, to those whose agency sustains them.

Since social institutions vary in kind and importance, so, too, do the reasons their members have for granting authority to their conventions, rules, and norms. Sometimes reasons are as thin as "this is how we have always done things (and we see no reason to change)"; sometimes they rest on the perceived utility of the practices in question: "Without the rules governing money, economic life would be severely impaired." Moreover, Hegel in particular makes clear that, given certain basic anthropological facts, if *human* social life is to be possible, there must be some institutions, those bound up most intimately with social reproduction, in which individuals' allegiance to norms and rules runs deeper than mere custom or utility, extending to the sense that those institutions realize the ethical good and that participating in them provides members with a social framework within which they can live meaningful, affirmable, "particular" lives. This is one point behind Hegel's claim that the being characteristic of human society must be understood as the living, self-(re)producing good. For Hegel, the family and civil society are paradigm examples of such core institutions, not only because the reproductive tasks carried out in them are demanding and time-consuming, but also because the labor and familial love their functioning depends on are potential sources of deep, identity-constituting satisfaction.

A further difference between animal and human social life is bound up with the fact that self-consciousness makes forms of internal division possible that are absent in merely living beings. Marx expresses this idea by saying that whereas "the animal is immediately one with its life activity, … the human being makes his life activity itself into an object of his willing and consciousness. … It is not a determination with which he immediately merges; … his own life is an object for him" (MER: 76/MEGA: XL.516). The self-conscious reflexivity of human subjects means that humans, in contrast to mere animals, not only participate in life but are also capable of taking their social life to be good (or not). This capacity for evaluative distance makes not only self-affirmation possible but also the relation to oneself and one's activities on which social critique depends. Questioning the goodness of the world within which one is situated is a possibility built into the nature of the human subject and an essential ingredient of healthy social life.

The reflexivity of human subjectivity also makes possible a more robust kind of freedom – a more substantive form of *self-determination* – than the "freedom" available to nonhuman animals. Rousseau alludes to this animal freedom in saying that just "as an untamed steed ... struggles impetuously at the very sight of a bit, ... so barbarous man ... prefers the most tempestuous freedom to a tranquil subjection" (DI: 177/OC: 181). Whereas freedom in this sense – the absence of external impediments to an animate being's strivings – might pertain even to living beings lacking self-consciousness, the internal division characteristic of human subjects enables them to distinguish who they presently are (or what they happen to desire) from some conception of who they "really" are that can function as a criterion for distinguishing actions that are truly their own – consonant with or expressive of who they take themselves essentially to be – from actions incompatible with their deepest sense of self. Such beings, in other words, can determine their wills in accordance with a self-conception, or practical identity. While this self-determination can take many guises – acting as a parent, a kindergarten teacher, a citizen of one's country, or a Kantian moral agent – its most fundamental form (for Hegel and Marx) involves conceiving of oneself as a free, spiritual being that both has essential relations to nature and is at the same time "higher" (because freer) than living beings that are immediately merged with life and unable to make their immersion in nature an object of their own consciousness. This means that the various practices of social life resist comprehension if one fails to take account of the distinction its participants draw between themselves as spiritual beings (and what is appropriate to them) and animate beings lacking in self-consciousness. Joining this point with those above, one could say that human societies consist in cooperative life practices that are imbued with ethical significance because of their potential to be consciously self-determined and to achieve the good of all participants – that is, they are self-conscious life processes aimed at realizing the human good.

A further distinguishing characteristic of human societies is that they can realize the good appropriate to them only through the awareness, will, and actions of their members, who, as individuated bearers of consciousness, are themselves spiritual beings and therefore unlike the cells or organs of biological organisms. That the processes of social life are accomplished only through the subjectivities of its individual participants has both theoretical and normative implications. In the first case, comprehending the practices of social life requires taking into account (but without simply accepting as true) social members' own understanding of what they do in social life; in the second, the tasks of social life must be carried out in a way that does justice to their spiritual nature. Members of a human society enjoy greater existential and normative independence from the whole to which they belong than do the parts of a biological organism: human individuals, once produced, can live longer in isolation than a heart torn from its animal body, and, more importantly, their good is not simply identical with – conceptually determined by – the good of their society. The latter claim means that certain ends of a social whole – material reproduction, for example – might in principle be achievable in ways that take no account of the freedom and well-being of its individual members,[1] whereas such a scenario is not a conceptual possibility for biological organisms. This implies that acceptable theories of social pathology must avoid normative holism in its strongest form, where the good of the whole can be specified without reference to the good of its individual members.

Finally, this vision of human social life has implications for understanding social transformation as a response not merely to narrowly functional deficiencies but to problems at once ethical and functional. If social life is to be conceived of as the living good – as self-reproducing, functionally organized practices in which the ends of material reproduction and realizing freedom are inextricably interwoven – then the problems that beset it will (typically) be ethical failures, and rational responses to those problems will (typically) involve ethical progress, not merely improvements in social engineering. Ethical crises demand ethical remedies, which, if successful, provide a kind of ethical justification for the conventions, rules, and norms that govern the newly transformed social practices.

[1] I mean here only that this is conceptually possible; I have argued throughout that, given certain basic anthropological facts, it is not a real possibility.

Bibliography

Alznauer, Mark. 2016a. "Hegel's Theory of Normativity," *Journal of the American Philosophical Association* 2: 196–211

Alznauer, Mark. 2016b. "Rival Versions of Objective Spirit," *Hegel Bulletin* 37: 209–31

Aristotle. 2009a. *The Nicomachean Ethics*. Trans. David Ross. Oxford University Press

Aristotle. 2009b. *Politics*. Trans. Ernest Barker. Oxford University Press

Aron, Raymond. 1970. *Main Currents in Sociological Thought*, Vol. 2. Trans. Richard Howard and Helen Wever. Routledge

Boorse, Christopher. 1975a. "Health as a Theoretical Concept," *Philosophy of Science* 44: 542–73

Boorse, Christopher. 1975b. "On the Distinction between Disease and Illness," *Philosophy and Public Affairs* 5: 49–68

Boorse, Christopher. 1976. "Wright on Functions," *Philosophical Review* 85: 70–86

Boyle, Matthew. 2016. "Additive Theories of Rationality: A Critique," *European Journal of Philosophy* 24: 527–55

Brandom, Robert B. 2007. "The Structure of Desire and Recognition: Self-Consciousness and Self-Constitution," *Philosophy and Social Criticism* 33: 127–50

Brudney, Daniel. 1998. *Marx's Attempt to Leave Philosophy*. Harvard University Press

Canguilhem, Georges. 1978. *On the Normal and the Pathological*. Trans. Carolyn R. Fawcett. D. Reidel

Canivez, Patrice. 2011. "Pathologies of Recognition," *Philosophy and Social Criticism* 37: 851–87

Carré, Louis. 2013. "Die Sozialpathologien der Moderne: Hegel und Durkheim im Vergleich," *Hegel-Jahrbuch* 19: 312–17

Cohen, G. A. 1978. *Karl Marx's Theory of History: A Defence*. Princeton University Press

Cohen, G. A. 1982. "Functional Explanation, Consequence Explanation, and Marxism," *Inquiry* 25: 27–56

Cohen, Joshua. 1997. "The Arc of the Moral Universe," *Philosophy & Public Affairs* 26: 91–134

Comte, Auguste. 2015 [1844]. *A General View of Positivism*. Trans. J. H. Bridges. Routledge
Comte, Auguste. 1875 [1851]. *System of Positive Polity, or Treatise on Sociology*, Vol. 2. Trans. J. H. Bridges. Burt Franklin
Corning, Peter A. 1982. "Durkheim and Spencer," *The British Journal of Sociology* 33: 359–82
Descombes, Vincent. 1994. "Is There an Objective Spirit?" in J. Tully and D. Weinstock (eds.), *Philosophy in an Age of Pluralism: The Philosophy of Charles Taylor in Question*. Cambridge University Press: 96–119
Descombes, Vincent. 2014. *The Institutions of Meaning: A Defense of Anthropological Holism*. Trans. Stephen Adam Schwartz. Harvard University Press
Durkheim, Émile. 1966 [1892]. *Montesquieu et Rousseau: Précurseurs de la sociologie*. Librairie Marcel Rivière et Cie
Durkheim, Émile. 1984. *The Division of Labor in Society*. Trans. W. D. Halls. Free Press
[1893. *De la division du travail social*. Presses universitaires de France]
Durkheim, Émile. 1982. *Rules of Sociological Method*. Trans. Sarah A. Solovay and John H. Mueller. Free Press
[1937 [1895]. *Les règles de la méthode sociologique*. Presses universitaires de France]
Durkheim, Émile. 1997. *Suicide: A Study in Sociology*. Trans. John A. Spaulding and George Simpson. Free Press
[1930 [1897]. *Le suicide: Étude de Sociologie*. Presses universitaires de France]
Durkheim, Émile. 1969. "Individualism and the Intellectuals," *Political Studies* 17: 14–30. Trans. Steven Lukes
[1970 [1898]. "L'individualism et les Intellectuels," in *La science sociale et l'action*. Presses universitaires de France]
Durkheim, Émile. 1995 [1912]. *The Elementary Forms of Religious Life*. Trans. Karen E. Fields. Free Press
Durkheim, Émile. 2014. *Sociology and Philosophy*. Trans. J. G. Peristiany. Free Press
[2014 [1924]. *Sociologie et philosophie*. Presses universitaires de France]
Durkheim, Émile. 1961. *Moral Education*. Trans. Everett K. Wilson, Herman Schnurer. Free Press
[2012 [1925]. *L'éducation morale*. Presses universitaires de France]
Engelhardt, H. Tristram. 1976. "Ideology and Etiology," *Journal of Medicine and Philosophy* 1: 256–68
Forster, Jeremy James. 2015. "Nietzsche and the Pathologies of Meaning," PhD diss., Columbia University
Foster, John Bellamy. 1999. "Marx's Theory of Metabolic Rift: Classical Foundations for Environmental Sociology," *American Journal of Sociology* 105: 366–405
Fox, R. E. 1988. "Proceedings of the American Psychological Association," *American Psychologist* 43: 508–31
Freyenhagen, Fabian. 2015. "Honneth on Social Pathologies: A Critique," *Critical Horizons* 16: 131–52

Freyenhagen, Fabian. 2019. "Critical Theory and Social Pathology," in P. E. Gordon, E. Hammer, and A. Honneth (eds.), *Routledge Companion to the Frankfurt School*. Routledge: 410–23

Gane, Mike. 2006. *August Comte*. Routledge

Gangas, Spiros. 2009. "Hegel und Durkheim: Sittlichkeit and Organic Solidarity as Political Configurations," *Hegel-Jahrbuch* 2009: 222–6

Geuss, Raymond. 2001. *History and Illusion in Politics*. Cambridge University Press

Geuss, Raymond. 2021. *A Philosopher Looks at Work*. Cambridge University Press

Giddens, Anthony. 1973. *Capitalism and Modern Social Theory*. Cambridge University Press

Giddens, Anthony. 1978. *Durkheim*. Fontana Press

Giddens, Anthony. 1984. *The Constitution of Society*. Polity Press

Gilbert, Margaret. 1989. *On Social Facts*. Routledge

Gilbert, Margaret. 1983. "Notes on the Concept of a Social Convention," *New Literary History* 14: 225–51

Ginsborg, Hannah. 2006. "Kant's Biological Teleology and Its Philosophical Significance," in G. Bird (ed.), *A Companion to Kant*. Blackwell: 455–69

Gould, Carol. 1978. *Marx's Social Ontology: Individuality and Community in Marx's Theory of Social Reality*. MIT Press

Greene, Amanda R. 2017. "Legitimacy without Liberalism: A Defense of Max Weber's Standard of Political Legitimacy," *Analyse and Kritik* 39: 295–324

Haase, Matthias. 2013. "Life and Mind," in T. Khurana (ed.), *The Freedom of Life: Hegelian Perspectives*. August Verlag: 69–109

Haase, Matthias. 2017. "Geist und Gewohnheit," in A. Kern and C. Kietzmann (eds.), *Selbstbewusstes Leben: Texten zu einer transformativen Theorie der menschlichen Subjektivität*. Suhrkamp

Habermas, Jürgen. 1975. *Legitimation Crisis*. Trans. Thomas McCarthy. Beacon Press

Habermas, Jürgen. 1987. *Theory of Communicative Action*, Vol. 2. Trans. Thomas McCarthy. Beacon Press

Hahn, Songsuk Susan. 2007. *Contradiction in Motion: Hegel's Organic Concept of Life and Value*. Cornell University Press

Hardimon, Michael. 1994. *Hegel's Social Philosophy: The Project of Reconciliation*. Cambridge University Press

Harris, Jonathan Gil. 1998. *Foreign Bodies and the Body Politic*. Cambridge University Press

Hartmut, Rosa. 2015. *Social Acceleration: A New Theory of Modernity*. Trans. Jonathan Trejo-Mathys. Columbia University Press

Harvey, David. 2010. *A Companion to Marx's* Capital. Verso

Hedrick, Todd. 2018. *Reconciliation and Reification: Freedom's Semblance and Actuality from Hegel to Contemporary Critical Theory*. Oxford University Press

Hegel, Georg Wilhelm Friedrich. 1975 [1802–03]. *Natural Law: The Scientific Ways of Treating Natural Law, Its Place in Moral Philosophy, and Its Relation to the Positive Sciences of Law*. Trans. T. M. Knox. University of Pennsylvania Press

Hegel, Georg Wilhelm Friedrich. 2018 [1807]. *The Phenomenology of Spirit*. Trans. Terry Pinkard. Cambridge University Press

Hegel, Georg Wilhelm Friedrich. 1969 [1812]. *Science of Logic*. Trans. A. V. Miller. Humanities Press

Hegel, Georg Wilhelm Friedrich. 1983 [1817–19]. *Die Philosophie des Rechts: Die Mitschriften Wannenmann (Heidelberg 1817/18) und Homeyer (Berlin 1818/19)*. Klett-Cotta

Hegel, Georg Wilhelm Friedrich. 1991a [1820]. *Elements of the Philosophy of Right*. trans. H. B. Nisbet. Cambridge University Press

Hegel, Georg Wilhelm Friedrich. 1991b [1830]. *The Encyclopedia Logic*. Trans. T. F. Geraets, W. A. Suchting, and H. S. Harris. Hackett

Hegel, Georg Wilhelm Friedrich. 1971 [1830]. *Philosophy of Mind: Part Three of the Encyclopedia of the Philosophical Sciences*. Trans. William Wallace. Oxford University Press

Hegel, Georg Wilhelm Friedrich. 2004 [1830]. *Philosophy of Nature: Part Two of the Encyclopedia of the Philosophical Sciences*. Trans. A. V. Miller. Oxford University Press

Heisenberg, Lars Thimo Immanuel. 2019. "Hegel on Social Critique: Life, Action, and the Good in Hegel's 'Philosophy of Right'," PhD diss., Columbia University

Heisenberg, Lars Thimo Immanuel. 2021. "Death in Berlin: Hegel on Mortality and the Social Order," *British Journal for the History of Philosophy* 29: 871–90

Hesslow, Germund. 1993. "Do We Need a Concept of Disease?" *Theoretical Medicine* 14: 1–14

Hobbes, Thomas. 1994 [1651]. *Leviathan*. Hackett

Hofstadter, Richard. 1959. *Social Darwinism in American Thought*. George Braziller

Hollis, Martin. 1994. *The Philosophy of Social Science*. Cambridge University Press

Honneth, Axel. 2007. "Pathologies of the Social: The Past and Present of Social Philosophy," in *Disrespect: The Normative Foundations of Critical Theory*. Polity Press: 3–48. Trans. Joseph Ganahl

Honneth, Axel. 2009. *Pathologies of Reason: On the Legacy of Critical Theory*. Trans. James Ingram. Columbia University Press

Honneth, Axel. 2010. *The Pathologies of Individual Freedom: Hegel's Social Theory*. Trans. Ladislaus Löb. Princeton University Press

Honneth, Axel. 2014a. "The Diseases of Society: Approaching a Nearly Impossible Concept," *Social Research. An International Quarterly* 81: 683–703. Trans. Arvi Särkelä

Honneth, Axel. 2014b. *Freedom's Right: The Social Foundations of Democratic Life*. Trans. Joseph Ganahl. Columbia University Press

Honneth, Axel. 2016. "The Depths of Recognition: The Legacy of Jean-Jacques Rousseau," in A. Lifschitz (ed.), *Engaging with Rousseau: Reaction and Interpretation from the Eighteenth Century to the Present*. Cambridge University Press: 189–206

Honneth, Axel. 2021. "Hegel and Durkheim: Contours of an Elective Affinity," in N. Marcucci (ed.), *Durkheim and Critique*. Palgrave: 19–42

Humber, James. (ed.) 1997. *What Is Disease?* Humana
Hume, David. 1896. *A Treatise of Human Nature*. Oxford University Press
Jaeggi, Rahel. 2019. *Critique of Forms of Life*. Trans. Ciaran Cronin. Harvard University Press
Jung, Carl G. 1964. *Man and His Symbols*. Trans. Marie-Luise von Franz. Doubleday
Kandiyali, Jan. 2014. "Freedom and Necessity in Marx's Account of Communism," *British Journal for the History of Philosophy* 22: 104–123
Kandiyali, Jan. 2020. "The Importance of Others: Marx on Unalienated Production," *Ethics*: 555–587
Kant, Immanuel. 1987 [1790]. *Critique of Judgment*. Trans. Werner S. Pluhar. Hackett
Karsenti, Bruno. 2006a. *La société en personnes: Études durkheimiennes*. Economica
Karsenti, Bruno. 2006b. *Politique de l'esprit: Auguste Comte et la naissance de la science sociale*. Hermann
Karsenti, Bruno. 2012. "Durkheim and the Moral Fact," in D. Fassin (ed.), *A Companion to Moral Anthropology*. Blackwell: 21–36
Keohane, Kieran, and Anders Petersen (eds.). 2013. *The Social Pathologies of Contemporary Civilization*. Ashgate
Kern, Andrea, and Kietzmann, Christian (eds.). 2017. *Selbstbewusstes Leben: Texten zu einer transformativen Theorie der menschlichen Subjektivität*. Suhrkamp
Khurana, Thomas. 2013. "Life and Autonomy," in T. Khurana (ed.), *The Freedom of Life: Hegelian Perspectives*. August Verlag: 155–93
Khurana, Thomas. 2017. *Das Leben der Freiheit: Form und Wirklichkeit der Autonomie*. Suhrkamp
Kincaid, Harold. 1990. "Assessing Functional Explanations in the Social Sciences," *PSA: Proceedings of the Biennial Meeting of the Philosophy of Science Association*, Vol. 1: 341–354
Kitcher, Philip. 1997. *The Lives to Come: The Genetic Revolution and Human Possibilities*. Simon & Schuster
Kitcher, Philip. 2014. *The Ethical Project*. Harvard University Press
Kitcher, Philip. 2021. *Moral Progress*. Oxford University Press
Knapp, Peter. 1985. "The Question of Hegelian Influence upon Durkheim's Sociology," *Sociological Inquiry* 55: 1–15
Korsgaard, Christine M. 1996. *The Sources of Normativity*. Cambridge University Press
Kreines, James. 2005. "The Inexplicability of Kant's 'Naturzweck': Kant on Teleology, Explanation and Biology," *Archiv für Geschichte der Philosophie* 87: 270–309
Kreines, James. 2013. "Kant and Hegel on Teleology and Life from the Perspective of Debates about Free Will," in T. Khurana (ed.), *The Freedom of Life: Hegelian Perspectives*. August Verlag: 155–93
Krugman, Paul. 2019. "Notes on Excessive Wealth Disorder," *The New York Times*, June 22
LaCapra, Dominick. 1972. *Emile Durkheim: Sociologist and Philosopher*. Cornell University Press

Laitinen, Arto, and Särkelä, Arvi. 2019. "Four Conceptions of Social Pathology," *European Journal of Social Theory* 22: 1–23

Lenoir, Timothy. 1982. *The Strategy of Life: Teleology and Mechanics in Nineteenth Century German Biology*. University of Chicago Press

Lukes, Steven. 1985. *Emile Durkheim, His Life and Work: A Historical and Critical Study*. Stanford University Press

Machiavelli, Niccolò. 1950 [1532]. *The Prince and the Discourses*. Trans. Max Lerner. Modern Library

Marshall, Eugene. 2013. *The Spiritual Automaton: Spinoza's Science of the Mind.* Oxford University Press

Marx, Karl. 1992 [1867]. *Capital*, Vol. 1.Trans. Ben Fowkes. Penguin

Marx, Karl. 1978. *The Marx-Engels Reader*. Ed. Robert C. Tucker. W. W. Norton

Marx, Karl. 1975. *Karl Marx, Friedrich Engels Gesamtausgabe*. Dietz Verlag

Mau, Steffen. 2019. *Lütten Klein: Leben in der ostdeutschen Transformationsgesellschaft.* Suhrkamp

Menke, Christoph. 2018. *Autonomie und Befreiung*. Suhrkamp

Mills, C. Wright. 1943. "The Professional Ideology of Social Pathologists," *American Journal of Sociology* 49: 165–180

Moran, Richard. 2016. "Williams, History, and the 'Impurity of Philosophy,'" *European Journal of Philosophy* 24: 315–30

Murphy, Dominic. n.d. "Concepts of Disease and Health," *Stanford Encyclopedia of Philosophy*, http://plato.stanford.edu/entries/health-disease/

Neuhouser, Frederick. 1990. *Fichte's Theory of Subjectivity*. Cambridge University Press

Neuhouser, Frederick. 2003. *Actualizing Freedom: Foundations of Hegel's Social Theory*. Harvard University Press

Neuhouser, Frederick. 2008. "Desire, Recognition, and the Relation between Bondsman and Lord," in K. Westphal (ed.), *The Blackwell Guide to Hegel's Phenomenology of Spirit*. Blackwell: 37–54

Neuhouser, Frederick. 2010. *Rousseau's Theodicy of Self-Love: Evil, Rationality, and the Drive for Recognition*. Oxford University Press

Neuhouser, Frederick. 2011. "The Concept of Society in 19th Century Thought," in A. W. Wood and S. S. Hahn (eds.), *Cambridge History of Philosophy in the 19th Century (1790–1870)*. Cambridge University Press: 651–75

Neuhouser, Frederick. 2012. "Rousseau und die Idee einer 'pathologischen' Gesellschaft," *Politische Vierteljahresschrift* 53: 628–45

Neuhouser, Frederick. 2015. *Rousseau's Critique of Inequality: Reconstructing the Second Discourse.* Cambridge University Press

Neuhouser, Frederick. 2017a. "Nietzsche on Spiritual Health and Cultural Pathology," in P. Katsafanas (ed.), *The Nietzschean Mind.* Routledge: 355–68

Neuhouser, Frederick. 2017b. "The Method of the 'Philosophy of Right'," in D. James (ed.), *Hegel's Elements of the Philosophy of Right*. Cambridge University Press: 16–36

Neuhouser, Frederick. 2021. "Anomie: On the Link between Social Pathology and Social Ontology," in N. Marcucci (ed.), *Durkheim and Critique*. Palgrave: 131–62

Ng, Karen. 2015. "Ideology Critique from Hegel and Marx to Critical Theory," *Constellations* 22: 393–404
Ng, Karen. 2017. "From Actuality to Concept in Hegel's Logic," in D. Moyar (ed.), *The Oxford Handbook of Hegel.* Oxford University Press: 269–90
Ng, Karen. 2020. *Hegel's Concept of Life: Self-Consciousness, Freedom, Logic.* Oxford University Press
Nietzsche, Friedrich. 1989 [1887]. *On the Genealogy of Morals.* Trans. Walter Kaufman and R. J. Hollingdale. Random House
Novakovic, Andreja. 2017. *Hegel on Second Nature in Ethical Life.* Cambridge University Press
Parsons, Talcott. 1971. *The System of Modern Societies.* Prentice-Hall
Pettit, Philip. 1996. "Functional Explanation and Virtual Selection," *The British Journal for the Philosophy of Science* 47: 291–302
Pettit, Philip. 1998. "Defining and Defending Social Holism," *Philosophical Explorations* 1: 169–84
Pettit, Philip. 2014. "Three Issues in Social Ontology," in *Rethinking the Individualism-Holism Debate: Essays in the Philosophy of Social Science*, J. Zahle and F. Collin (eds.). Springer International: 77–96
Piketty, Thomas. 2017. *Capital in the Twenty-First Century.* Trans. Arthur Goldhammer. Harvard University Press
Piketty, Thomas. 2020. *Capital and Ideology.* Trans. Arthur Goldhammer. Harvard University Press
Pinkard, Terry P. 2012. *Hegel's Naturalism: Mind, Nature, and the Final Ends of Life.* Oxford University Press
Pippin, Robert B. 2008. *Hegel's Practical Philosophy: Rational Agency as Ethical Life.* Cambridge University Press
Plato. 1992. *Republic.* Trans. G. M. A. Grube. Hackett
Poggi, Gianfranco. 2000. *Durkheim.* Oxford University Press
Polanyi, Karl. 1957 [1944]. *The Great Transformation: The Political and Economic Origins of Our Time.* Beacon Press
Postone, Moishe. 1996. *Time, Labor, and Social Domination.* Cambridge University Press
Quante, Michael. 2011. *Die Wirklichkeit des Geistes: Studien zu Hegel.* Suhrkamp
Radcliffe-Brown, A. R. 1935. "On the Concept of Function in Social Science," *American Anthropologist* 37: 394–402
Radcliffe-Brown, A. R. 1957. *A Naturalist Science of Society.* University of Chicago Press
Ramos, Cesar Augusto. 2009. "O conceito hegeliano de liberdade como estar junto de si em seu outro," *Filosofia Unisinos* 10: 15–27
Rand, Sebastian. 2015. "What's Wrong with Rex? Hegel on Animal Defect and Individuality," *European Journal of Philosophy* 23: 68–86
Rawls, John. 1955. "Two Concepts of Rules," *The Philosophical Review* 1955: 3–32
Rawls, John. 1999. *A Theory of Justice (Revised Edition).* Harvard University Press
Rawls, John. 2001. *Justice as Fairness: A Restatement.* Harvard University Press
Renault, Emmanuel. 2010. "A Critical Theory of Social Suffering," *Critical Horizons* 11: 221–41

Renault, Emmanuel. 2017. *Social Suffering: Sociology, Psychology, Politics*. Trans. Maude Dews. Rowman & Littlefield

Richards, Robert J. 2002. *The Romantic Conception of Life: Science and Philosophy in the Age of Goethe*. University of Chicago Press

Rigdon, Susan M. 1988. *The Culture Facade: Art, Science, and Politics in the Work of Oscar Lewis*. University of Illinois Press

Rousseau, Jean-Jacques. 1964. *Oeuvres Complètes*, Vol. 3, (eds.) B. Gagnebin and M. Raymond. Gallimard

Rousseau, Jean-Jacques. 1997a [1755]. *Discourse on the Origin of Inequality*, in V. Gourevitch (ed. and trans.), *The Discourses and Other Early Political Writings*. Cambridge University Press

Rousseau, Jean-Jacques. 1997b [1755]. *Discourse on Political Economy*, in V. Gourevitch (ed. and trans.), *The Social Contract and Other Later Political Writings*. Cambridge University Press

Rousseau, Jean-Jacques. 1979 [1762]. *Emile, or On Education*. Trans. Allan Bloom. Basic Books

Rousseau, Jean-Jacques. 1997c [1762]. *The Social Contract*, in V. Gourevitch (ed. and trans.), *The Social Contract and Other Later Political Writings*. Cambridge University Press

Sandkaulen, Birgit. 2010. "Die Seele ist der existierende Begriff," *Hegel-Studien* 45: 35–50

Sandkaulen, Birgit. 2019a. "Der Begriff des Lebens in der Klassischen Deutschen Philosophie – eine naturphilosophische oder lebensweltliche Frage?," *Deutsche Zeitschrift für Philosophie* 67: 911–929

Sandkaulen, Birgit. 2019b. *Jacobis Philosophie: Über den Widerspruch zwischen System und Freiheit*. Meiner

Särkelä, Arvi. 2018. *Immanente Kritik und soziales Leben: Selbsttransformative Praxis nach Hegel und Dewey*. Klostermann

Schmidt, Alfred. 2014. *The Concept of Nature in Marx*. Trans. Ben Fowkes. Verso

Searle, John. 1995. *The Construction of Social Reality*. The Free Press

Searle, John. 2008. "Language and Social Ontology," *Theory and Society* 37: 443–59

Searle, John. 2010. *Making the Social World: The Structure of Human Civilization*. Oxford University Press

Sedgwick, Sally. 2001. "The State as Organism: The Metaphysical Basis of Hegel's Philosophy of Right," *The Southern Journal of Philosophy* 34: 171–188

Sedgwick, Sally. 2004. "Hegel on Kant's Idea of Organic Unity: The Jenaer Schriften," in S. Doyé, M. Henz, and U. Rameil (eds.), *Metaphysik und Kritik: Festschrift für Manfred Baum zum 65. Geburtstag*. De Gruyter: 285–298

Smith, Adam. 1977. *An Inquiry into the Nature and Causes of the Wealth of Nations*. University of Chicago Press

Sontag, Susan. 1978. *Illness as Metaphor*. Farrar, Straus and Giroux

Spencer, Herbert. 1969 [1898]. *The Principles of Sociology*, Vol. 1. Archon Books

Stahl, Titus. 2013. *Immanente Kritik: Elemente einer Theorie sozialer Praktiken*. Campus Verlag

Taylor, Charles. 2010. *Human Agency and Language: Philosophical Papers 1*. Cambridge University Press

Testa, Italo. 2008. "Selbstbewusstsein und zweite Natur," in K. Vieweg and W. Welsch (eds.), *Hegels Phänomenologie des Geistes*. Suhrkamp: 286–307

Testa, Italo. 2020. "Embodied Cognition, Habit, and Natural Agency," in M. Bykova and K. Westphal (eds.), *The Palgrave Hegel Handbook*. Palgrave: 269–95

Thompson, Michael. 2008. *Life and Action: Elementary Structures of Practice and Practical Thought*. Harvard University Press

Tuomela, Raimo. 2013. *Social Ontology: Collective Intentionality and Group Agents*. Oxford University Press

Turner, Jonathan H., and Maryanski, Alexandra R. 1979. *Functionalism*. Benjamin/Cummings

Turner, Jonathan H., and Maryanski, Alexandra R. 1988. "Is 'Neofunctionalism' Really Functional?" *Sociological Theory* 6: 110–121

Turner, Stephen P. 1986. *The Search for a Methodology of Social Science*. Springer

Weber, Max. 1992 [1905]. *The Protestant Ethic and the Spirit of Capitalism*. Trans. Talcott Parsons. Routledge

Whitebook, Joel. 2011. "Sigmund Freud: Ein 'philosophischer Arzt,'" in M. Leuzinger-Bohleber and R. Haubl (eds.), *Psychoanalyse: interdisziplinär-international-intergenerationell*. Vandenhoeck & Ruprecht: 120–42

Wilkinson, Richard G. 1996. *Unhealthy Societies: The Afflictions of Inequality*. Routledge

Wood, Allen W. 1981. *Karl Marx*. Routledge

Wood, Allen W. 1990. *Hegel's Ethical Thought*. Cambridge University Press

Wright, Frank Lloyd, and Pfeiffer, Bruce Brooks. 2008. *The Essential Frank Lloyd Wright: Critical Writings on Architecture*. Princeton University Press

Zurn, Christopher. 2011. "Social Pathologies as Second-Order Disorders," in D. Petherbridge (ed.), *Axel Honneth: Critical Essays*. Brill: 345–70

Index